REEFS AND BANKS OF THE NORTHWESTERN GULF OF MEXICO

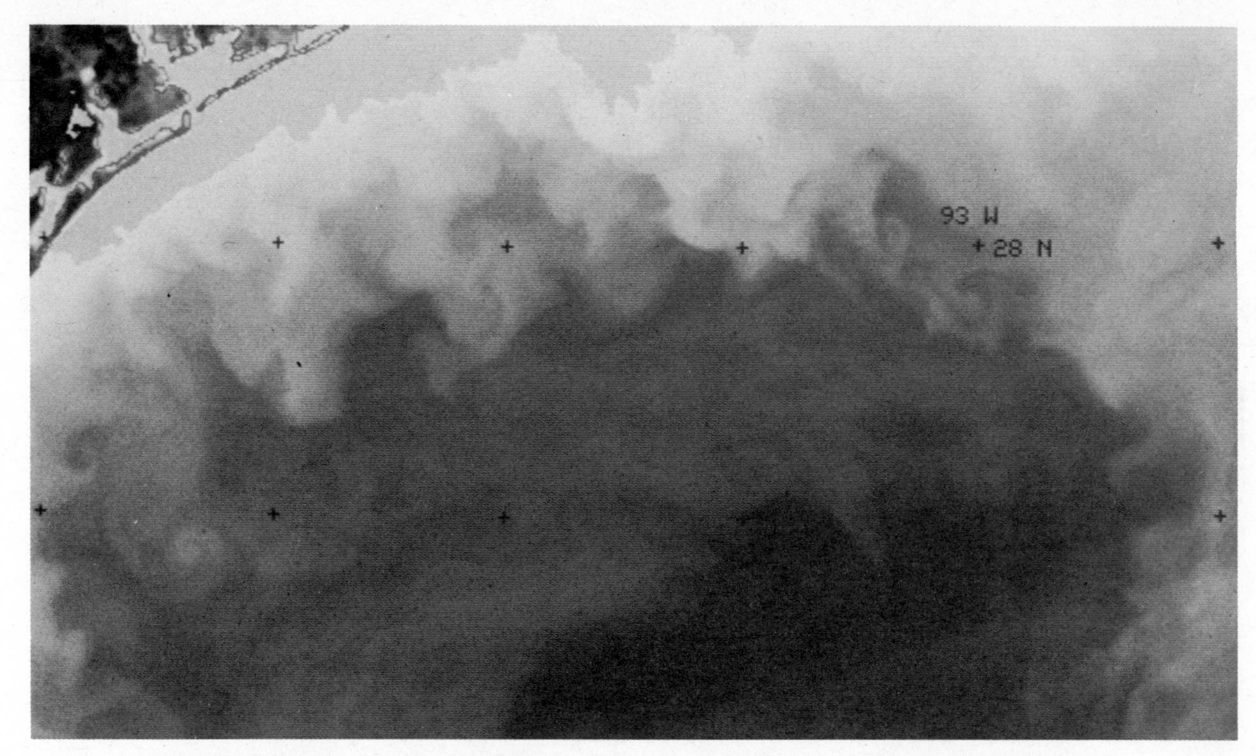

Figure. 11 Dec. NOAA-7 AVHRR image of the northwestern Gulf of Mexico taken on December 11, 1983. Light areas denote cooler water in comparison with dark, warmer waters. A pattern of alternating cold and warm water on the Texas shelf indicates surface expressions of wavelike perturbations found in hydrographic studies (Fig. 5.26). The average wavelength for these waves, measured on the images, is 80 km.

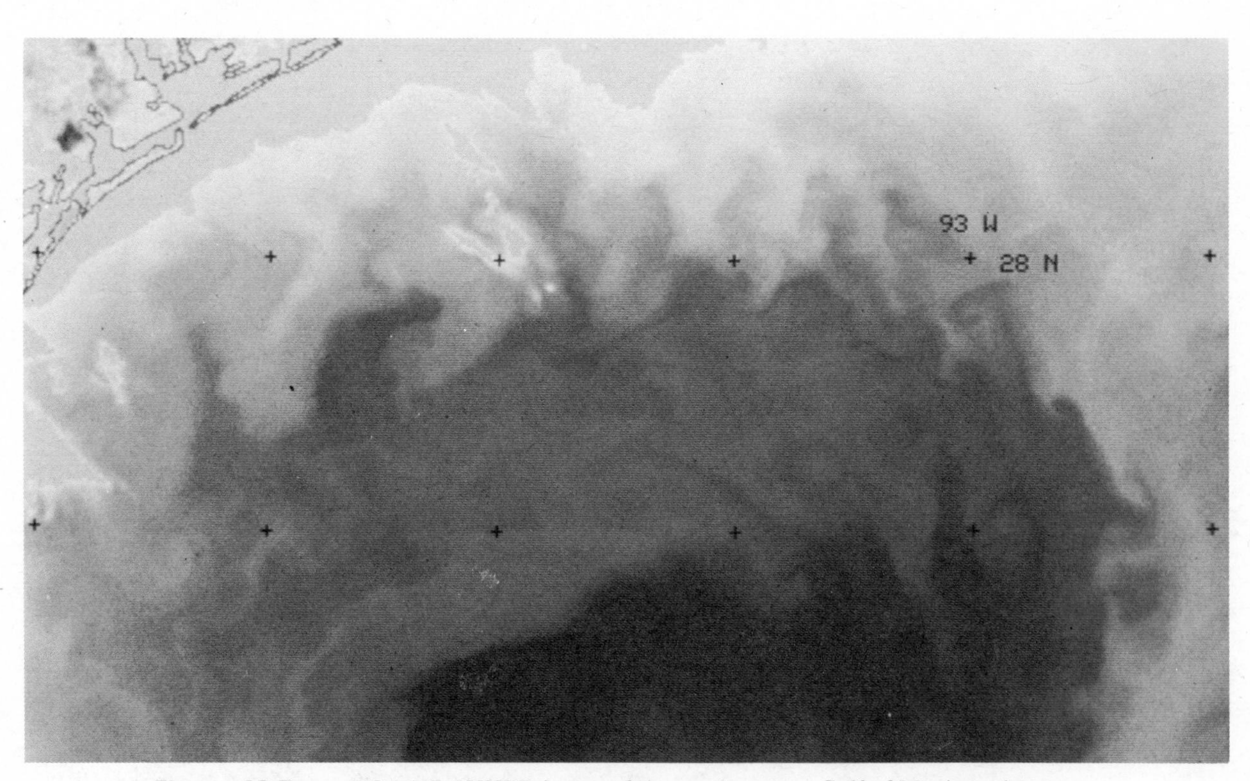

Figure. 12 Dec. NOAA-7 AVHRR image of the northwestern Gulf of Mexico taken on December 12, 1983. Shelf waves on the Texas shelf 24 hours after the image shown in Figure 11 Dec. As the text of this book was completed in August 1983, these images substantiate David McGrail's speculation in the caption for Figure 5.26 that these are shelf waves.

REEFS AND BANKS OF THE NORTHWESTERN GULF OF MEXICO

THEIR GEOLOGICAL, BIOLOGICAL, AND PHYSICAL DYNAMICS

RICHARD REZAK
THOMAS J. BRIGHT
DAVID W. McGRAIL

Department of Oceanography
Texas A&M University
College Station, Texas

A Wiley-Interscience Publication
JOHN WILEY & SONS
New York / Chichester / Brisbane / Toronto / Singapore

Library of Congress Cataloging in Publication Data:

Rezak, Richard, 1920–
Reefs and banks of the northwestern Gulf of Mexico.

"A Wiley-Interscience publication."
Bibliography: p.
Includes indexes.
1. Reefs—Texas. 2. Reefs—Louisiana. 3. Continental shelf—Texas. 4. Continental shelf—Louisiana. 5. Coral reef ecology—Texas. 6. Coral reef ecology—Louisiana. I. Bright, Thomas J. II. McGrail, David W. III. Title.
GB465.T4R49 1985 551.4′24 84-26989
ISBN 0-471-89379-X

Printed in the United States of America

10 9 8 7 6 5 4 3 2 1

To Billy, Ken, Mike, and DIAPHUS,
who conveyed us to undersea places
defiant of proper description
in scientific prose

DAVID WAYNE McGRAIL

David Wayne McGrail, husband of Susan and father of Harlan, died on September 24, 1984.

David McGrail was awarded a Bachelor of Arts in Geology from the College of Wooster in 1967. He also earned a Master of Science with Geology as a major field in 1972 and a Doctor of Philosophy with Oceanography as a major field in 1976, both from the University of Rhode Island. While working on his advanced degrees, David McGrail found time to serve with distinction as an Officer of the United States Coast Guard and as a researcher with the Coastal Resource Center of the State of Rhode Island.

Upon receiving his Ph.D., Dr. McGrail joined the faculty of the Department of Oceanography at Texas A&M University. At Texas A&M, he had a distinguished career as an administrator, researcher, and teacher. As an administrator in the Department of Oceanography, McGrail served as Deputy Department Head, Manager of the Gulf of Mexico Topographic High Program, and Director of Technical Services. As a researcher, David McGrail combined his considerable skills in both geological oceanography and physical oceanography to produce more than thirty manuscripts. In particular, his work on the currents, long waves, and sediment transport near the shelf edge of the northwestern Gulf of Mexico stands out as the major effort to understand this complex oceanographic region.

Research and graduate teaching are complementary endeavors. This was so true with David McGrail. He was loved and respected as a teacher and graduate advisor. His teaching ability was formally recognized through his receipt of the Distinguished Teaching Award given by the Association of Former Students of Texas A&M University. Moreover, he was designated the most outstanding faculty member by the Oceanography Graduate Student Council. David McGrail was the only individual to receive this unique award decided solely by a vote of the students.

David McGrail used his considerable administrative and research skills to produce many of the ideas and results which appear in this manuscript. Perhaps of even more importance was the example he set with his wise counsel, good cheer, and genuine zeal for his work. Indeed, during his long illness, David continued to work enthusiastically with his colleagues and his students. He was an inspiration then. Memories of him inspire us now.

William Merrell
College Station, Texas
September, 1984

DAVID WAYNE McGRAIL
1944–1984

FOREWORD

In this book oceanographers Richard Rezak, Thomas J. Bright, and David W. McGrail, aided by the editorial skills of technical writer Rose Norman, have attempted to describe the complex natural environment of the Texas–Louisiana shelf in terms readily understandable by decision makers in goverment and industry and by interested citizens. Such writing represents a formidable challenge to scientists. Yet it is of considerable importance to them as well as to the reader because it forces them to examine and clarify scientific arguments to the point at which they can be explained in simple, straightforward terms.

I believe that the benefits and occasional frustrations of such writing are described best by Thomas Huxley in the preface to his *Discourses: Biological and Geological,* published in 1894. Huxley wrote:

> I found that the task of putting the truths learned in the field, the laboratory and the museum, into language which, without bating a jot of scientific accuracy shall be generally intelligible, taxed such scientific and literary faculty as I possessed to the uttermost; indeed my experience has furnished me with no better corrective of the tendency to scholastic pedantry which besets all those who are absorbed in pursuits remote from the common ways of men, and become habituated to think and speak in the technical dialect of their own little world, as if there were no other.

I hope that you will agree with me that the authors of this book have lived up to Huxley's high standards by producing a document that is both scientifically accurate and easily readable.

WILLIAM J. MERRELL

College Station, Texas
January 1985

PREFACE

This book has been written to provide a synthesis of scientific information regarding the geology, biology, and physical oceanography of the Texas–Louisiana Outer Continental Shelf. A considerable portion of the data collected on the shelf is the result of a series of studies funded by the U.S. Department of the Interior, Bureau of Land Management (BLM; now Minerals Management Service, MMS) and conducted principally by the present authors. Our conclusions rely primarily on data generated during these investigations, which started in 1974. It also incorporates pertinent scientific information generated by other studies before and during these BLM/MMS-sponsored investigations.

BACKGROUND

Initial interest in the banks on the outer continental shelf was expressed more than 50 years ago in a paper on the Mississippi Delta (Trowbridge, 1930) in which Trowbridge recorded the presence of a bathymetric prominence with a relief of 33 m and a covering of coarse sediments. Six years later the U.S. Coast and Geodetic Survey carried out a detailed survey of the Gulf of Mexico west of the Mississippi River. Analysis of these data led Shepard (1937) to suggest that the numerous pinnacles present in that area were due to the intrusion of salt into the sediments. Later Carsey (1950) reported 164 "topographic features" that occur along the shelf off the coast of Texas and Louisiana.

Parker and Curray (1956) dredged dead corals, but Pulley (1963) was the first to report that the Flower Garden Banks were indeed flourishing coral reefs. In 1961, with the assistance of scuba divers, he photographed and collected live corals from both banks. At about the same time Richard Rezak, while employed by the Shell Development Company, Houston, Texas, was invited by Texas A&M University to participate in a short cruise to the Flower Garden Banks. Bathymetric profiles and Van Veen grabs confirmed Stetson's description of living coral and coralline algae (1953). Levert and Ferguson (1969) further substantiated Stetson's and Pulley's findings.

The Department of Oceanography at Texas A&M University conducted several cruises to the Flower Garden Banks, led primarily by Rezak and Bright during the period from 1968 to 1974. Their work began as separate geological and biological investigations but very early on became a joint effort funded by Texas A&M University and the Flower Garden Ocean Research Center at the University of Texas Marine Biomedical Institute. The goal of the geological effort was to develop a conceptual model of a coral reef growing on a terrigenous shelf for hydrocarbon exploration. Biological efforts sought to document the flora and fauna of the reef and bank environments.

Edwards (1971), in his Ph.D. dissertation, presented a most comprehensive report on the West Flower Garden Bank. He used 3.5-kHz and air-gun seismic profiles for the first time to illustrate the salt diapir subbottom structure of that bank. Edwards also described very accurately the sediment distribution and the biota associated with each sediment facies.

The initial Texas A&M investigation for BLM (now Minerals Management Service) began in late 1974 (Bright and Rezak, 1976). This study provided baseline biological and geological information to facilitate judgments regarding the need for and the nature of protective regulations to be imposed on drilling operations near these banks. Seventeen banks were mapped by using precision navigation, precision depth recorder, and side-scan sonar. Three of the areas mapped showed no bathymetric relief and were not studied further. Six of the 17 banks were examined and sampled from the Texas A&M submersible DRV DIAPHUS. All seafloor observations were documented by 35-mm color still photography and black and white videotape recordings. Surface samples of sediments (grabs and cores) were taken at five of the banks. An important discovery made in 1975 was the existence of a layer of turbid water that blankets the continental shelf and surrounds and/or covers all of the banks examined on the South Texas Outer Continental Shelf. The layer of turbid water is associated with increased sedimentation and limited light penetration, both of which have a profound influence on the biota of the banks and the sediments on and around them. This turbid layer is known as the nepheloid layer.

A second investigation, initiated in 1976 (Bright and Rezak, 1978a), extended the mapping program to three more banks and included additional submersible work on four banks. At four of the seven banks investigated, studies included postdrilling environmental assessments, and at two banks the quantitative ecological relationship between the nepheloid layer and epibenthic community population dynamics was investigated. An important discovery during this investigation was the presence of a high-salinity brine lake at the East Flower Garden Bank. The brine lake has provided a unique opportunity for the study of the effects of natural brine discharges. Its presence also provided information on the nature of salt tectonism at the East Flower Garden Banks and other banks near the shelf break.

Studies of the nepheloid layer at various banks and studies of the brine lake at the East Flower Garden Bank were continued by Bright and Rezak in 1977 (Bright and Rezak, 1978b). Additional mapping studies in 1977 provided physiographic and subbottom data on eight more banks, seven of which were observed and sampled from the submersible. A biological monitoring study was also initiated for the first time within the living coral facies of the East Flower Garden Bank. Biological monitoring of the East Flower Garden coral reef was continued in 1978, as were studies of the nepheloid layer at selected banks (Rezak and Bright, 1981a). Mapping and submersible observations were undertaken at nine banks not previously studied, among which was the West Flower Garden Bank, where monitoring studies identical to those at the East Flower Garden Bank were initiated.

During the 1978–1980 study (Rezak and Bright, 1981a) several technological changes were made. Provision for seismic and side-scan sonar equipment on mapping cruises made possible the preparation of a series of seafloor roughness maps and structure/isopach maps for several of the banks. Color video cassette recordings were made of the submersible observations, and the biological monitoring study at the Flower Garden Banks instituted several experimental techniques that permitted quantitative statistical analysis.

Significant advances were also made in the instrumentation used for hydrographic studies of the nepheloid layer. The deployment of current meter moorings and the development of a sophisticated new system for simultaneous hydrographic measurements have created a large data base for the measurement of turbidity, current velocities, temperature, and salinity in the region of the Flower Garden Banks.

Investigations from 1979 through 1981 focused on biological and hydrologic monitoring at the East and West Flower Garden Banks, and on the geological analysis and interpretation of subbottom, side-scan, and sedimentological data from these two banks (McGrail, Rezak, and Bright, 1982). Subbottom and side-scan data for seven selected banks were also analyzed and interpreted; with one exception, these data had been acquired in previous studies.

Geological studies at the Flower Garden Banks produced a sediment-distribution map for the Flower

Garden region and examined the relationship between sediment facies and biotic zones. Use of the EG&G Seafloor Mapping System made it possible to prepare a side-scan sonar mosaic for the West Flower Garden Bank. This mosaic is for all intents and purposes a photographic representation of the seafloor.

Biological investigations continued to be directed toward assessment of the health of biotic communities at the two banks. Biotic zonation maps were developed from direct observations and data were gathered in the course of the continuing monitoring program. Identical coral ecology studies were carried out at the East and West Flower Garden coral reefs and the results were compared.

Investigations of water and sediment dynamics at the Flower Garden Banks had three goals: (1) to study the hydrographic climate (salinity, temperature, turbidity, and currents) in which the banks exist; (2) to develop an understanding of the dynamics of the nepheloid layer, particularly as it impinges on the shelf-edge banks; and (3) to ascertain the nature of the shelf-edge flow, including the driving mechanisms.

The present study is a synthesis of data from the five previous Texas A&M studies and the published record. Chapters 1–3 provide a regional setting for the geology, biology, and hydrography of the Texas–Louisiana Outer Continental Shelf. Chapters 4–6 describe the geology, biology, and hydrology of the Flower Garden Banks, which have been most intensively studied and are used as a model to which other banks on the outer continental shelf are compared in the final part of the report. Chapter 7 is an attempt to categorize the many banks that we have examined and to establish their relationship to the dynamic system that we have described at the Flower Garden Banks.

Richard Rezak
Thomas J. Bright
David W. McGrail
College Station, Texas
January 1985

Acknowledgments

We wish to express our gratitude to the U.S. Department of the Interior, Minerals Management Service (formerly Bureau of Land Management) for funding the investigations that have resulted in the writing of this book. Funding was provided under the following contracts: 08550-CT5-4, AA550-CT6-18, AA550-CT7-15, AA550-CT8-35, AA851-CT0-25, and AA851-CT1-55. More than 80 graduate students contributed to the successful completion of this study. Their efforts, both at sea and in the laboratory, are gratefully acknowledged. They are listed as contributors in the final reports for the above contracts, which are available from the National Technical Information Service (NTIS).

Thanks are also due Arnold Bouma, William Bryant, T. K. Treadwell, Thomas Hilde, George Sharman, Eric Powell, and Linda Pequegnat for their contributions to the study. Rose Norman served ably as managing editor for the project. The captain and crew of the R/V GYRE provided exceptionally professional seamanship in support of our field work. The marine and electronics technicians and personnel of the Marine Operations facility at Texas A&M University also lent excellent support to our field efforts. Michael Carnes and Doyle Horne provided computer programming data analysis, numerical modeling, and the writing of software for the PHISH System. James Stasney designed and constructed the PHISH System's hardware, calibrated instrumentation, and maintained all electronics equipment. Sandra Drews provided word-processing skills to reformat and make numerous text corrections on this manuscript.

The crew of the research submersible DIAPHUS, to whom the book is dedicated, deserve special thanks for their professional execution of a most critical logistical component of our program.

Last but not least we wish to acknowledge the efforts of William J. Merrell, Jr., Program Manager, and Sylvia C. Herrig, Assistant Program Manager of the MMS Synthesis Project in their attempts to keep three wayward scientists on schedule.

R.R.
T.J.B.
D.W.M.

CONTENTS

REEFS AND BANKS OF THE NORTHWESTERN GULF OF MEXICO

1

REGIONAL GEOLOGIC SETTING

To establish an understanding of the processes active on the outer continental shelf (OCS) this chapter is devoted to a general description of the Gulf of Mexico, with special emphasis on the physiography, sediments, and structure of the shelf in the northwestern Gulf. The Gulf of Mexico is a small ocean basin which, with the four basins of the Caribbean, constitutes the American Mediterranean (Harding and Nowlin, 1966). The Gulf covers an area of about 1.5 million km^2 and has a maximum depth of about 3700 m. Figure 1.1 illustrates the bathymetry and major physiographic provinces of the Gulf.

PHYSIOGRAPHY OF THE GULF OF MEXICO

Continental Shelf

The continental shelf is the submerged portion of the continental platform that slopes gradually to depths between 100 and 200 m. It is widest off southern Florida (about 300 km) and narrowest off the modern Mississippi Delta (about 10 km). The West Florida and Yucatan shelves are characterized by carbonate sediments, whereas the East Mexico and Texas–Louisiana shelves are composed primarily of terrigenous sediments (Figure 1.2). Local areas of carbonate sediment production exist from Veracruz to the Mississippi Delta (Rezak and Edwards, 1972).

Continental Slope

The continental slope is variable in degree from place to place in the Gulf. In general, the upper slope shows a slight increase in declivity from the shelf. Throughout most of the Gulf, however, the lower part of the slope is an escarpment. The West Florida Escarpment, one of the steepest submarine slopes known, attains a declivity of about 39° (810 m/km). The Campeche continental slope is somewhat less steep, whereas the slope off Texas and Louisiana is considerably less so. The Campeche Escarpment forms the lower part of the Campeche continental slope and the Sigsbee Scarp forms the base of the Texas–Louisiana continental slope. The irregular topography of the Texas–Louisiana continental slope and the Gulf of Campeche (Figure 1.1) is due to salt diapirism; the Sigsbee Escarpment is believed to be a wall of salt (Amery, 1978). The slope off the East Mexico shelf is characterized by a ridge system that is believed to be due to shale ridges at depth (Watkins et al., 1978).

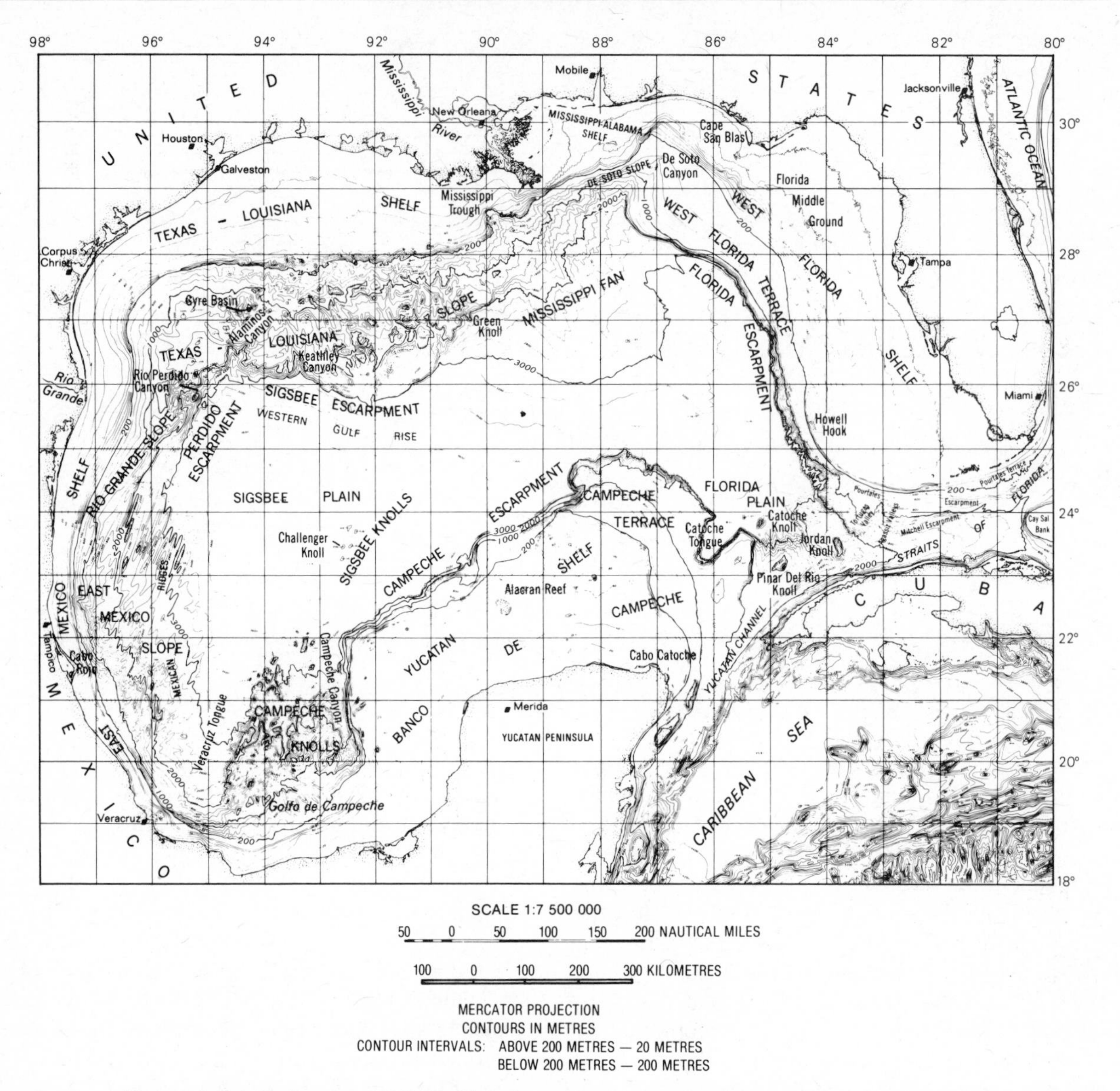

Figure 1.1. Bathymetry of Gulf of Mexico region showing physiographic provinces and major topographic features. Contour intervals: 0 to 200 m in 20-m isobaths; > 200 m in 200-m isobaths. Reprinted with permission from Martin and Bouma (1978).

Mississippi Fan

The Mississippi Fan is a broad feature that covers about 160,000 km^2 in the bathyal and abyssal depth ranges. The fan has developed since Pleistocene time by the transport of sediment from the Mississippi River through the Mississippi Trough by slumping, debris and turbidity flows.

Continental Rise and Sigsbee Abyssal Plain

The continental rise in the Western Gulf descends gradually from the base of the continental slope and merges with the Sigsbee Abyssal Plain. The Sigsbee Plain is a flat surface with a maximum depth of 3700 m. It is underlain by turbidites and interbedded pelagic oozes.

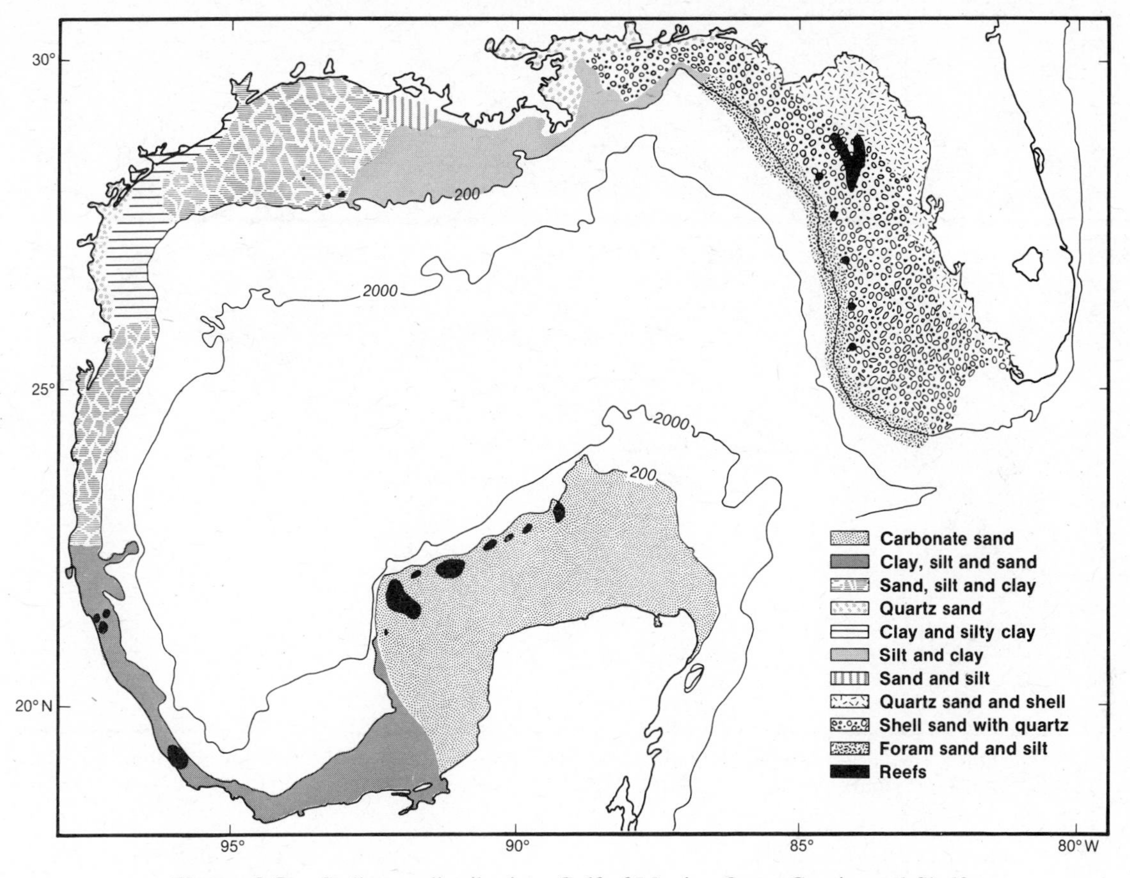

Figure 1.2. Sediment distribution, Gulf of Mexico Outer Continental Shelf.

GEOLOGIC HISTORY OF THE GULF OF MEXICO

Mesozoic Era

The basic structure and stratigraphic framework of the Gulf of Mexico was formed by events that took place during the Late Triassic and Jurassic periods (Salvador, 1980). Late Triassic and Early Jurassic time saw continental conditions in the area now occupied by the Gulf of Mexico. As the North American Plate began to move away from the African and South American plates, tensional grabens filled with red beds and volcanics began to form in the area. Because of continued subsidence, Pacific waters began to encroach by way of Central Mexico in the latter part of the Middle Jurassic, during which time and in the early part of the Late Jurassic the area was intermittently covered by shallow seawater that evaporated and produced the extensive salt deposits known today as the Louann Salt. Connection with the Atlantic Ocean was finally established late in the Kimmeridgian (Late Jurassic). At that time evaporite conditions ceased and the deposition of shallow marine limestones began. It continued into the Middle Cretaceous (Aptian to Cenomanian) and culminated in an extensive shelf-edge reef complex composed mainly of rudistids, corals, algae, and foraminifers. The reef complex occurs in the subsurface of south Texas, trends northeastward into southern Louisiana, then southeastward to the West Florida Escarpment, where it has been sampled by rock dredging (Antoine et al., 1967; Bryant et al., 1969).

During Late Cretaceous time there was a marked rise in sea level and the sediments of the northern

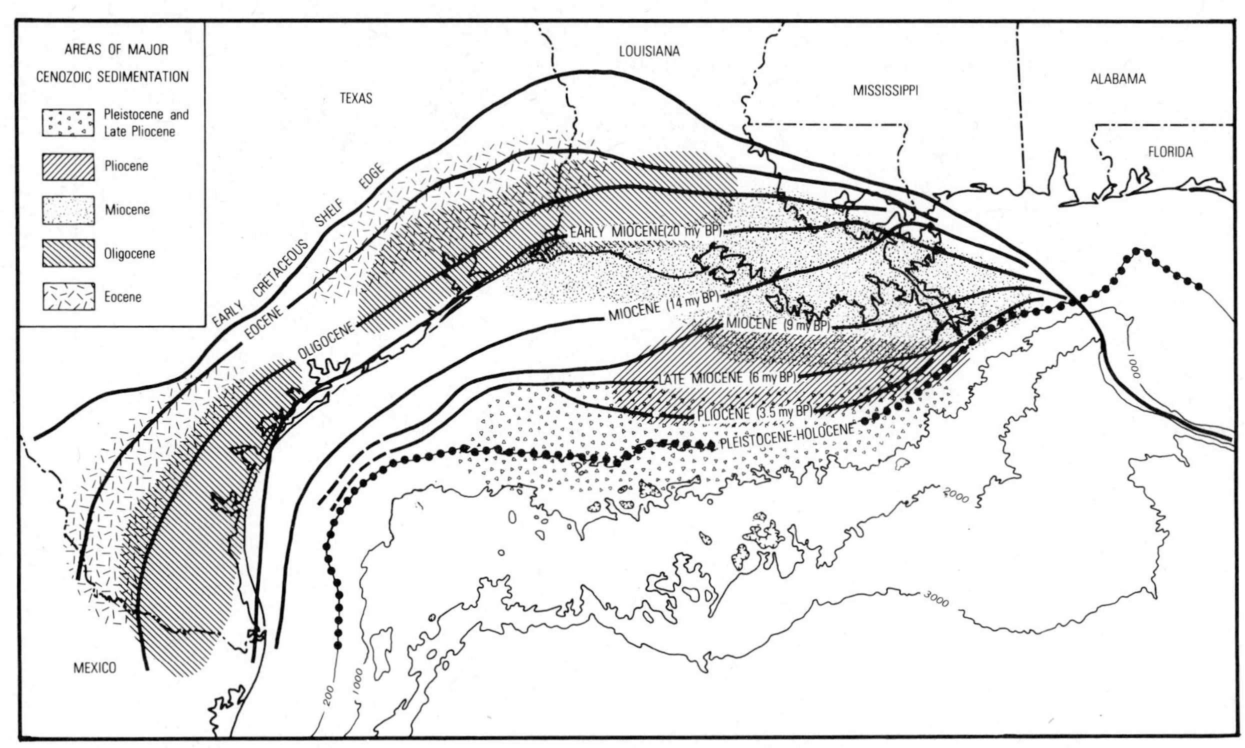

Figure 1.3. Sketch map showing paleoshelf edges in Gulf Coast basin and distribution of major Tertiary depocenters. Reprinted with permission from Martin (1978).

Gulf became mainly sandstones, shale, marl, and chalk.

Cenozoic Era

The most significant feature of post-Mesozoic history of the Gulf has been the tremendous seaward growth of the Texas–Louisiana Continental Shelf. Early in Cenozoic time the Gulf began to receive the detritus from the Laramide orogeny to the west and northwest. The major transporters of this sediment were the Mississippi and Rio Grande Rivers. The supply of sediment was so great that the rate of subsidence could not accommodate the great volume of material. As a consequence, the shelf edge prograded by as much as 400 km from the edge of the Cretaceous shelf to the present shelf break (Figure 1.3). This great wedge of sediment is illustrated in Figure 1.4. Note that the thickness of the sediments near the present coastline is approximately 15 km.

TEXAS–LOUISIANA CONTINENTAL SHELF

Physiography

The northwestern Gulf of Mexico shelf is dominated by one major (Mississippi) and two minor (Brazos–Colorado and Rio Grande) deltas. Other streams in the area are building deltas in bays and estuaries behind barrier islands (Figure 1.5). The sediments on the continental shelf reflect the environments of the adjacent coasts. This is to be expected because sea level has risen approximately 130 m during the past 16,000 to 18,000 years and these same environments have migrated across the shelf to their present positions. Consequently, we have adopted the areas delineated by Curray (1960) to subdivide the sediments of the shelf and adjacent coastal areas (Figure 1.5).

Central Louisiana Area. Among the four areas Curray (1960) defines, the Central Louisiana Area

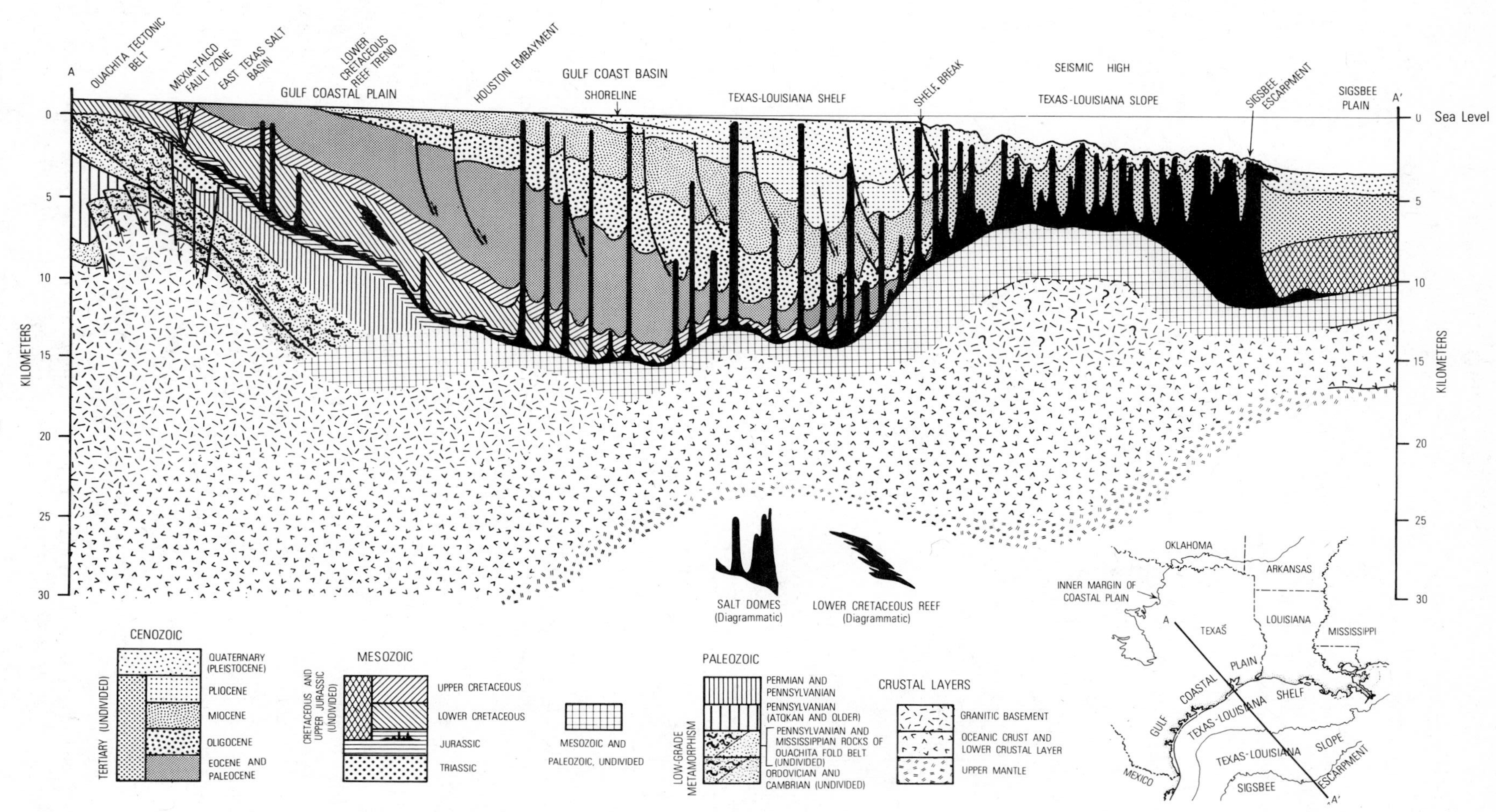

Figure 1.4. Generalized cross section of northern Gulf of Mexico margin. Reprinted with permission from Martin (1978).

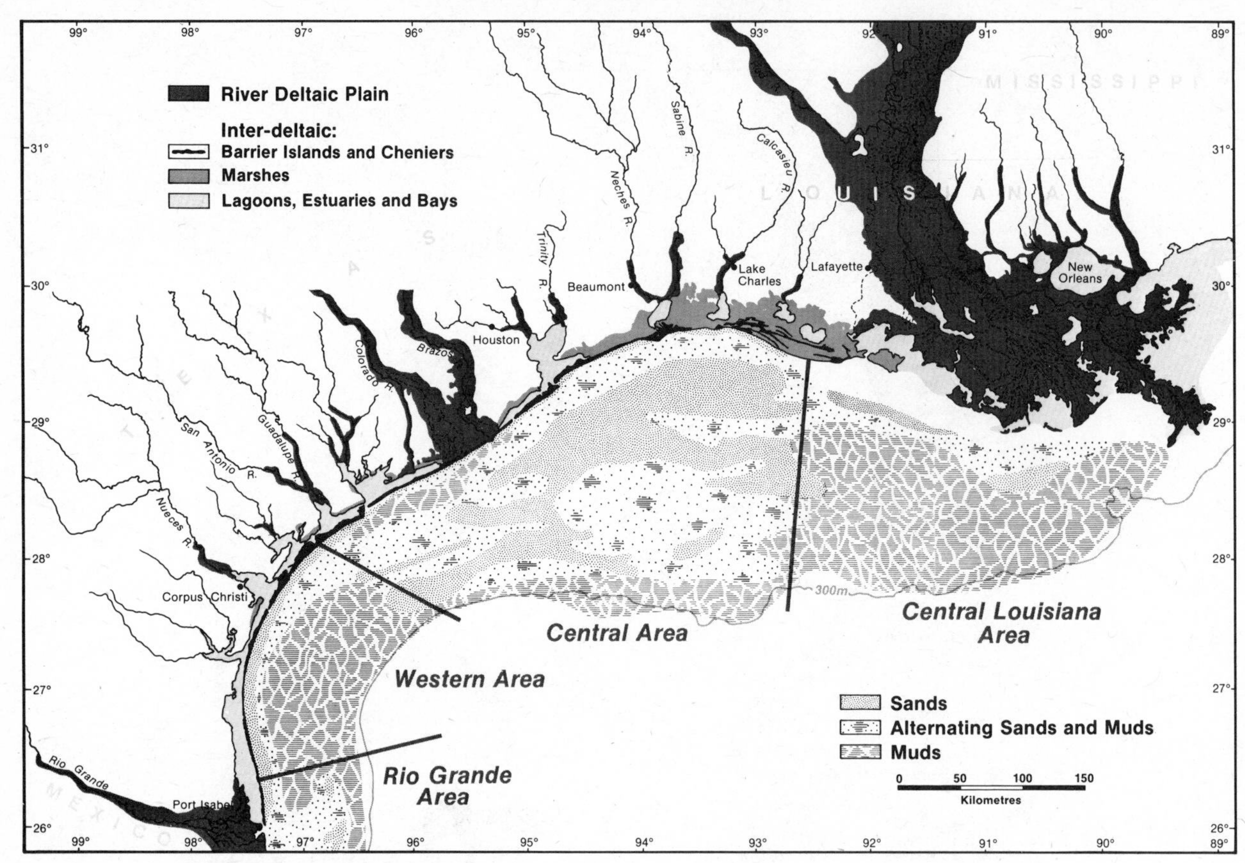

Figure 1.5. Coastal environments and shelf sediments, Texas–Louisiana Continental Shelf. Sediment distribution, after Curray (1960).

is most affected by the Mississippi River outflow (Figure 1.5). The western edge of this area is White Lake, just west of Vermilion Bay. The outflow of the Mississippi has migrated back and forth along this portion of the Gulf coast since Early Tertiary time. During the Miocene the outflow was close to Sabine Lake; since that time it has migrated eastward to Mississippi Sound and then back toward the west and its present position. Subbottom profiles in these areas show an abundance of sand-filled distributary channels with interdistributary swamp and lake deposits between them. Close to the shelf break, particularly near the head of the Mississippi Trough, large slumps and debris flows are common. Shoal areas, such as Tiger and Trinity off Vermilion Bay and Ship Shoal farther to the east, are probably remnants of Late Pleistocene deltas.

Central Area. To the west of the present delta, from just west of Vermilion Bay westward to Sabine Lake, lies the area known as the Chenier Plain (Figure 1.5). This plain consists of low ridges of sand separated by marshy swales underlain by sandy muds. The sand ridges (cheniers) are beach deposits that have been formed in a rather unusual manner. LeBlanc (1972) presented a conceptual development model for these beach ridges. Shoreline accretion is mainly by massive influxes of mud and sand from the Mississippi River Delta, eroded and transported westward into the Chenier Plain, where they are deposited as muddy tidal flats. Repeated minor fluctuations in sea level over long periods of time eroded the margin of the mud flats and winnowed the fine sediment particles, leaving a lag of fine sand and silt, with coarser shell debris along the shoreline. Later, renewed influx of muds moved the shoreline farther seaward and the cycle was repeated. The rapid rate of sedimentation in this area has effectively sealed streams like the Calcasieu and Mermentau Rivers and created vast coastal marshes and lakes.

The coast between Sabine Pass and Port Isabel

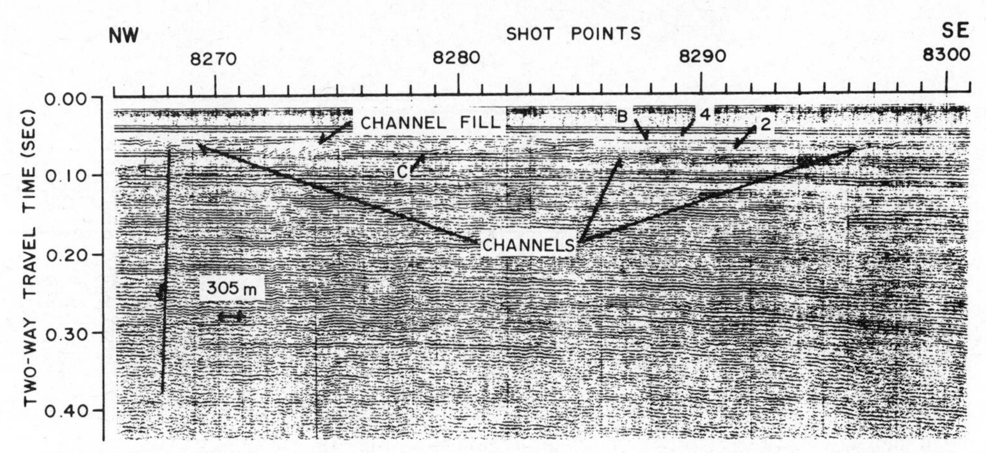

Figure 1.6. Sparker profile showing channeling in Holocene sediments off Matagorda Bay (Pyle, 1977). Vertical exaggeration, 12.5X.

is characterized by barrier islands that separate lagoons and embayments from the main body of the Gulf of Mexico. Most of the streams that flow into the Gulf along this stretch of coastline do not drain large areas, and the climate varies from subhumid in the neighborhood of Galveston to semiarid south of Corpus Christi.

The Central Area extends to the southwestern tip of Matagorda Island. The Brazos–Colorado deltaic plain has filled in the lagoon behind the barrier island from the east end of Matagorda Bay to the west end of Galveston Bay. This process may occur rapidly; for example, the Colorado River built its delta across the eastern area of Matagorda Bay to the Matagorda peninsula in 25 to 30 years, between approximately 1930 and 1956. Since 1956 it has been building a delta outside the barrier bar, but strong longshore currents carry a large amount of this sediment southwestward along the coast.

Western Area. The Western Area, which extends from the southern tip of Matagorda Island to a point about 10 miles north of Port Isabel, is characterized by an almost continuous barrier island and the absence of any major streams. The semiarid nature of its climate precludes any significant transport of sediment to the shelf along this stretch of coastline. This area, which lies between the deltas of the Brazos–Colorado complex and the Rio Grande River, could be designated as the South Texas interdeltaic plain. Shideler (1977) shows that the shelf sediments range from silty sands to muds. In the central part, off Corpus Christi and Baffin Bay, the sediments are primarily clayey silts. The coarser sediments are close to the northern and southern limits of Shideler's study area, near the deltas. Subbottom profiles also show distinctive features of the interdeltaic plain. Pyle (1977) illustrates the subbottom structure by using minisparker records. Figure 1.6 (Pyle's Figure 16), a line across part of the ancestral Brazos–Colorado Delta, indicates abundant channeling in the youngest seismic unit. His Figures 15 and 19 show lines within the interdeltaic plain that indicate no channeling in the youngest seismic unit. Figure 1.7 is a reproduction of Pyle's (1977) Figure 29 which shows a minisparker profile across Baker Bank. This line reveals the absence of channels and the presence of regular reflectors typical of an interdeltaic area.

Rio Grande Delta Area. The Rio Grande Delta was actively built during Pleistocene and Holocene time. Until recent times the Rio Grande River supplied a significant amount of sediment to the continetal shelf, but after the construction of numerous dams on the river, the amount of sediment transported by the Rio Grande decreased significantly. Berryhill (1981) illustrates a major stream valley (possibly the Nueces River) that flowed southward along the inner shelf during the latest Wisconsin low-sea-level stand. This stream joined with the Rio Grande and contributed sediment to the ancestral Rio Grande Delta.

Hard Banks. Hard banks are distributed over the entire shelf from the Mississippi Delta to Port Isabel,

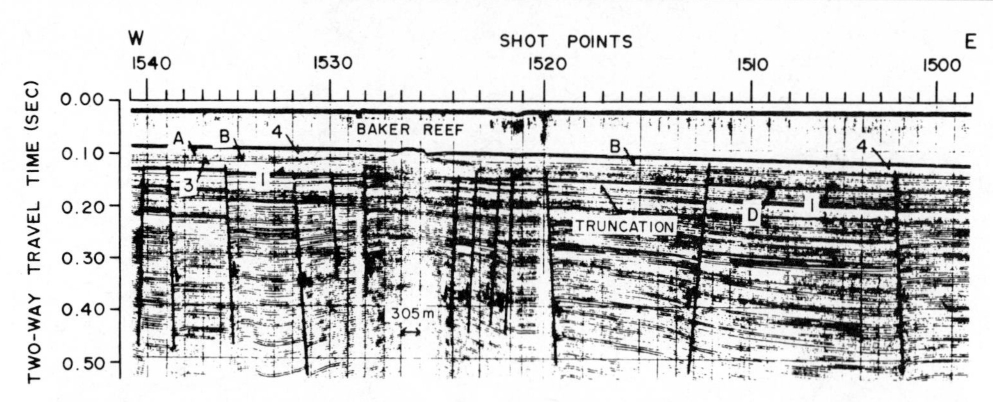

Figure 1.7. Sparker profile showing Baker Bank, a typical South Texas Outer Continental Shelf bank (Pyle, 1977). Note the continuous reflectors beneath the bank. Vertical exaggeration 11X.

Texas. Several hundred hard banks are known to exist on the continental shelf and upper continental slope. Most of the banks are associated with salt diapirs, but many are not. Some are bare Tertiary and Cretaceous bedrock, whereas others are heavily encrusted by organisms associated with coralgal reefs (reefs built from coral and coralline algae; Bathurst, 1975).

The coralgal reefs may be living, as they are at the Flower Garden Banks, or they may be drowned, as they are on the South Texas shelf, due to the presence of turbid bottom water. A line drawn from Port O'Conner (near Matagorda Bay) toward the shelf break, in a southeasterly direction, separates two major hard-bank areas. To the south lie numerous mid-shelf banks that are not associated with salt domes; to the north and east are mid- and outer-shelf banks that are mostly involved with salt tectonics, although some may be growing on shale diapirs.

Several classifications of banks have been based on combinations of three factors: (1) position on the shelf (nearshore, mid- and outer shelves); (2) total relief on the bank; and (3) water depths adjacent to the bank. Aside from the fact that there are no high relief features in nearshore, shallow water and that some of the banks in deeper water have high relief, there is little relationship between any of these factors. Location on the shelf is a key factor however: (1) it determines the maximum amount of surface relief on the bank; (2) it strongly affects maximum and minimum water temperature; and (3) relief and temperature control the distribution of the biotic assemblages on the banks. Position on the shelf also reflects the maturity of the underlying salt diapir (see Chapter 8).

Recent Sediments

The normal sediments on the Texas–Louisiana Continental Shelf are land-derived sandy muds with varying amounts of locally produced skeletal material. The sources of the land-derived sediment are the major streams that flow into the northwestern Gulf of Mexico. The annual discharge of the Mississippi river, the major contributor of sediment, is 497 billion kg, of which 45% is clay, 36% silt, and 19% is very fine to fine sand (Everett, 1971). The other rivers of the region contribute minor amounts to the Gulf because their deltas are mostly in estuaries behind the barrier islands and beaches that are more or less continuous from the Calcasieu River to the Rio Grande. Curray (1960), in describing the distribution of sediments on the northwest Gulf of Mexico Continental Shelf, divided the shelf into four areas: (1) the Rio Grande, (2) the Western, (3) the Central, and (4) the Central Louisiana (Figure 1.5.). The Rio Grande and Central areas are characterized by sandy sediments. The Central Louisiana and Western areas are underlain primarily by muds. A significant conclusion that resulted from API Project 51 (Curray, 1960) is that most of the transport of sand is only in areas close to the shoreline and that sand is not being transported from the rivers to any part of the mid- and outer shelves between the Mississippi and Rio Grande Deltas. Silt and clay, on the other hand, are distributed independently of the sand. Sands on the middle and outer shelves are relict Pleistocene and Holocene sediments.

Much has been written about the sources of sediments on the Texas–Louisiana Shelf (Van Andel, 1960; Van Andel and Poole, 1960; Davies and Moore, 1970; Shideler, 1978; Berryhill, 1975), and the consensus is that most of the sediment is derived from the Mississippi and Rio Grande Rivers, with some contribution from the Brazos–Colorado Delta (Berryhill et al., 1976; Davies and Moore, 1970). Heavy mineral studies cannot determine unambiguously the extent to which a certain source has contributed to a sediment or if indeed it has at all (Hawkins, 1983). In his study of the sediments in the Central area, Hawkins (1983) used quartz-grain-shape analysis to differentiate among the sources that contribute to sediments on the shelf. The purpose of his study was to compare the results of heavy mineral counts and quartz-grain-shape analysis (Ehrlich and Weinberg, 1970) to determine sources of fine sand in the area. Hawkins concluded that much of the sand in the Central area is a mixture of Colorado River, Brazos River, Trinity River, and Red River sands, with little influence in the fraction from the Mississippi River. This conclusion that the sediments in this area are the result of the Pleistocene progradation of deltas across the shelf agrees with those of Winker (1982) and Curray (1960).

There has been much speculation concerning the rates of sedimentation on the Texas–Louisiana Continental Shelf. Van Andel and Curray (1960) suggested that rates will vary with time, depending on the state of equilibrium between the rate and type of sediment supplied by the river and the rate of winnowing and redistribution, by marine agents. Among the deltas formed by streams flowing onto the Texas–Louisiana Shelf, the Mississippi River area has the highest rate of sedimentation and progradation because of the dominance of fluviatile over marine processes. The Rio Grande and Brazos–Colorado Rivers, on the other hand, do not transport the volume of sediment that the Mississippi does; consequently marine processes dominate. Most of the sediment is removed from the area, leaving a narrow zone of littoral deposits along a nonprograding delta front. Other streams, like the Nueces, San Antonio, Guadalupe, Trinity, Neches, and Sabine Rivers, flow into bays that are separated from the Gulf of Mexico by bay-mouth bars and barrier islands. These streams have little or no influence on continental shelf sedimentation. As a consequence, the large region designated as the Central Area (Figure 1.5) by Curray (1960) has low sedimentation rates, and on some parts of the shelf no sediments are being deposited at the present time. Curray's Central Area is underlain by the relict Pleistocene and Early Holocene deltaic deposits described earlier in this chapter.

Recent geochemical studies of sediments on the Texas Continental Shelf between Mexico and Matagorda Bay have used the ^{210}Pb method to determine sedimentation rates (Berryhill and Trippet, 1980, 1981a, 1981b; Holmes, 1982). Unfortunately, the Berryhill and Trippet maps do not include text, and the Holmes paper is rather cryptic in that his maps are inconsistent with one another and with his text. Holmes (1982) states that among the 36 cores analyzed, 16 showed a constant decrease in the log activity of ^{210}Pb versus depth, thus indicating a constant rate of sedimentation. Fifteen cores had no excess ^{210}Pb, which indicates that no sediment has accumulated in the area of the 15 cores for at least the past 150 years. Yet Holmes mid-shelf minimum sedimentation rate (his Figure 5) occurs in the same area as his high-surface ^{210}Pb activity (his Figure 6). Holmes states that the ^{210}Pb activity in the sediments of the continental shelf ranged from 25 to 1.0 dpm/g (decays per minute per gram). Holmes Figure 6, however, shows isopleths that range from only 6 to 20 dpm/g.

A disturbing feature of the maps in the Berryhill and Trippett publications and the paper by Holmes is the lack of conformity of the sedimentation-rate isopleths to bathymetric contours. One would expect such high rates of sedimentation to be reflected in the bathymetry, but for some unexplained reason they are not (see Berryhill and Trippet, 1980, 1981a, 1981b; Holmes, 1982, figure 5).

Classification of Sediments

At this point we need to discuss the principles of classification and how they apply to the sediments of the Texas–Louisiana Shelf. These sediments range in composition from terrigenous to mixtures of terrigenous and carbonate skeletal sediments to pure carbonate sediments. Each of the three kinds has its own classification problems.

Sediments may be classified according to texture, mineralogy, or genesis. A textural classification is used to describe terrigenous sediments because they have been subject to transport from the continent by moving fluids. Determination of the particle-size distribution in these sediments allows interpretation of the process of transportation and the velocities required to transport them. A greater flow velocity is required to transport a sand grain than a grain of silt.

The classification of terrigenous sediments in general use by sedimentologists today is that of Folk

(1974). In his classification scheme Folk used the grade scale devised by Wentworth (1922). According to this grade scale, the diameters of the sediment particles are as follows:

Gravel	> 2.0 mm	
Sand	0.0625–2.0 mm	
Silt	0.0020–0.0625 mm	Mud
Clay	< 0.0020 mm	Mud

Folk places major emphasis on the presence of even minute quantities of gravel because he regards the proportion as a function of the highest current velocity at the time of deposition. Consequently, even a trace (0.01%) is enough to term the sediments "slightly gravelly." This emphasis on the importance of gravel creates a problem when dealing with sediments that are mixtures of land-derived detritus and locally produced skeletal matter; for example, if an echinoid living on the bottom dies and its skeleton is buried by mud, sampling at that site will yield a sediment that consists of mud and the dissociated plates of the echinoid skeleton. In the analyis, these plates could conceivably amount to 5 or 6% of the sediment, which would require it to be classified as gravelly mud. Yet the presence of 6% gravel is in no way related to the current velocities at the time the sediment was deposited. Present studies indicate that the amount of gravel in the sediment on the OCS is not a function of the highest current velocities at that site but rather proximity to a reef, either living or drowned. This concept has not been understood by those who cite the presence of large amounts of gravel at depths of 60 to 100 m as an indication of strong bottom currents. The consequences of this erroneous reasoning have great bearing on the theorized fate of pollutants introduced into the bottom boundary layer by shunting cuttings and mud from drilling platforms.

In carbonate sediments, which are produced by biological activity and accumulate more or less *in situ*, textural analysis is of little value in the interpretation of their origin. Carbonate particles, either whole or fragmented skeletons, give carbonate sediments a clastic texture that may be described in terms of the terrigenous sediment classification; however, the interpretation based on that classification may be completely erroneous because it is impossible to distinguish between mechanically deposited corbonates (calcarenites) and sediments that accumulate by *in situ* deposition and fragmentation of skeletons (bioclastics) (Logan, 1969). The only textural measurement useful for *in situ* accumulations of carbonate sediments is the ratio between the amount of sediment finer than and coarser than 0.0625 mm, which is indicative of the current regime at the depositional site, a clean skeletal sand that suggests currents strong enough to remove silt- and clay-sized particles from the site, and a muddy sand that indicates currents too weak to remove silt- and clay-sized particles. A knowledge of the nature of the constituent particles is basic to an understanding of the origin of carbonate sediments. All sediments are intimately related to the fauna and flora from which they derive and the name of the carbonate sediment facies is taken from its dominant skeletal component.

Late Quaternary Sea-Level Fluctuations

Poag (1973) reviewed the published record of Late Quaternary sea levels and compiled a list of 26 still stands and their ages in the Gulf of Mexico. Poag concluded that "it is fully expectable that the 'post glacial' sea level rose in cyclical pulses." However, eustatic changes in sea level due to the growth and withdrawal of the polar ice caps are complicated by isostatic and tectonic movements of the seafloor. Isostatic adjustments are made by two mechanisms. Differential loading of the shelf by seawater during high stands of sea level causes depression of the outer shelf. Lowering of sea level removes the differential load, and the outer shelf begins to rebound.

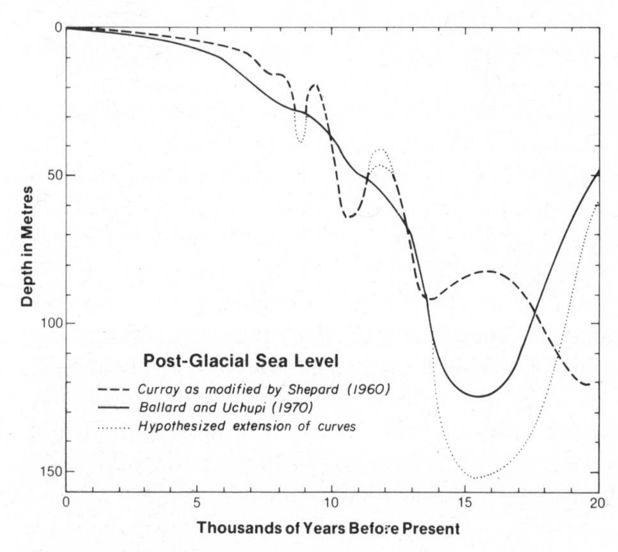

Figure 1.8. Postglacial sea-level curves.

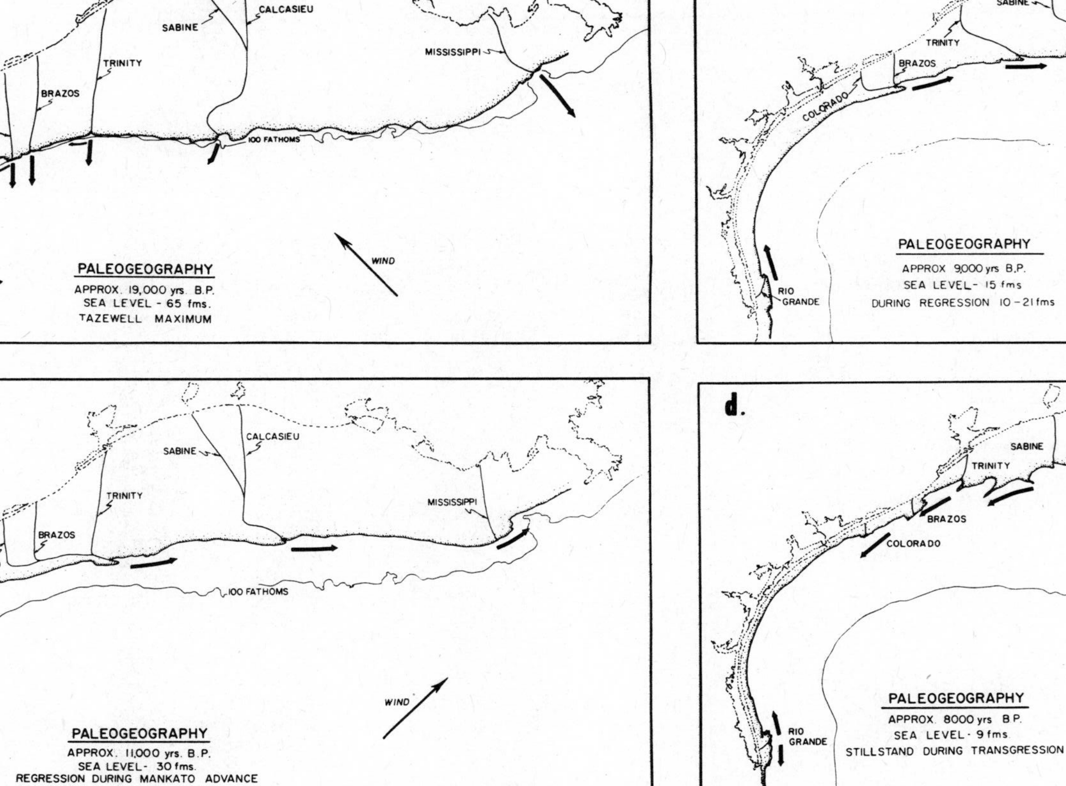
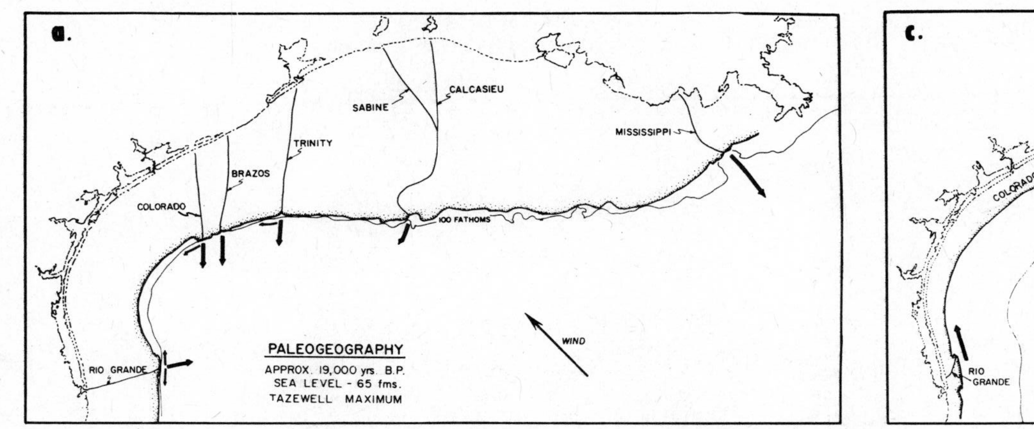

Figure 1.9. (*a*) Paleogeography of the northwest Gulf of Mexico about 19,000 years BP, when sea level is believed to have been about −65 fathoms. This correlates with the maximum of the Tazewell glaciation. (*b*) Paleogeography of the northwest Gulf of Mexico about 11,000 years BP, when sea level is believed to have been about −30 fathoms. This stage was during the regression from −22 to −35 fathoms during the advance of the Mankato glaciers of North America, following the Two Creeks interval. Wind was from the southwest during this period. (*c*) Paleogeography of the northwest Gulf of Mexico about 9000 years BP, when sea level was about −15 fathoms. This stage was during the temporary regression from about −10 to −21 fathoms. Wind was from the southwest during this period. (*d*) Paleogeography of the northwest Gulf of Mexico about 8000 years BP, when sea level was about −9 fathoms. This was a period of stillstand of sea level before the final stage of transgression to the present level. Reprinted with permission from Curray (1960).

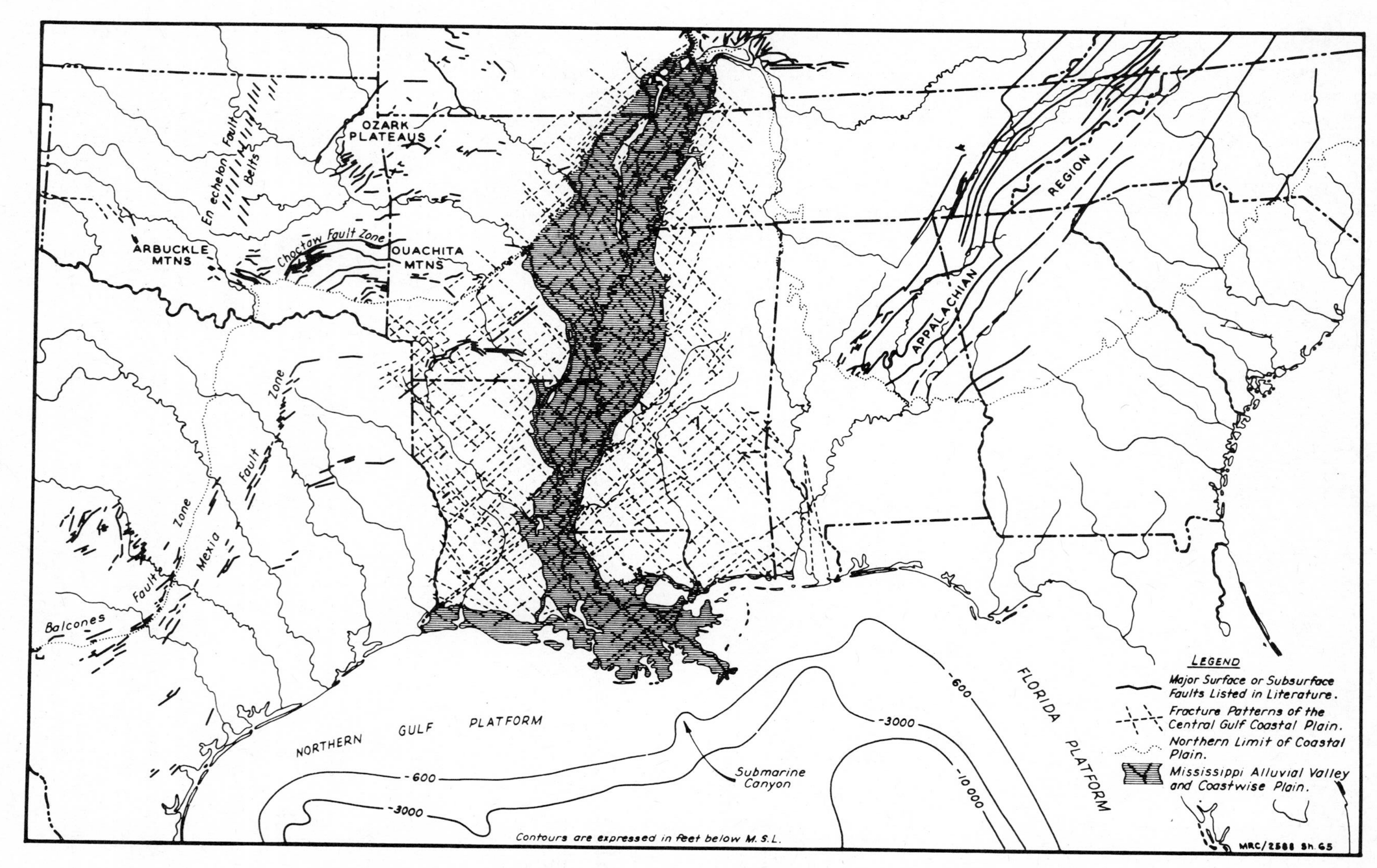

Figure 1.10. Regional fracture pattern of southern United States. Reprinted with permission from Murray (1961).

A more permanent isostatic adjustment is made by sediment loading. The Mississippi River has been disgorging its sediment load onto the Louisiana Shelf and slope since long before the beginning of the Pleistocene, and more than 3048 m (10,000 ft) of shelf sediments have been accumulated since then in the vicinity of the Mississippi Canyon. The tremendous mass of sediment has caused a major amount of subsidence in the area. McFarlan (1961), who devised a graphic method for determining the amount of subsidence due to sediment loading, estimated that the maximum structural subsidence following the deposition of approximately 270 m of "postglacial" sediment was 41 m in the vicinity of the head of the Mississippi Canyon. Structural movement decreased to zero shoreward (approximately 30°N latitude) and laterally away from the axis of the Mississippi Canyon (approximately 200 km east and west of the axis).

Tectonic movements on the Texas–Louisiana Shelf are caused by salt diapirism. Tectonism may cause local shoaling (upthrusting of salt) or local deepening (dissolution of salt at the crest of a diapir and subsequent collapse of overlying sediments). Tectonism is probably the most important factor in creating errors in sea-level curves. Broecker (1961) casts serious doubt on the validity of McFarlan's (1961) sea-level curve. At best, sea-level curves

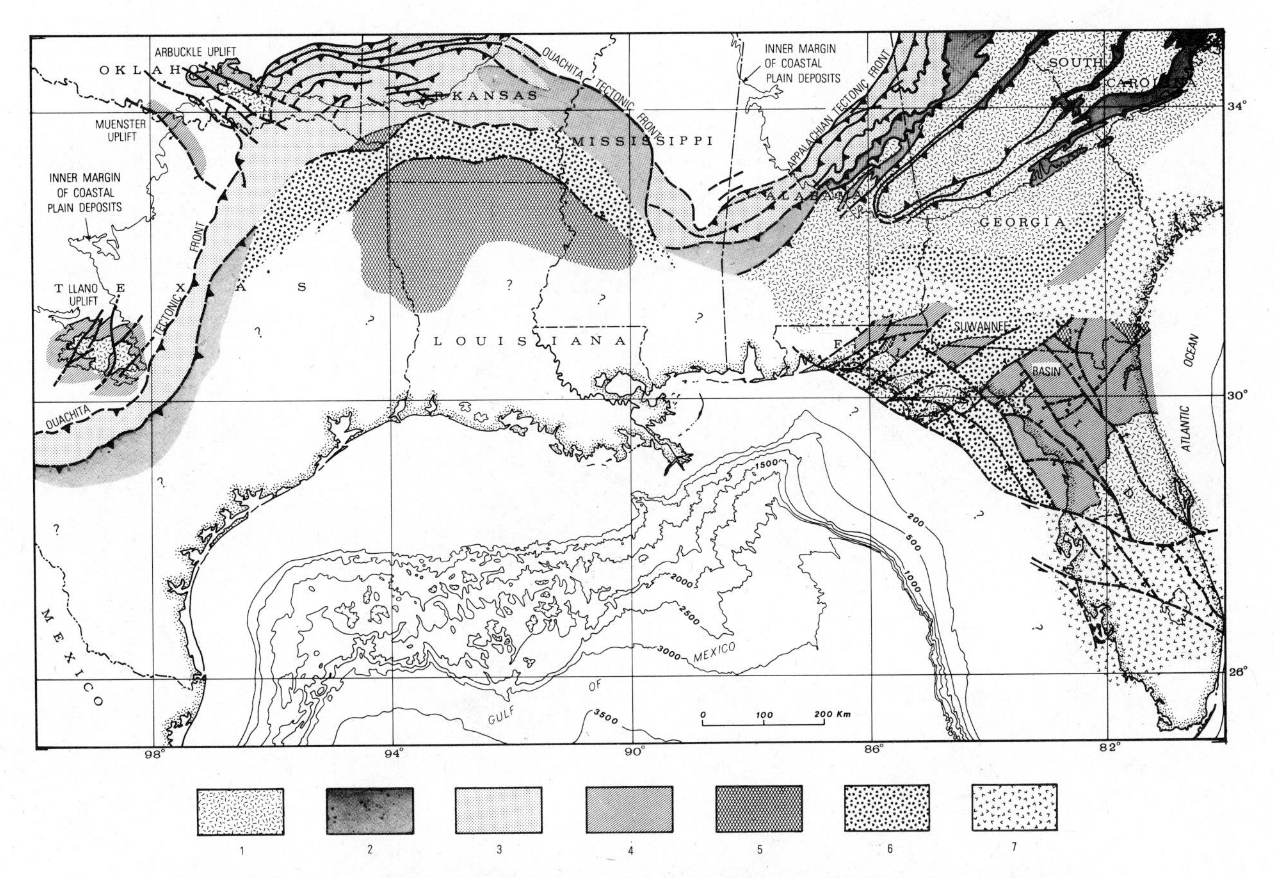

Figure 1.11. Distribution of pre-Middle Jurassic (Callovian) rocks in northern and eastern Gulf of Mexico regions. Explanation of patterns and symbols: (1) Precambrian and Paleozoic metamorphic and plutonic basement rocks; (2) Precambrian and Paleozoic metasedimentary rocks: slate, quartzite, marble, and schist; (3) Paleozoic sedimentary rocks deformed by Paleozoic orogenies; (4) Lower Paleozoic platform deposits—undeformed; (5) Upper Paleozoic platform deposits—undeformed; (6) Triassic and Jurassic graben deposits (red beds and diabase), Eagle Mills Formation, Newark Group, and equivalent; (7) Triassic and Lower Jurassic volcanic and plutonic complex, mainly volcanic rocks of Early Jurassic age. Bar and ball on downthrown side of normal fault; sawteeth on overthrust plate of reverse fault. Reprinted with permission from Martin (1978).

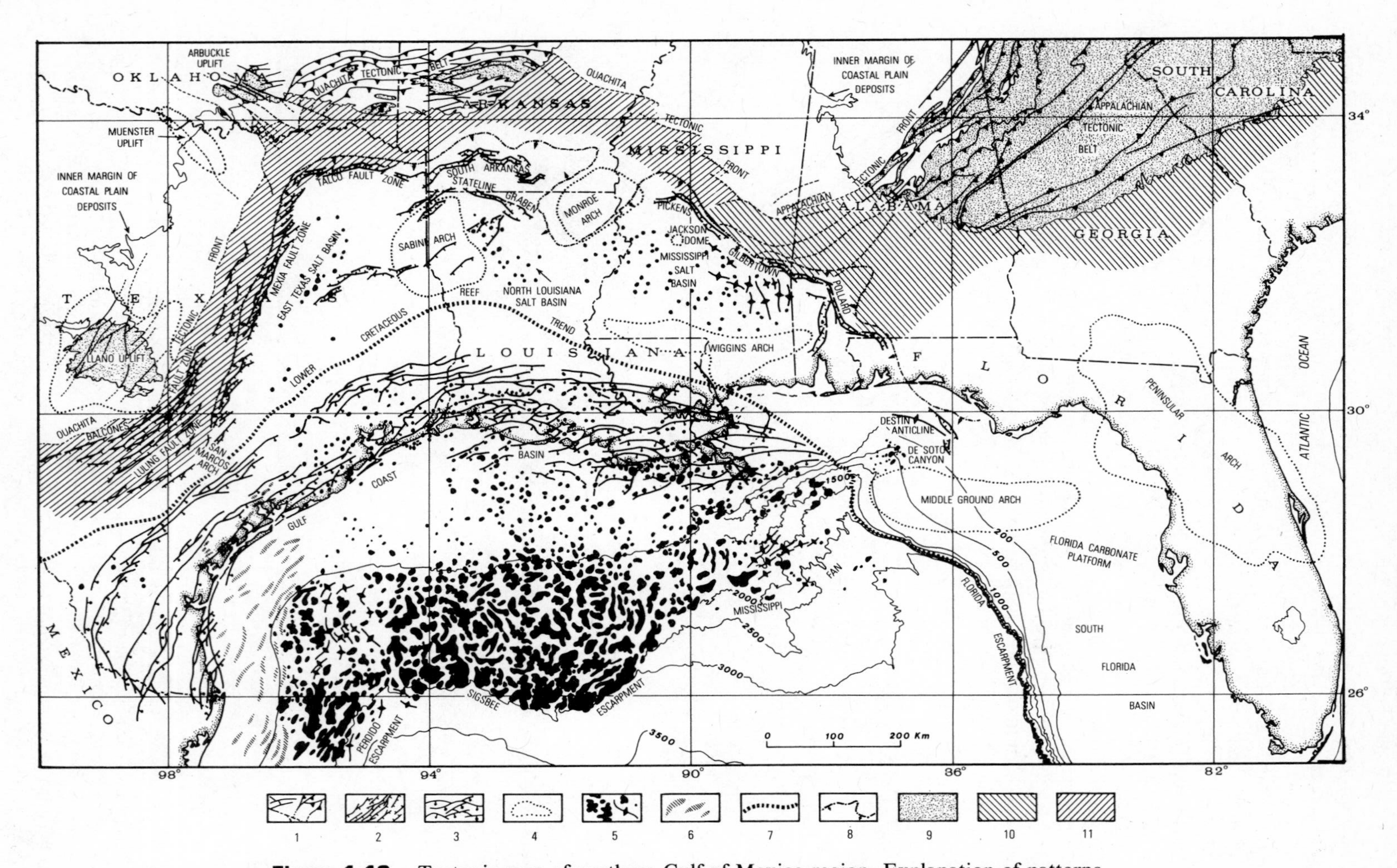

Figure 1.12. Tectonic map of northern Gulf of Mexico region. Explanation of patterns and symbols: (1) reverse fault, sawteeth on overthrust plate; (2) normal fault, hachures on downthrown side; (3) fault of undetermined movement; (4) broad anticline, or arch, of regional extent; (5) salt diapirs and massifs, indicating relative size and shape; (6) shale anticlines and swells (nondiapiric), showing general trend; (7) Lower Cretaceous reef trend; (8) updip limits of Louann Salt; (9) uplifts of exposed Paleozoic strata and crystalline basement rocks; (10) buried Ouachita tectonic belt; (11) Blue Ridge and piedmont. Scale: 1° latitude equals 110 km. Reprinted with permission from Martin (1978).

based on ^{14}C age dating alone are tenuous. Much more detailed information is needed on the regional nature and ages of uncomformities in addition to the diagenetic history of the sediments involved.

The sea-level curves of Ballard and Uchupi (1970) and Curray (1960) seem to be fairly close except for the time between about 14,000 and 20,000 years BP. Ballard and Uchupi (1970) propose a low stand of sea level of -150 m at 15,000 years BP, whereas Curray's curve bottoms out at slightly less than -120 m at 20,000 years BP (Figure 1.8). We have observed drowned reefs at about -150 m at the West Flower Garden Bank, which lends credence to that part of Ballard and Uchupi's curve. Changes in paleogeography as interpreted by Curray (1960) are shown in Figure 1.9.

Regional Structure

The major structural features on the outer continental shelf are gravity faults and salt diapirs that penetrate a thick monoclinal accumulation of Mesozoic and Cenozoic sediments (Figure 1.4). Gravity faults may be locally controlled by salt diapirs or they may be regional in extent like some growth faults that parallel the shelf break (see Figure 1.12). Faulting associated with salt diapirs develops patterns that are controlled by regional stresses (Withjack and Scheiner, 1982). Martin (1978) proposed five significant causative factors as explanations of Gulf Coast growth faults: (1) crustal loading and basement tectonics; (2) slumping along shelf edges and flexures as a result of rapid sediment accumulation; (3) salt and shale flow into local structures and systems of regional extent; (4) gravitational creep and sliding; and (5) differential compaction due to abrupt changes in sediment thickness and facies. Examination of bathymetric charts like the National Oceanographic Survey charts NH 15-11 (Bouma Bank), NH 15-12 (Ewing Bank), NG 15-2 (Garden Banks), and NG 15-13 (Green Canyon) reveals reentrants at the shelf break that coincide more or less with the alignment of salt structures on the shelf and upper slope. The reentrants are real; they have been mapped by several different investigators. The orientation of these lineaments approximates the orientation of fractures in the Mississippi Embayment (Figure 1.10) and the Triassic and Jurassic grabens in northern and northwestern Florida (Figure 1.11). The 200- and 500-m isobaths in Figure 1.11 between 90 and 94°W longitude show the orientation of the reentrants.

The similarity between these patterns and those of the pre-Middle Jurassic fault basins is striking. The modern fault systems are most likely inherited from the Early Jurassic structures in areas between the Middle Jurassic salt basins, where little or no salt was deposited. See Figure 1.12 for locations of salt basins in East Texas, Louisiana, and Mississippi. The same patchy distribution of salt could occur beneath the continental shelf. There are areas of higher and lower concentrations of salt diapirs on the continental shelf and slope off Louisiana and Texas (Figure 1.12). Lateral movement of the salt was caused by the overburdern of the thick wedge of Tertiary sediments. The salt flowed into the faulted areas underlying the Late Cretaceous sediments and rose along post-Late Cretaceous–reactivated Early Jurassic faults, which were readily available avenues, to form the salt diapirs and ridges that we now see on the continental shelf and slope.

SUMMARY

The Gulf of Mexico is a small ocean basin that originated approximately 160 million years ago (late Jurassic time) due to the rifting of the North American, African, and South American plates. The shallow, primordial Gulf has gradually evolved into its present configuration—enlarged by continued spreading, deepened by subsidence, and slowly filled by sedimentation. The spreading phase ended during the Cretaceous Period (approximately 80 to 100 million years ago). Sedimentation and subsidence will continue as long as streams flow into the Gulf from the continent and lime-secreting organisms continue to thrive in areas of low stream outflow.

The Western Gulf, from the Mississippi Delta to the Campeche Canyon on the west side of the Yucatan Shelf, illustrates clearly the influence of deltaic sedimentation on the continental shelf of the Gulf. The Mississippi River, because of its great drainage area, has been the major contributor of deltaic sediments throughout Cenozoic time and as a consequence has built an extremely large coastal plain–continental shelf complex in the northwestern Gulf of Mexico. The narrowing of the coastal plain–continental shelf complex southward into Mexico is due to the limited drainage areas of the streams that flow into that part of the Gulf.

The West Florida and Yucatan shelves are broad, shallow areas that are composed primarily of carbonate sediments produced by lime-secreting organisms. These areas of little or no continental sediment influx are underlain by thousands of meters of shallow-water carbonate sediments, which indicate that accumualtion of carbonate skeletons on the seafloor in these areas has kept pace with subsidence for many tens of millions of years.

The final structural complication in the northwestern Gulf is the formation of salt diapirs on the outer continental shelf and continental slope. It is on these structures that most of the banks and reefs discussed in this report occur.

2

CIRCULATION, HYDROGRAPHY, AND DISTRIBUTION OF SUSPENDED SEDIMENT ON THE CONTINENTAL SHELVES IN THE GULF OF MEXICO

A firm knowledge of the behavior of the flow field over the continental shelves in the Gulf of Mexico is critical to an understanding of modern sedimentary processes and the zoogeography on those shelves. Unfortunately, the quantity of high-quality direct measurements of current velocities on these shelves is woefully small. The locations of current meter moorings from which data or interpretation of data are available are shown in Figure 2.1. A substantially greater amount of hydrographic data is available, but it is clustered geographically (see Figure 2.1) and the station spacings are rather coarse for resolving anything but large scale, long-term differences. As shown in Figure 2.1, observations of suspended sediment are exceedingly sparse and are reported in such a variety of units that they are comparable only in a relative sense.

With that necessary caveat it is also appropriate to state that it is possible to deduce a great deal of useful information from the data set available in spite of its shortcomings.

FORCING MECHANISMS

A very limited number of mechanisms drive flow phenomena on continental shelves. All vary, however, with respect to the amplitude of their input, in both time and space, and all operate at the same time. Those mechanisms are astronomical forcing (tides), atmospheric forcing (primarily wind), differential heating, river runoff, and interaction with shallow flow of the deep basin.

Tides

Tides are a good place to start a discussion of forcing and flow responses because they are nearly alone

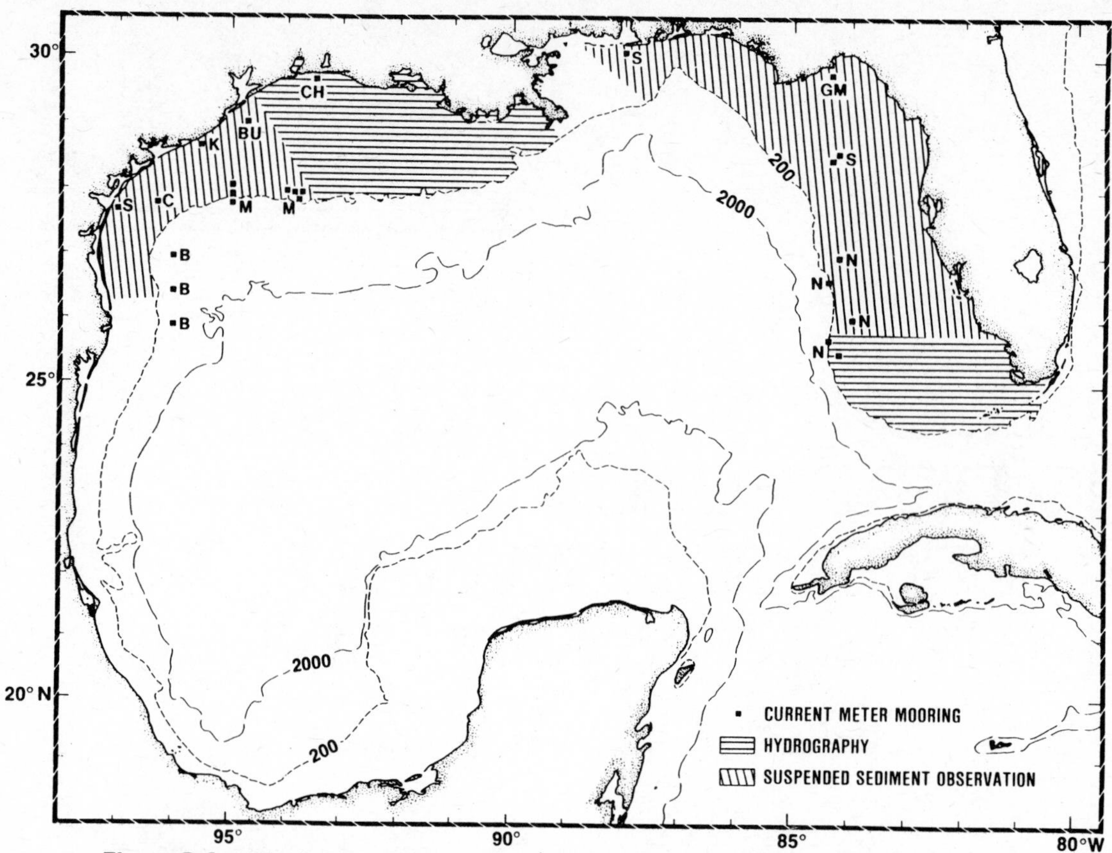

Figure 2.1. Distribution of data on which this analysis of circulation, hydrography, and sediment dynamics on the continental shelves of the Gulf of Mexico was based. The diagonal pattern indicates areas for which there is both hydrographic and suspended sediment data. The horizontal line pattern indicates areas for which only hydrographic data were available. The deployment periods for the current meters and the sources of these current meter data appear in Table 2.1.

TABLE 2.1. Deployment Periods of Current Meters in Gulf of Mexico, 1973–1983, Keyed to Notation in Figure 2.1

Deployment Period		Sources
B	= July 1980 to February 1981	Brooks and Eble (1982)
BU	= September 3 to 4, 1973	Forristal et al. (1977)
C	= March 1978 to March 1981	Continental Shelf Associates (1982)
CH	= January 1979 to August 1979	Crout and Hamiter (1981)
GM	= August 1978 to September 1978	Marmorino (1982)
K	= 1977 to 1982	Cochrane and Kelly (1982)
M	= March 1982 to May 1982, 95°W transect	This report
M	= January 1979 to July 1981, Flower Gardens region	McGrail and Carnes (1983)
N	= August 1973 to April 1974	Niiler (1976)
S	= October 1978 to December 1979, Florida Middle Ground and Anderson Reef	Hopkins and Schroeder (1981)
S	= 43-day deployment, 1977, South Texas Shelf	Smith (1980)

in being truly periodic. Therefore, one can predict tide heights, phase, and currents for any station merely by observing the tides for a sufficiently long time without recourse to determining how the tide propagates or what its amplitude is in deep water.

Tides are, of course, caused by the gravitational attraction of the moon and sun. Out of this duo we are able to derive six major tidal constituents: (1) the principal diurnal lunar tide (O_1); (2) the principal diurnal solar tide (P_1); (3) the diurnal luni-solar tide (K_1); (4) the principal semidiurnal lunar tide (M_2); (5) the principal semidiurnal solar tide (S_2); (6) and the largest lunar elliptic (N_2). Among these six components the most important (largest amplitudes) are the O_1, K_1, M_2, and S_2. In addition, there are fortnightly variations (*Mf*) in the amplitudes of these consituents as the moon passes through its phases in relation to the sun. Maximum amplitudes occur when the sun and moon are aligned (in phase) and minimum amplitudes occur when the moon and sun are 90° apart (in quadrature). These changes are known as spring (maximum) and neap (minimum) tides.

In any given basin the tide may be directly forced by the variation in gravity or it may be driven indirectly by entry of the tidal wave through a port or entrance. The latter is known by the term co-oscillating tide. It is also possible, of course, for the tide to be a combination of the two.

The Gulf of Mexico is essentially a small (1.555 $\times$ 10^6 km^2) ocean basin with two ports, one located at the Florida Straits, the other, at the Straits of Yucatan. The tides of the Gulf vary from diurnal (one high and one low tide per day) to mixed. The mixed tide is semidiurnal (two high and two low tides per day) but with a large inequality between the heights of the succeeding highs and/or lows. Representative tide records are shown in Figure 2.2. Note the contrast between the tides of the Gulf and those of Miami, which lies on the Atlantic seaboard, where semidiurnal tides are the rule. The records also show that the Gulf of Mexico has a microtidal environment with a maximum range of less than 1 m.

Reid and Whitaker (in press) have shown, rather persuasively, that the diurnal portion of the tide is primarily a co-oscillating tide driven by that in the Caribbean Sea and Atlantic Ocean. They show that the contribution due to direct forcing is only on the order of 15%. For the M_2 and S_2 components, direct forcing accounts for 65% of the signal. From the results of their model Reid and Whitaker (in press) suggest that the semidiurnal tide rotates counter-

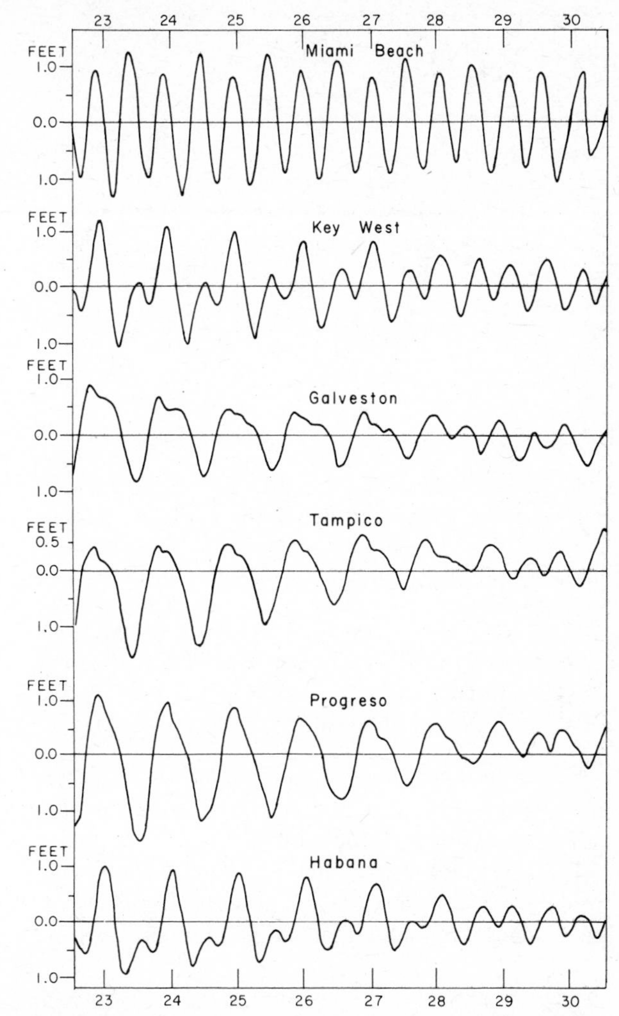

Figure 2.2. Tide curves at various places in the Gulf of Mexico, 23–30 June 1948 (Marmer, 1954).

clockwise about an amphidromic point at the tip of the Yucatan Peninsula. The tidal ellipses (major and minor axes) for the $K_1$and M_2 tides from the Reid and Whitaker (in press) model are shown in Figures 2.3 and 2.4.

Atmospheric Forcing

The wind blowing over the sea surface produces three types of response in the sea, two direct, one indirect. First, the wind deforms the sea surface into the bane of *mal de mer*, surface gravity waves. It also induces surface currents by direct frictional drag (stress) on the sea surface. The frictionally driven currents redistribute water, producing horizontal pressure gradients between areas where the water

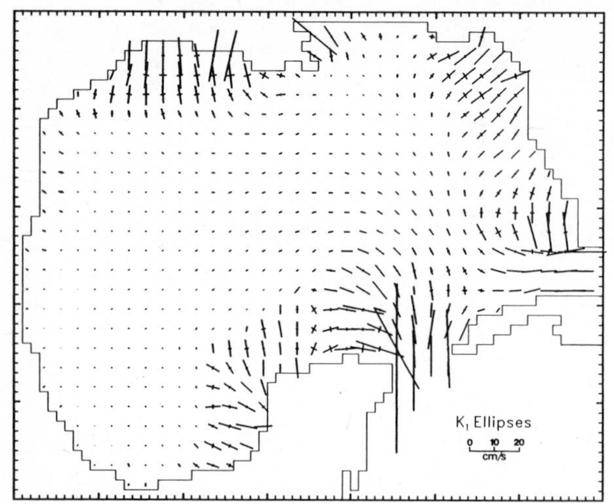

Figure 2.3. Ellipses for the K_1 diurnal tidal currents. The lines are the major and minor axes of the current ellipses in their proper orientation. The lengths of the axes are proportional to the speeds in cm/s. Plots are courtesy of R. O. Reid and R. Whitaker, Texas A&M University (unpublished), based on a model described by Reid and Whitaker (in press).

is blown out and where it is blown in. The resulting pressure gradients, in turn, drive flow that is only indirectly coupled to the wind.

Blumberg and Mellor (1981) computed wind stress over the Gulf of Mexico on a monthly basis at a 1° grid spacing from more than a million reports of wind velocity from ship observations. They did this as input for a numerical model of wind-driven circulation in the Gulf. The average wind-stress vectors for January, March, June, and September from Blumberg and Mellor (1981) are shown in Figure 2.5. In January the regional wind stress over the Gulf is, in general, from northeast to southwest. By March the stress vectors have swung more westerly and are, in general, oriented east to west. In the summer months, represented by June, the Gulf has come under the influence of the Bermuda High Pressure system and the stress vectors point primarily to the northwest. Where Blumberg and Mellor wind-stress vectors can be compared with those calculated by Cochrane and Kelly (1982) from shore stations along the northwestern Gulf they agree rather well, at least in orientation.

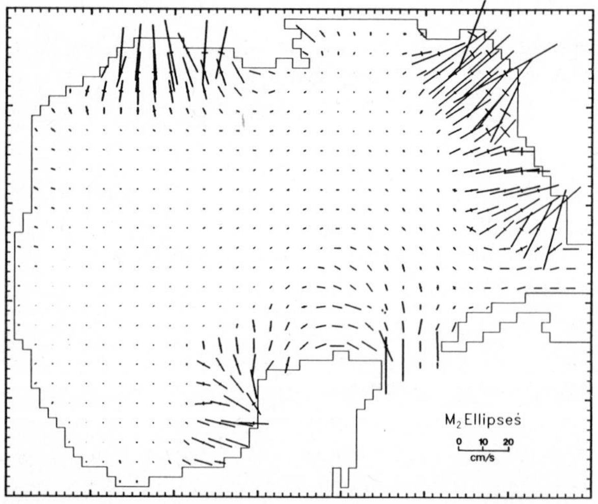

Figure 2.4. Ellipses for the M_2 semidiurnal tidal currents. The lines are the major and minor axes of the current ellipses in their proper orientation. The lengths of the axes are proportional to the speeds in cm/s. Plots are courtesy of R. O. Reid and R. Whitaker, Texas A&M University (unpublished), based on a model described by Reid and Whitaker (in press).

These monthly mean-stress fields are perturbed in the winter by the intrusion of polar air masses into the Gulf. This intrusion takes the form of a frontal passage. Along the south side of the front there is usually a low-pressure trough. Strong southerly winds flow into the trough, thus heralding the approach of the cold front. As the front passes strong north winds take the polar air out over the Gulf.

In the summer and fall, tropical cyclones may migrate into the Gulf to produce large anticlockwise wind systems that alter the local flow field significantly.

Differential Heating

Inequalities in insolation caused by latitudinal variations across the Gulf produce horizontal variances in water density. The variances in the pressure field that results, in turn cause circulation. One might think of it as convection on a very large scale.

On the shelf, winter cold-air outbreaks can extract large amounts of heat from the shelf waters, thereby altering their density structure. These changes can happen on very short time scales.

Runoff

The Louisiana Continental Shelf receives the outflow of the Mississippi, one of the world's largest rivers. At peak discharges in April it can deliver more than 10^4 m^3/s to the Gulf. This volume of water poured onto the surface of the Gulf produces strong local pressure gradients that induce currents that would not otherwise exist. The outflow also pro-

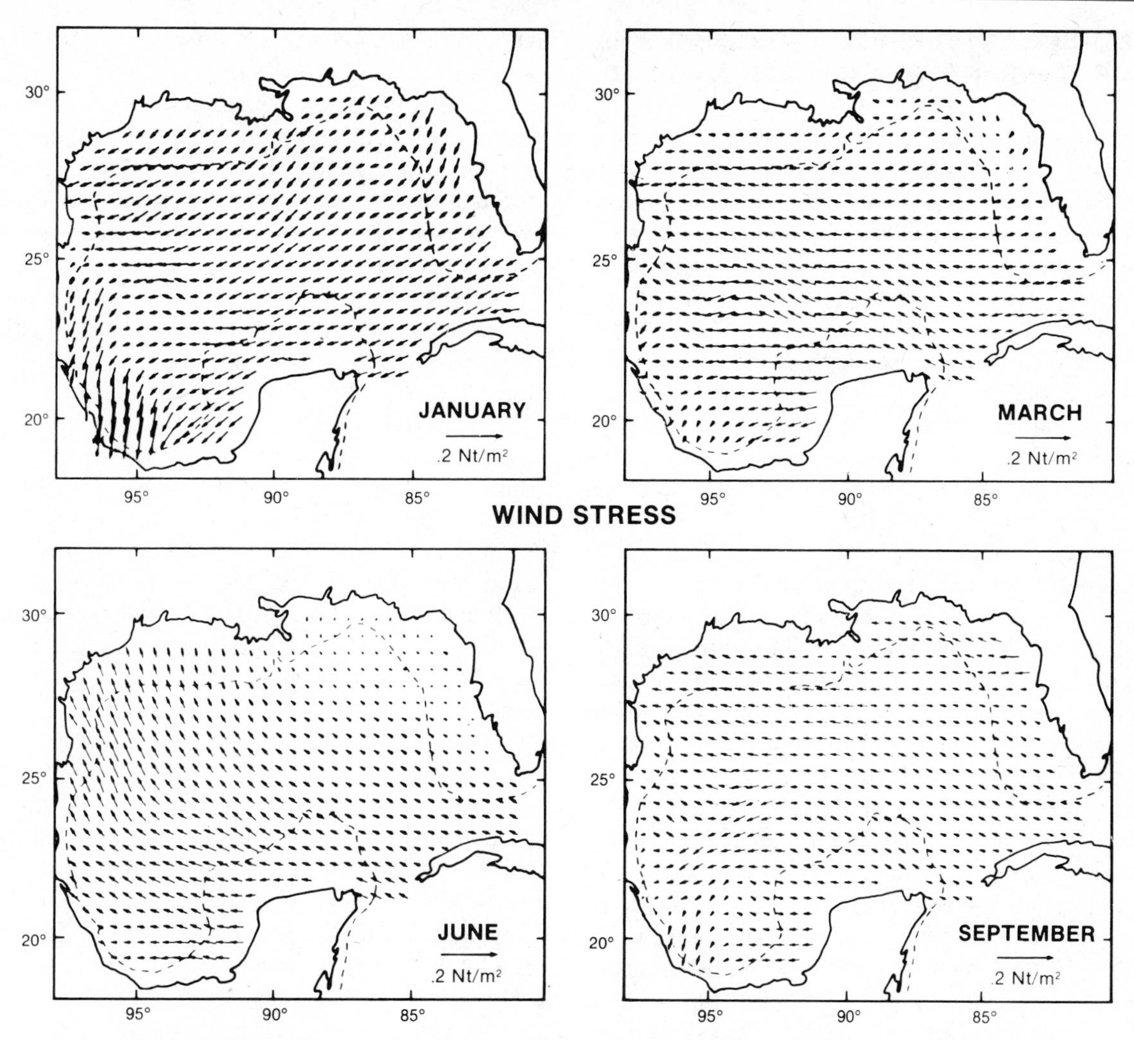

Figure 2.5. Mean monthly wind stress vectors for the Gulf of Mexico (after Blumberg and Mellor, 1981).

duces the classic estuarine responses, with surface flow away from the river mouth and bottom flow toward the river mouth.

Cochrane and Kelly (1982) show that the outflow hugs the Texas–Louisiana coast throughout much of the year and is augmented by the outflow from rivers to the west of the Mississippi. They suggest that this stand of brackish water along the coast sets up a pressure gradient that adds to the wind-driven, west-to-southwest mean flow along the Texas coast.

Interaction with Shallow Flow of the Deep Gulf

In addition to locally produced circulation, the Gulf is intruded by the Loop Current, a jet of swift-flowing water. This current enters through the Yucatan Straits and exits through the Florida Straits after "looping" up toward Alabama. The northward extension of the Loop Current in the Gulf varies considerably. Its descending limb courses along the steep Florida continental slope. The fluctuation of this flow causes cross-shelf variances in the horizontal pressure and, one would suspect, a flow field. It also should entrain some of the shelf-edge water and drag it toward the south. As the Loop Current enters the Gulf it often becomes unstable and develops meanders which may pinch off and form rings that propagate into the western Gulf (see Elliott, 1982, and Merrell and Morrison, 1981). These rings may also influence flow on the outer shelf and slope off Texas as they spin down.

WIND-DRIVEN SURFACE FLOW AND THE LOOP CURRENT IN THE DEEP BASIN

One of the simplest methods of studying surface flow is the use of drifters. Unfortunately, surface drifters provide only two data points (drop position and time and location and time of retrieval) and the path from start to end cannot be traced. Some conclusions, however, can be drawn about wind-driven surface flow.

The Office of Naval Research supported a project by researchers at Texas A&M University to deploy a large number of woodhead drifters from research vessels and ships of opportunity passing through the Caribbean Sea and Gulf of Mexico. The project ran from 1975 through 1978.

Woodhead drifters are yellow plastic discs, 18 cm in diameter and red plastic stems 52 cm in length. A serial number, offer of reward for information regarding the date and location of the drifter's recovery, and address of the researcher are imprinted on the disk. In the period from October 1975 through December 1978, 15,684 of these drifters were released. The statistics of the releases and recoveries were reported by Parker et al. (1979). Of the 15,684 drifters launched, reports on 2127, or 14%, were returned to Texas A&M University.

A remarkable observation which developed from the study is that drifters released anywhere in the Gulf of Mexico west of a line from the Mississippi Delta to the middle of the Yucatan Strait washed ashore, almost exclusively, on the Texas shore. The greatest number of drifters were found between the southern tip of Padre Island and Galveston Bay. Drifters released to the east of the abovementioned line were found, almost exclusively, on the east coast of Florida in the area between Miami and Cape Canaveral. This includes drifters released on the West Florida Shelf.

This distribution fits well with expectations based on the wind-stress fields of Blumberg and Mellor (1981) (Figure 2.5) and a knowledge of the Loop Current. In general, the wind stress should produce a significant westward drift in the surface waters of the Gulf of Mexico. It is no particular suprise, then, that surface drifters accumulate at the western end of the basin. The convergence on Padre Island, as demonstrated later, is a function of the shape of the coastline in the western Gulf. Similarly, the absence of drifter returns from the west Florida coast is consistent with the expectation that surface waters should be blown offshore by the westward wind stress.

The Loop Current shows up in this data set as the only obvious agent that could entrain drifters in the eastern Gulf of Mexico and move them out through the Straits of Florida. It then carries them up the east coast of the peninsula where the westward wind stresses drive the surface waters, and the drifters, onshore. It is the entrainment of most drifters in the eastern Gulf by the Loop Current that accounts for the paucity of drifters from that area that find their way to the Texas shelf.

One problem with drifters is that one knows only where they were released and where they were found, not the path they took between those two points. The general pattern they reveal however, is useful as a control of the possibilities offered by theoretical models (Figure 2.6).

Blumberg and Mellor's (1981) model of circulation in the Gulf of Mexico is of interest because it suggests how the flow field behaves over the whole Gulf. It is possible therefore to compare its predictions in areas for which observations exist to test its validity and to gain some insight into circulation patterns for which data are sparse. Their model is rather comprehensive in that it treats both barotropic and baroclinic modes, contains provisions for eddy (turbulent) vicosity, and uses realistic bathymetry and input through the boundaries to drive the model.

The horizontal velocity vectors for the level 6 m below the sea surface, generated by the model at

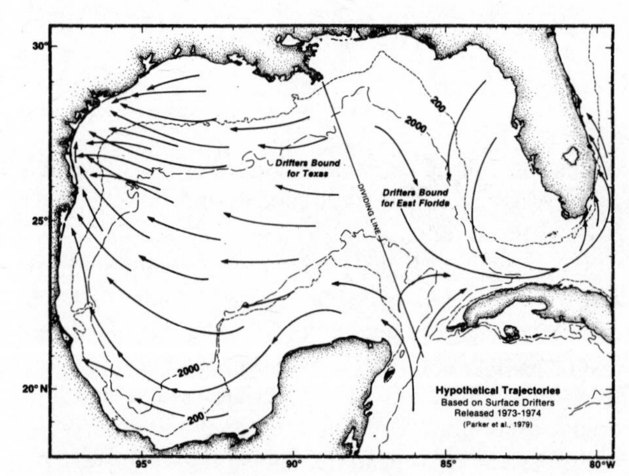

Figure 2.6. Idealized surface drift based on surface-drifter data from Parker et al., 1979. The dividing line is the demarcation between release points for which nearly all drifters went west and those for which the drifters exited from the Gulf through the Straits of Florida.

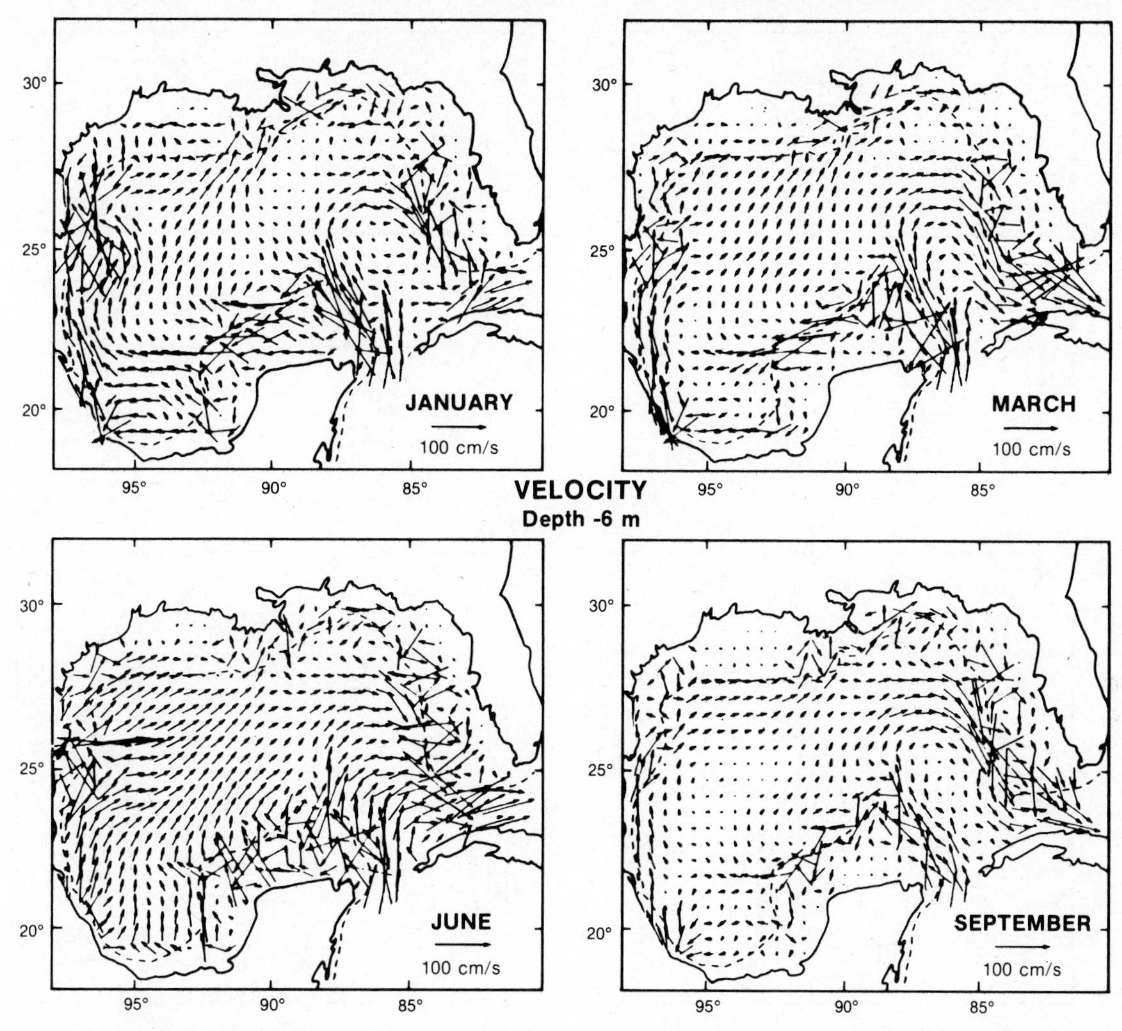

Figure 2.7. Mean monthly velocity vectors at 6-m depth for the Gulf of Mexico (after Blumberg and Mellor, 1981).

90-day increments, are shown in Figure 2.7. Similar vector plots for the 100-m level appear as Figure 2.8. The flow patterns on the shelves are consistent with the drifter data of Parker et al. (1979), but the flow patterns for the deep basin are not. The model does develop a Loop Current that would remove drifters from the eastern Gulf. The flow field the model portrays in the western Gulf would, however, carry the drifters to the north and northeast and not toward the western boundary of the basin. This may be related to a more disturbing failure of the model, which is that it does not shed eddies from the Loop Current. These eddies and their importance to the western Gulf of Mexico are well documented (see, e.g., Nowlin and McLellan, 1967; Behringer et al., 1977; Elliott, 1982; and Merrell and Morrison, 1981).

The models of Hurlbert and Thompson (1982) do generate variations in the intrusion of the Loop Current which lead to meanders that pinch off to form anticyclonic (clockwise rotating) rings which drift westward toward the western boundary of the basin. Variations in the Loop Current intrusion and a pinched-off ring were shown in the data of Ichiye et al. (1973) (Figure 2.9). Nowlin and McLellan (1967) suggested that a large anticyclonic gyre occupied a major portion of the western Gulf. This study was based on a rather coarse grid of hydrographic stations taken on two cruises. Blaha and Sturges (1981) put forth the hypothesis that the anticyclonic flow is generated by the curl of the wind stress. Elliott (1979), however, showed that when the wind stress curl was calculated on a finer scale than that used

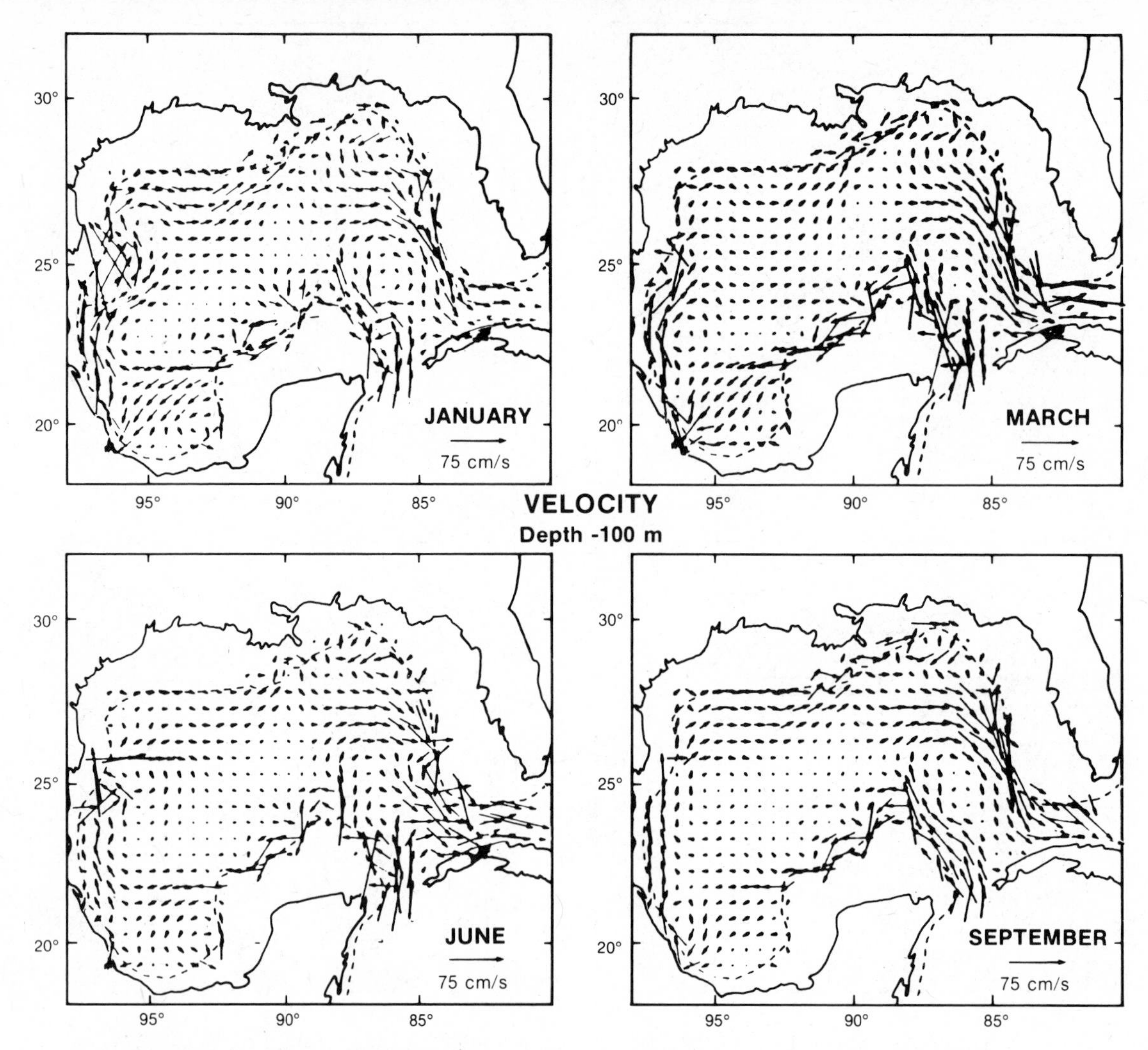

Figure 2.8. Mean monthly velocity vectors at 100-m depth for the Gulf of Mexico (after Blumberg and Mellor, 1981).

by Blaha and Sturges it yielded a cyclonic circulation in the Gulf of Campeche, with anticyclonic circulation to the north.

Elliott (1982) showed that eddies were shed from the Loop Current at a rate of slightly more than one per year, that they drifted to the west at about 2 km/day, and that they contributed significantly to the salt and heat transport into the western Gulf of Mexico. Merrell and Morrison (1981) drew on Nowlin and McLellan (1967), Elliott (1979), Blaha and Sturges (1981), and new hydrographic data from the western Gulf of Mexico to present a rather complicated picture of circulation at the western margin of the basin. They hypothesized that wind stress produces a cyclonic circulation in the Gulf of Campeche, that an anticyclonic gyre exists north of the Gulf of Campeche as a result of westward drifting anticyclonic rings shed by the Loop Current, and that a second cyclonic eddy may exist even farther north (above 24°30'N latitude). Merrell and Morrison (1981) postulate that the northern cyclone is also generated by the Loop Current. These can form only when the Loop Current is at its full northern extension. At that time a low pressure trough forms west of the northward-flowing limb of the Loop Current. A cyclonic eddy could then form if a meander enfolded the trough. They suggest therefore that the northern cyclone may not always be present because the Loop Current does not reach full extension every year.

In summary, then, circulation in the deep basin of the eastern Gulf of Mexico is dominated by the Loop Current, which is generated outside the Gulf. Judging from the drifter data of Parker et al. (1979),

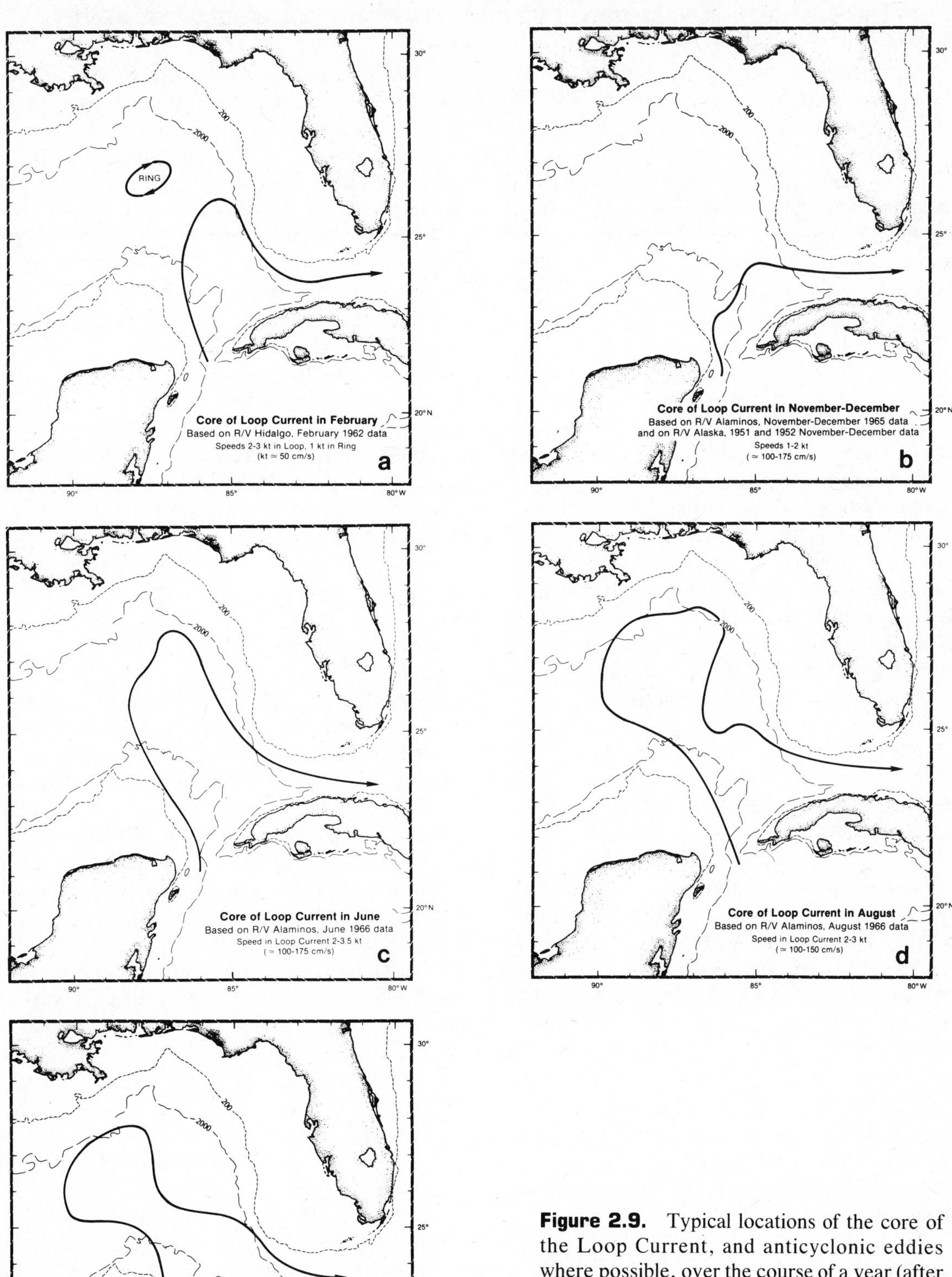

Figure 2.9. Typical locations of the core of the Loop Current, and anticyclonic eddies where possible, over the course of a year (after Ichiye et al., 1973).

this current entrains the surface waters of the eastern Gulf and sweeps them out through the Florida Straits. Though the Loop Current varies considerably in the extent of its northern penetration into the Gulf, it is clear that the current maintains a persistent anticyclonic circulation in the waters of the basin east of a line from the Mississippi Delta to Yucatan Straits.

Circulation in the deep basin west of the Mississippi Delta–Yucatan Strait line is much more complicated than in the eastern basin. The western basin appears to be dominated by large (200-km) anticyclonic and perhaps cyclonic eddies spun off by the Loop Current; these eddies drift to the western boundary. There they interact with the local wind-driven circulation in some complex and still incomprehensible fashion. Whatever the nature of the interaction, the circulation of the deep western Gulf should exhibit significant year-to-year variation because of the annual variance in the number of rings or eddies spawned. Also, the Parker et al. (1979) drifter data show unequivocally that the surface waters of the western Gulf converge on the Texas coast, whatever their circuitous path to that location might be.

LONG-TERM, LARGE-SCALE OBSERVATIONS

The distribution of observations on the continental margins around the Gulf of Mexico is uneven (see Figure 2.1). The shelf off Mexico, for example, is virtually devoid of reported observations. The only information available on which to base speculations is contained in the drifter data of Parker et al. (1979) and on the mathematical models. Similarly, there is little information on the segment of shelf from DeSoto Canyon west to the Mississippi Delta. Only the Texas–Louisiana Shelf and West Florida Shelf have had any large-scale hydrographic surveys or long-term current meter moorings from which data are available.

Mexican Shelf

The drifter returns from releases on the Mexican Shelf in October and November 1976 imply a rather complex circulation pattern. Drifters released on the Yucatan banks and throughout the Gulf of Campeche were found on Padre Island, Texas. On the other hand, only a few drifters from the western Gulf of Campeche went southeastward and landed at Veracruz. This matches up fairly well with the 6-m depth-velocity vectors from Blumberg and Mellor's (1981) model (Figure 2.7). Their velocity field on the Mexican shelf shows an anticyclonic drift with a convergence at Veracruz for all periods except midsummer.

From these data it appears that water which enters through the Yucatan Strait along the shelf sweeps westward across the broad carbonate platform of the Yucatan Peninsula, swings southwest along the outer shelf in the Bay of Campeche, then heads north to the Texas Shelf. This must be viewed as a long-term, general flow that would have many perturbations superimposed on it. It is not possible to make an estimate of the effect of the cyclonic and anticyclonic gyres of the deep basin on the shelf circulation. The surface waters of these gyres, however, do become entrained in the shelf flow somehow because the drifters released in them come to rest on the Texas coast (Figure 2.6).

Texas–Louisiana Shelf

Seasonal Circulation Patterns. The monthly averaged wind-stress field over the Texas–Louisiana Shelf is relatively uniform in magnitude and direction (Figure 2.5), but the orientation of the coastline varies from essentially north–south along the western boundary of the Gulf to east–west between Galveston, Texas, and the Mississippi Delta. Both Smith (1980) and Cochrane and Kelly (1982) have observed that the coastal currents are best correlated with the alongshore component of the wind stress. Smith (1980) suggests further that convergence in the alongshore flow ought to occur when the wind stress is normal on an arcuate coast because at that point the alongshore wind stress would be zero. Consider Figure 2.10, in which the wind stress is perpendicular to the coast at location 1 and there is no component of stress parallel to the coast. At position 2, however, the wind stress is not perpendicular to the coast; therefore, it possesses a component of stress in the down-coast (SW) direction. At position 3 the alongshore wind-stress component is directed up coast (NE). In shallow water the flow goes in the direction of the alongshore windstress and converges at location 1.

Using a combination of historical data (meteorological and oceanographic), long-term current meter records (see Figure 2.1 for locations), processed satellite imagery, and monthly hydrographic transects of the inner shelf, Cochrane and Kelly (1982) developed an interesting model of the mean monthly

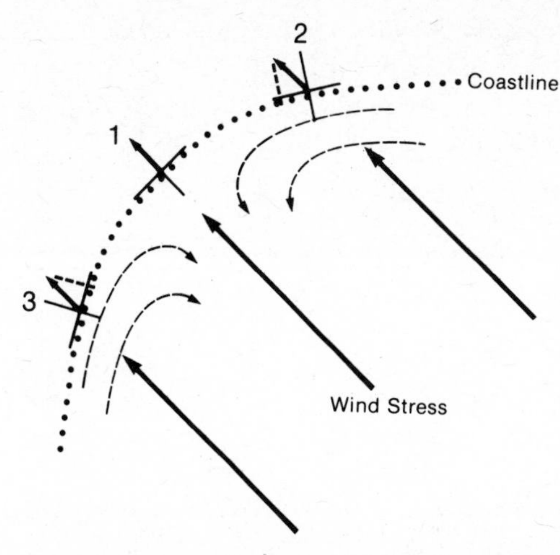

Figure 2.10. Parallel wind-stress vectors striking an arcuate coastline. At point 1 the stress vector is perpendicular to the coast and no alongshore current is generated. At point 2 the stress vector crosses the coast at an angle and the alongshore component of the stress vector is downcoast. At point 3 just the opposite is true: there is convergence of the alongshore flow at point 1.

flow on the Texas–Louisiana Shelf. The essence of their model appears in Figures 2.11 and 2.12. These charts show the mean geopotential anomaly at the sea surface (relative to the 70 db surface) with streamlines. With the exception of July, the coastal flow on the east-west segment of the shelf is downcoast, or westerly. In the spring a northerly flow develops on the north–south shelf segment. The convergence zone between these two flows migrates northeasterly very rapidly in the midsummer; therefore, in July the coastal currents are all up coast, with considerable cross-shelf exchange just west of the Mississippi Delta. Figure 2.5, the July wind stress of Blumberg and Mellor (1981), shows that the mean wind stress is normal to the shore at the Texas–Louisiana border.

Whereas the currents very near shore are driven directly by the alongshore component of the wind, the offshore flow tends to be directed to the right of the wind. This piles surface waters up against the coast along the east–west segment of the shelf during all periods when the wind stress is acting to the west and sets up the coastal waters so that the pressure gradient is normal to the shore. Under these conditions a geostrophic current should develop in which the Coriolis acceleration balances the pressure gradient as shown in equation 1:

$$u_f = \frac{-1}{\rho}\frac{\partial p}{\partial y} \quad (1)$$

where u = x component of flow (positive to the east)

$f = 2\omega \sin \phi$ = local angular velocity of the earth, in radians per second

ρ = density

p = pressure

y = north–south coordinate (positive to the north)

This means that pressure increasing to the north would drive a westerly flowing current.

During the periods of westerly flow at the coast, the discharge of the Mississippi River is tucked against the coast by the wind stress and by its density contrast augments the westerly flow. Cochrane and Kelly (1982) were able to show that decreases in salinity at their study site (Figure 2.1) occurred about six weeks after periods of discharge maxima in the Mississippi River.

In the offshore region the historical hydrography implies that the flow should be northward along the N–S shelf segment and eastward along the E–W segment. This is in good agreement with Blumberg and Mellor's (1981) model and the observations of Nowlin and McLellan (1967).

Cochrane and Kelly (1982) suggest that during the winter preferential cooling of the rather saline midshelf waters would create a dynamic low on the shelf which would then set up a cyclonic circulation on the outer shelf; that is, it would set up a counterclockwise flow. This low appears on the shelf for all periods except July. It is difficult, however, to account for the reestablishment of the low after July if recourse must be made to a remnant of cooler water at mid-shelf left over from winter cooling.

Continental Shelf Associates, Inc. (CSA) maintained a current meter mooring 2 km SSE of Baker Bank from March 1978 through March 1981 for Conoco, Inc. The mooring was set in 73 m of water with meters at depths of 12, 38, and 69 m. These data were released by Conoco, Inc., and provided by CSA to aid in the present study. The location of the mooring is shown in Figure 2.13 along with the location of the National Data Buoy Office (NDBO) buoy 42002 and the location of the moorings established by Texas A&M University researchers for the Minerals Management Services.

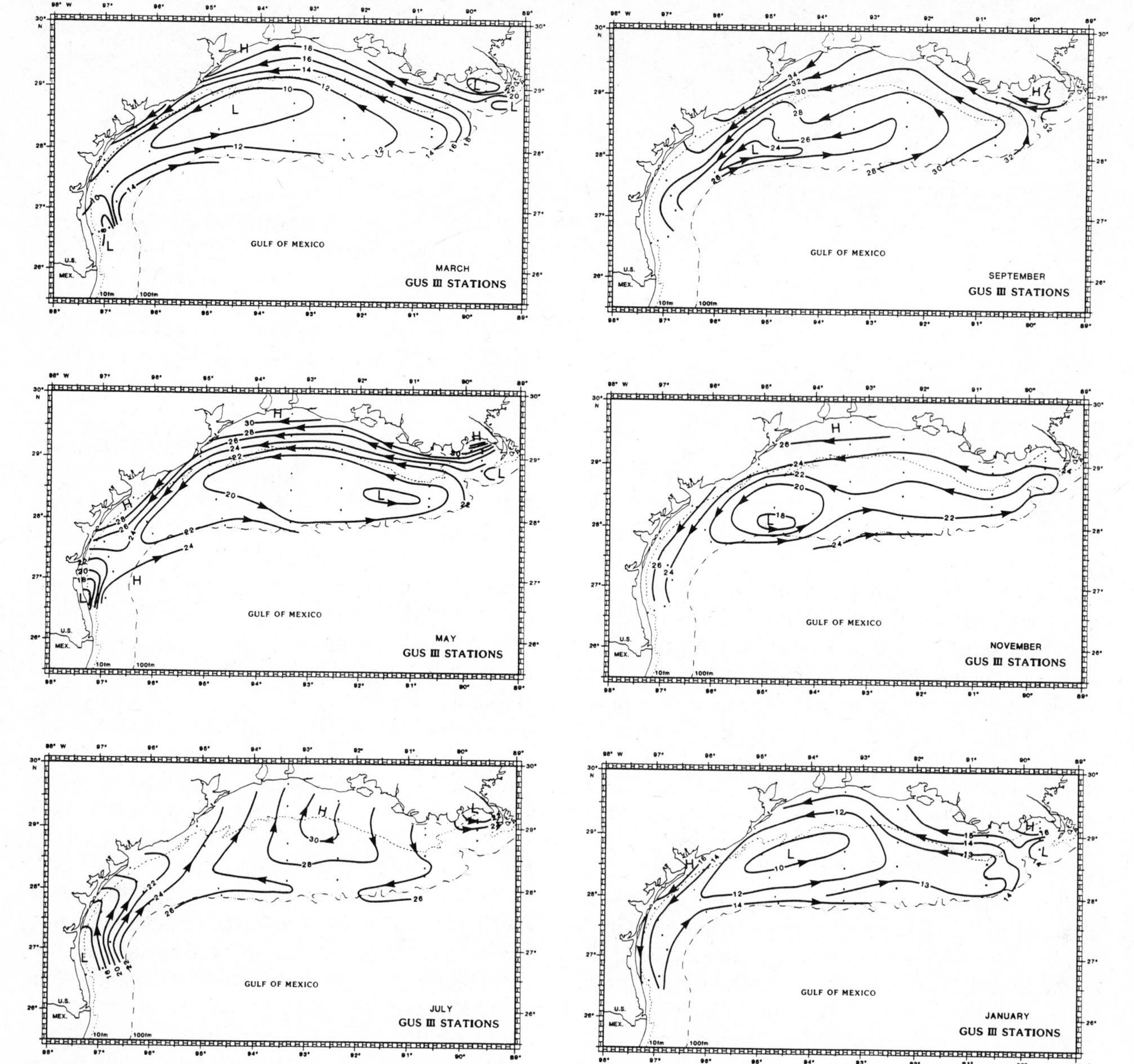

Figure 2.11. Mean geopotential anomaly (dyne/cm) of the sea surface relative to 70 decibars based on GUS III data for March, May, and July (from Cochrane and Kelly, 1982).

Figure 2.12. Mean geopotential anomaly (dyne/cm) of the sea surface relative to 70 decibars based on GUS III data for September, November, and January (from Cochrane and Kelly, 1982).

Progressive vector diagrams (PVD) for the currents at the CSA site and the East Flower Garden, as well as the surface winds recorded at NDBO 42002, were created from raw recorded data for the period from 23 April 1980 to 8 August 1980 (Figure 2.14). Before these PVDs are discussed it is essential that they be recognized for what they are. The current vector recorded at the beginning of each sample interval is multiplied by the time between samples. For the CSA data the interval is 30 minutes; it is 1 hour for the NDBO winds and 20 minutes for the Flower Garden currents. The resulting vector gives a distance and direction but it should not be viewed as a displacement. It is a representation of the time history of the flow through a point (the current meter). These vectors are plotted so that the origin for each new vector is the nose of the last vector; thus the name *progressive vector diagram*.

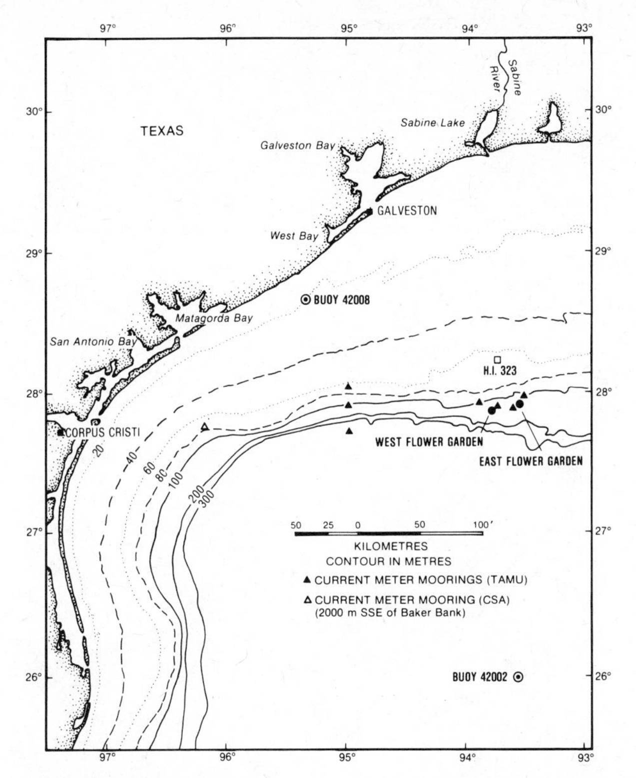

Figure 2.13. Shelf index map showing the locations of the CSA current meter mooring (near Baker Bank), NDBO buoys 42002 and 42008, and the TAMU current meter moorings along 95°W and near the Flower Garden Banks.

In the first two and a half weeks of the record, the wind oscillated between flow from the southeast and flow from the northeast. Thereafter the wind was out of the southeast except for brief periods of flow from the east at the end of the third week, beginning of the seventh week, and middle of the tenth week.

The average flow for the period at the CSA site, near Baker Bank, was toward the northeast, parallel to the local trend of the isobaths. During the first six weeks, the flow completed two counterclockwise rotations, which are not clearly related to the wind. The first oscillations may be related to the passage of the last weak cold fronts of the season appearing in the wind record. However, the southerly flow at the upper meter (12 m) between weeks 4 and 6 is hard to reconcile with the steady southeasterly wind. The record for the meter at a 38-m depth possesses much tighter loops during the first six weeks, thus indicating slower flow than at the surface. Directional differences also exist between the two, with more southerly flow at the surface. The near-bottom flow (69-m depth) exhibits reversals rather than rotational events during the first six weeks. It is apparent as well that if it were not for the large northerly flow between weeks 9 and 10 the net flow at the bottom would have been to the south.

Beginning about the middle of week 6, the flow at all depths turned to the north. The near-bottom flow reversed abruptly in the middle of week 7, then returned to northerly flow in the middle of week 8. This perturbation may be related to the wind shift during that period. The strongest flow at all levels is between weeks 9 and 10. The wind was not appreciably stronger but it did shift from southeasterly to southerly during that period.

The Flower Garden moorings were deployed in deeper water (about 100 m) than the mooring at the CSA site (73 m) (see Figure 2.13 for locations). Also, mooring 2 was set at the base of the East Flower Garden Bank; therefore, flow at that site is deformed by the bank.

The only current meter in the Flower Garden region as shallow as the top meter at the CSA site was the electromagnetic current meter (EMCM) on mooring 2. Unfortunately, the EMCM battery failed prematurely, leaving only a 2 week record. This very short piece does suggest that the surface waters were responding to the reversals in the wind which occurred at that time (Figure 2.14). The mid-depth current (50 to 60 m), however, was remarkably constant in its flow to the east. Ignoring the record from the bottom current meter on mooring 2 because of the strong topographic influence of the bank, we can see that the bottom flow in the Flower Gardens region was also to the east, much slower than the mid-depth flow, and oriented slightly more offshore.

Figure 2.15 is a PVD for the wind at the NDBO 42002 and the CSA site for 11 weeks, starting on St. Valentine's Day 1979. During that period flow at the level of the upper two meters responded much more directly to the wind. The bottom current also responded to variations in the wind, but it was flowing to the southwest with greater strength than mid-depth water was flowing to the northeast.

Monthly means for all moorings and all deployments. (January 1979 to July 1981) were computed for the Flower Gardens site (Figure 2.16a). The records were divided into mid-depth (32 to 64 m), near bottom (11 to 18 m above bottom), and very near bottom (4 to 8 m above bottom). The mid-depth currents were dominated by a rather strong easterly flow, except for two periods: mid-July to mid-August 1979 and December 1980. The cause of the anom-

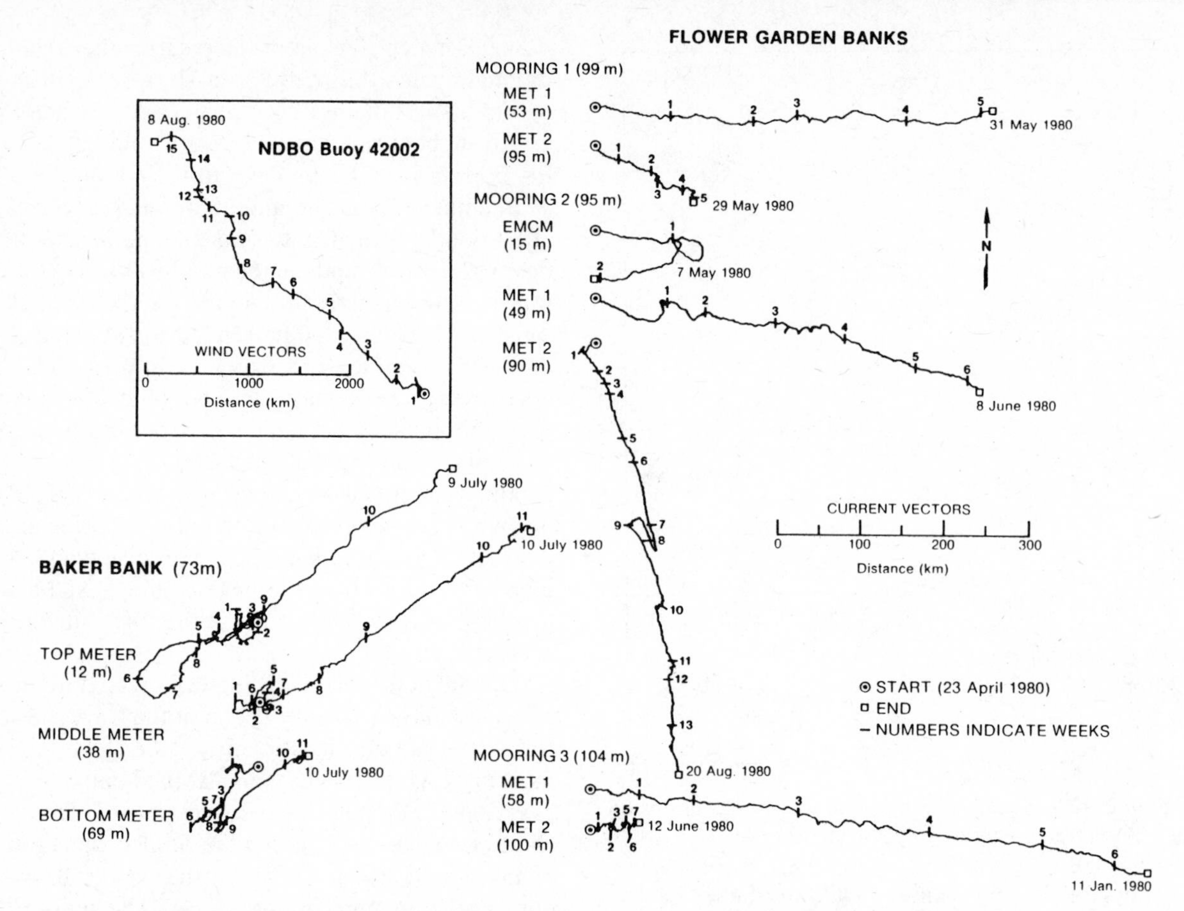

Figure 2.14. Progressive vector diagrams constructed from velocity time-series records beginning on 23 April 1980 for (*a*) the wind at NDBO buoy 42002 and the currents at the top, middle, and bottom meters on the CSA mooring near Baker Bank; (*b*) the currents at all available meters on moorings 1, 2, and 3 near the Flower Garden Banks; and (*c*) records beginning on 14 Feb 1979 for the wind at NDBO buoy 42002 and the currents at the top, middle, and bottom meters on the CSA mooring near Baker Bank. (See Figure 2.13 for locations.)

alous flow in 1979 was the passage of Tropical Storm Claudette. This storm passed over the Flower Gardens, traveling northward early on 25 July 1979. After passing over the region, the storm stalled over the Texas coast. The result was several days of high winds over the northwestern Gulf. In the second period of westerly currents several very strong, cold frontal passages produced a vector average wind out of the northeast.

The bottom currents were directed primarily toward the southeast. There was no apparent seasonal distribution to these vector means, but that may have been due to the relatively small sample period.

For the monthly vector averages the greatest variance was in the east–west, or u component, of flow. The cause of the variation was investigated by means of a multiple linear regression, using the north–south (τ_v) and east–west (τ_u) components of wind stress as the independent variables. The regression had the form

$$u = A + B\tau_u + C\tau_v$$

where A, B, and C were estimated from the data by using the best fit according to the least-squares criterion. The values for these constants were A = 7.89, ± 3.28, B = 3.9 ± 8.9, and C = 23.3 ± 6.7. This model will account for approximately 70% of the variance in the u component of the mid-depth current. The low value of B, its large standard error, and the large value of C imply that the variance of the east–west component of flow is driven by the

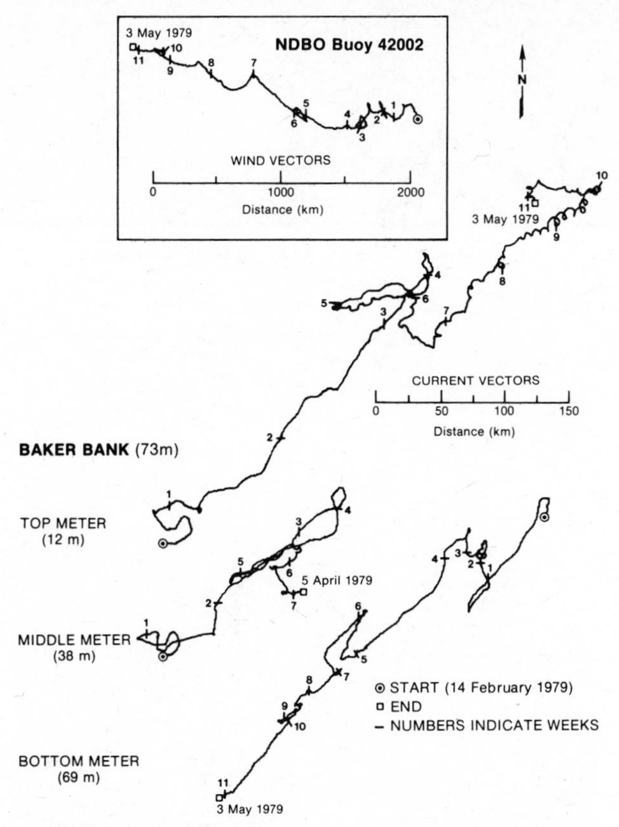

Figure 2.15. Progressive vector diagrams constructed from velocity time series records beginning on 14 February 1979 for the currents at the top, middle, and bottom meters on the CSA mooring near Baker Bank and from wind stress data at NDBO buoy 42002.

north–south component of the wind stress. The stronger the wind stress to the north, the greater the magnitude of the eastward flowing current.

The same regression was run for the very near bottom meters, excluding those from mooring 2. The values for the constants on this run were $A = 8.25 \pm 2.89$, $B = 16.1 \pm 6.1$, and $C = -4.4 \pm 4.4$. For the bottom flow it is obviously the east–west component of wind stress that is related to the variance. Again approximately 70% of the variance of u could be accounted for by the model. The latter result may be related to the fact that very strong wind stress to the west over the shelf would produce onshore-directed Ekman flow at the surface and southwesterly directed Ekman flow at the bottom. Convergence of the bottom flow from the inner shelf and the easterly bottom flow from the outer shelf would modulate the flow at the location of the Flower Garden Banks, as the model implies.

The set of monthly mean current vectors computed from all CSA mooring data are displayed in Figure 2.16*b*. The stronger currents (between 8 and 25 cm/s) at the top meter, 12 m from the surface, are directed primarily to the northeast, nearly parallel to the local shelf isobaths. For several cases with speeds less than 8 cm/s, however, the mean currents are directed nearly offshore or onshore. The few cases with a downcast (southwest) component to the flow are found only in the late fall and early winter and are perhaps related to the wind-stress patterns at this time of year. Over much of the year the winds are from the southeast but come from the northeast in late fall and early winter. The mid-depth currents (38-m depth) again show a predominance of northeast flow at higher speeds but with a greater percentage of occurrences with lower speeds and nearly random directions. The bottom current averages in Figure 2.16*b* show a predominance of southwesterly flow in the opposite direction to the surface currents and with less than half the speed. The few cases with upcoast flow at the bottom occur in the winter and spring.

Salinity Distribution. To belabor the obvious for a moment, the Mississippi River contributes an enormous volume of fresh water to the Texas–Louisiana Shelf. Elliott (1982) has provided an estimate that 371 km^3/yr are discharged into the western Gulf of Mexico by the Mississippi River. That is equivalent to approximately 5% of the water on the continental shelf between the Mississippi Delta and the 95°W meridian and on the order of 0.1% of the volume discharge of the Loop Current through the Yucatan Strait.

The discharge of the Mississippi River is not constant. In 1982 there were maxima in mid-February, early March, and late June, according to Cochrane and Kelly (1982). They further reported observations of corrsponding decreases in the salinities measured at the site of their current meter moorings (Figure 2.1) in late March and late May but that the brackish water of the June maximum did not reach that site because the flow was upcoast in late June and July.

The effects of the Mississippi effluent on the surface salinities of the Texas–Louisiana Shelf are shown for alternate months of 1964 in Figure 2.17. In the winter, with discharge of the river at a minimum, oceanic waters with salinities of 36 ppt intruded within a few kilometers of the coast. By May the nearshore salinities plunged to 20 ppt or less and the 36 ppt isohaline had been pushed off the shelf. Also notice the suggestion in these May surface sal-

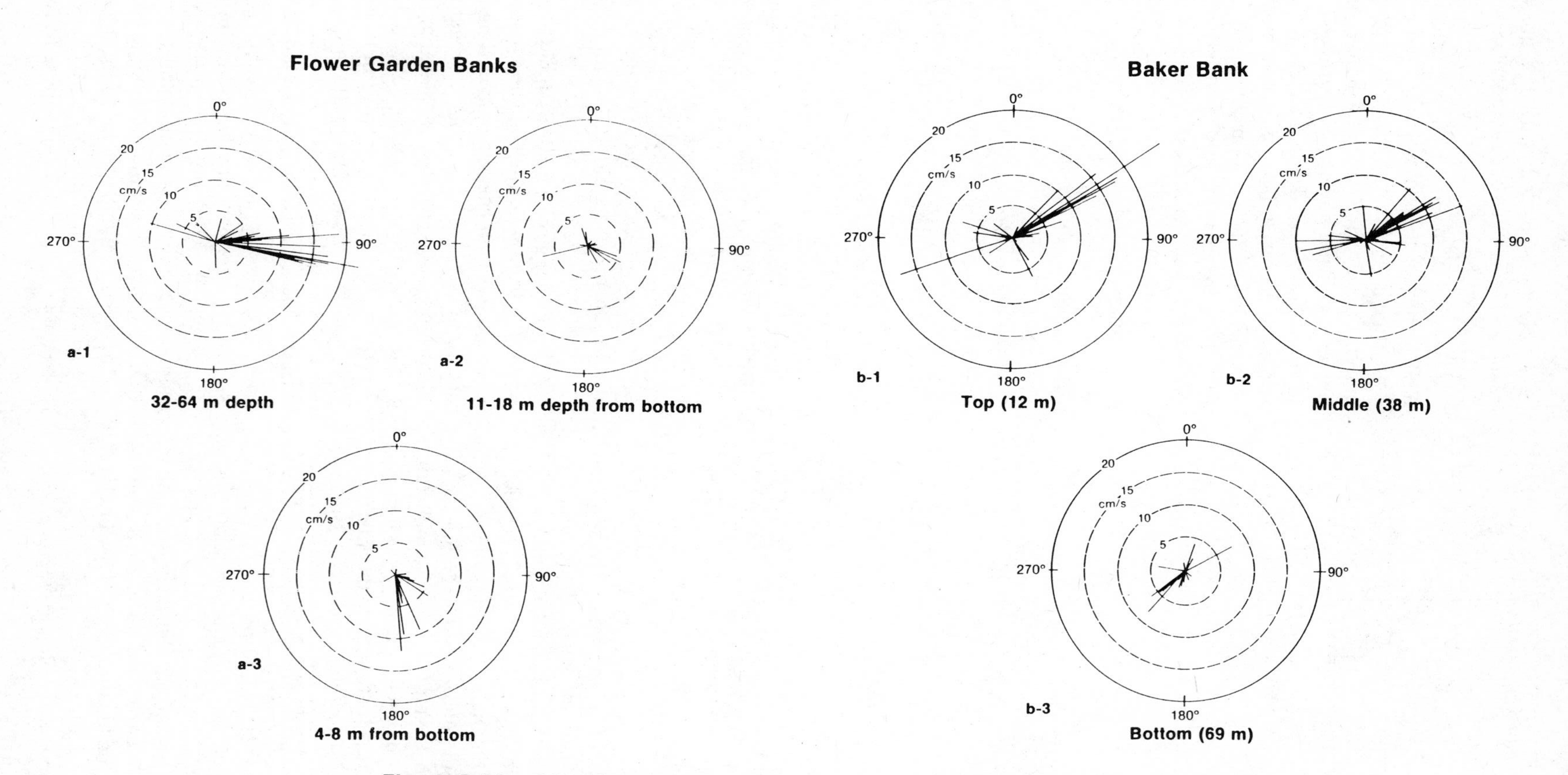

Figure 2.16. Monthly mean current vectors for (*a*) all deployments and all moorings near the Flower Garden Banks over the period January 1979 to July 1981; and (*b*) the CSA mooring near Baker Bank over the period March 1978 through March 1981.

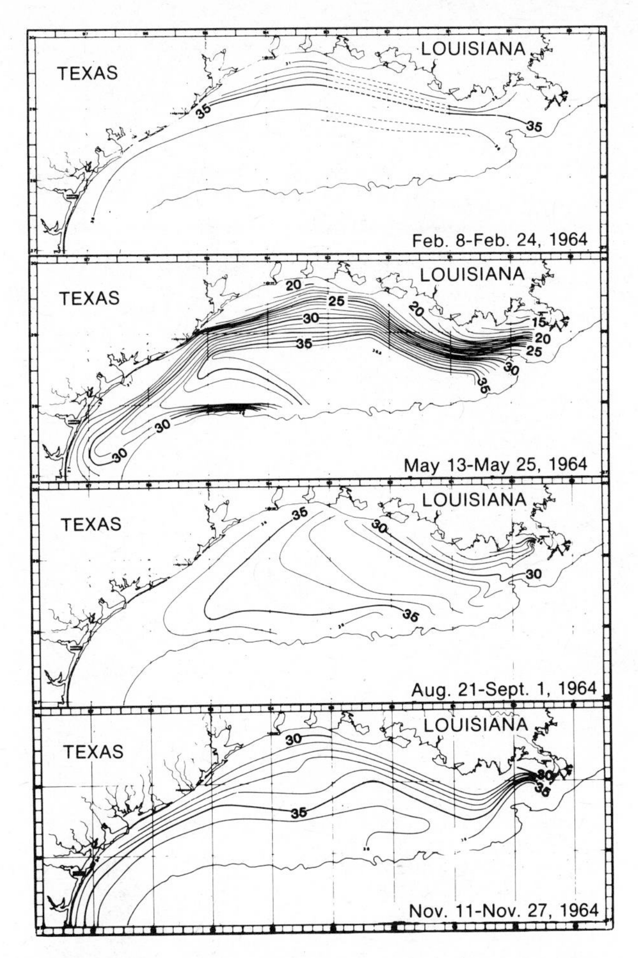

Figure 2.17. Sea-surface salinity from GUS III observations taken in February, May, August, and November 1964 (from Cochrane and Kelly, 1981).

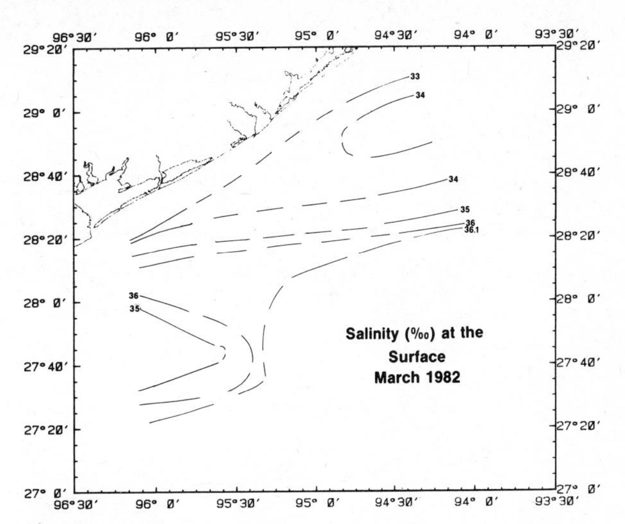

Figure 2.18. Sea-surface salinity from observations taken in March 1982 (Sahl et al., 1982).

inities that low salinity water was moving offshore and recurring to the east along the shelf break. A similar situation was found in March of 1982 (Sahl et al., 1982) when a thin veneer of water with a salinity of less than 36 ppt was found on the outer shelf between 95°30'W and 96°W (Figure 2.18). This is consistent with the model of Cochrane and Kelly (1982), which calls for a cross-shelf flow near the region of Matagorda Bay and entrainment in the eastward flowing current on the outer shelf (Figures 2.11 and 2.12).

The surface isohalines in August 1964 were not, as in other months, parallel to the coast (Figure 2.17). Rather there was a tongue of relatively low salinity over the central shelf as far west as 94°W. This distribution may result from blocking of the late June discharge of the Mississippi River by an eastward flow over the entire shelf in July and subsequent return to westward shelf flow in August, as suggested by Cochrane and Kelly (1982). By November 1964 the isohalines had once again become more or less parallel to the coast, with minimum nearshore salinities of 30 ppt or greater.

Temperature Distribution. Etter and Cochrane (1975) analyzed approximately 20 years of bathythermograph data collected on this segment of the shelf. They reported that surface and bottom temperatures were at a minimum in January and February (Figure 2.19) and at a maximum in July and August (Figure 2.20). The mean minimum surface temperatures show nearshore waters at approximately 14°C and offshore waters near 20°C, which can be compared with the sea-surface temperature maps of 17 February 1982 (Figure 2.21) and 23 February 1982 (Figure 2.22) produced from the NOAA-7 satellite Advanced Very High Resolution Radiometer (AVHRR) digital data by Richard M. Barazotto of NOAA-National Environmental Satellite Service (NESS). In the satellite data, one can see the complex crenulations of the isotherms due to the large-scale, turbulent flow of the shelf. It is also apparent that there is a strong nearshore gradient from about 13.5 to 16°C and another region of rather pronounced gradient between 17 and 19°C. Overall, the mean monthly data of Etter and Cochrane (1975) are consistent with the snapshot of the sea surface temperature provided by the satellite data.

The bottom temperatures for late winter range from less than 13°C nearshore to a maximum of 18.5°C near the shelf break (Figure 2.20).

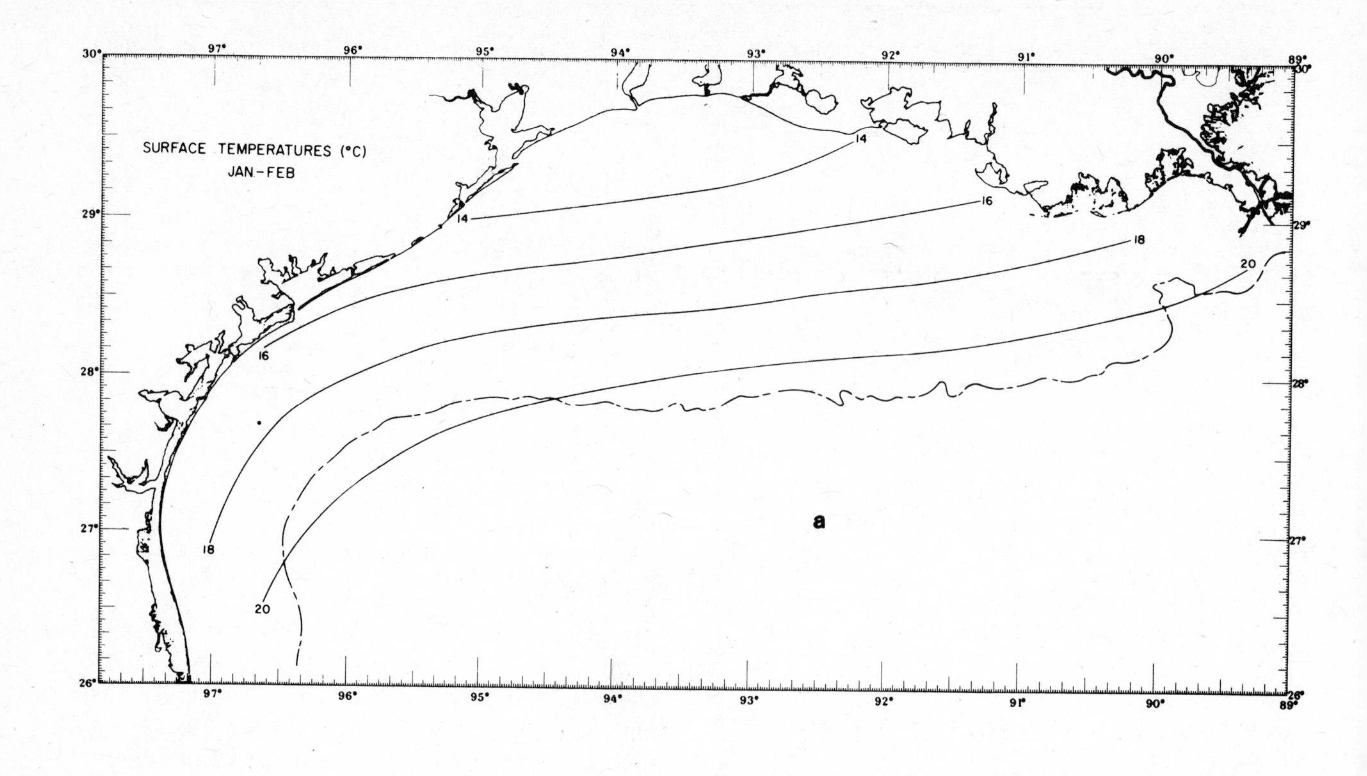

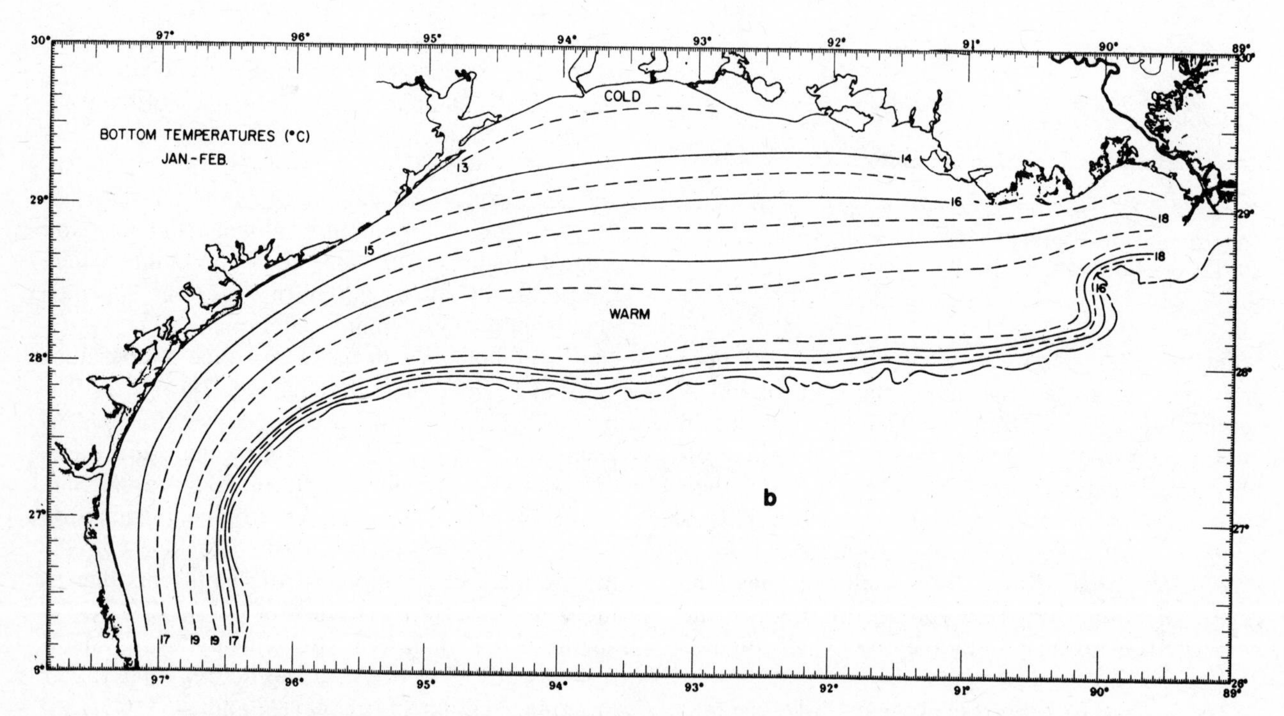

Figure 2.19. Map of (*a*) mean surface temperature (°C) for January–February and (*b*) mean bottom temperatures for January–February (from Etter and Cochrane, 1975).

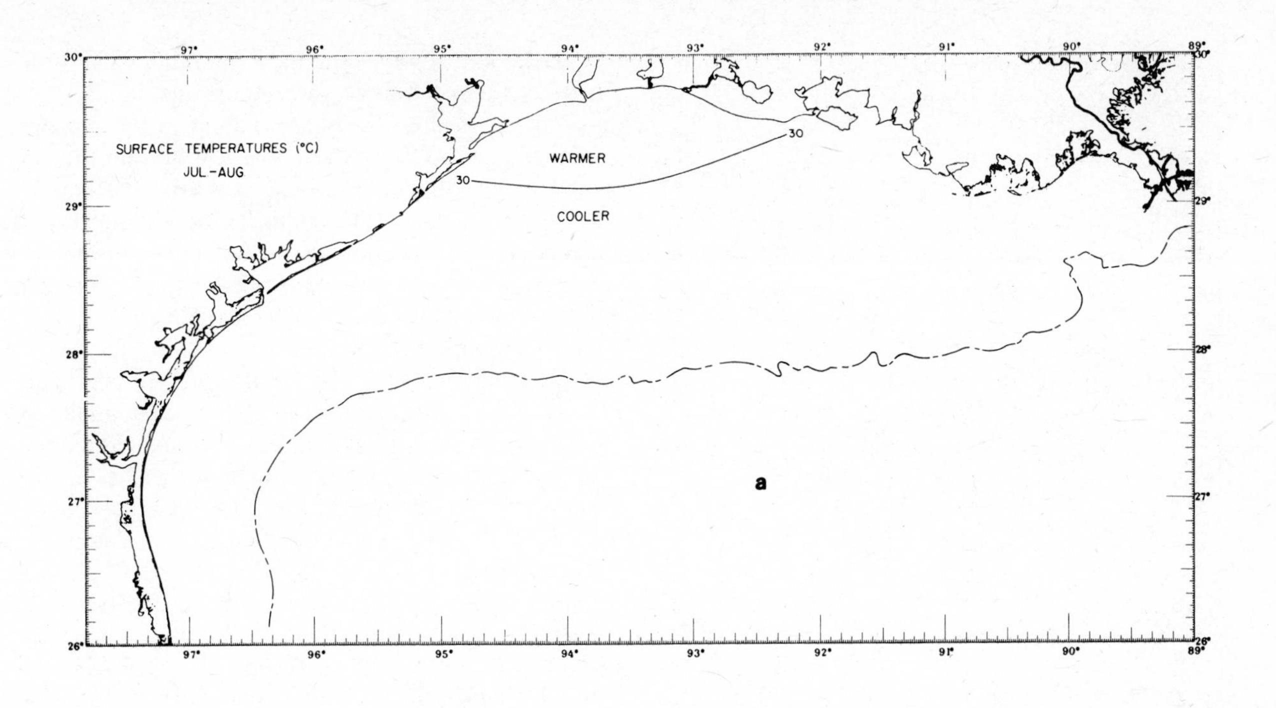

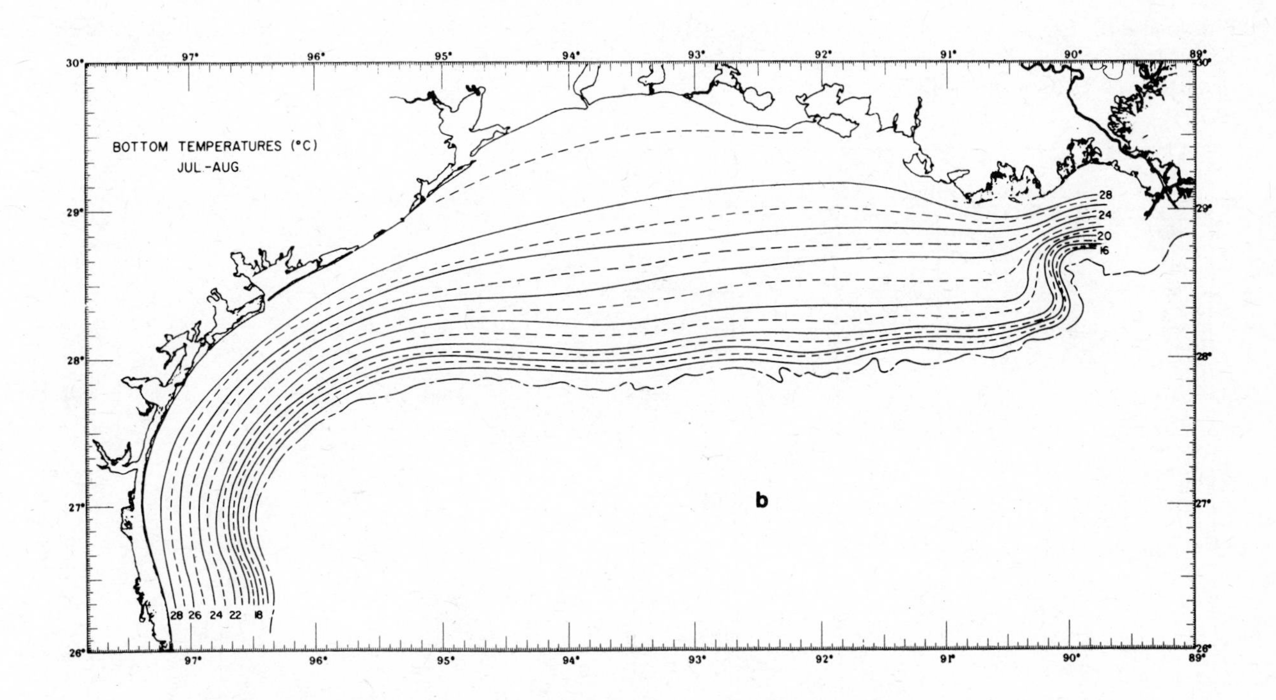

Figure 2.20. Maps of (*a*) mean surface temperature (°C) for July–August and (*b*) mean bottom temperatures for July–August (from Etter and Cochrane, 1975).

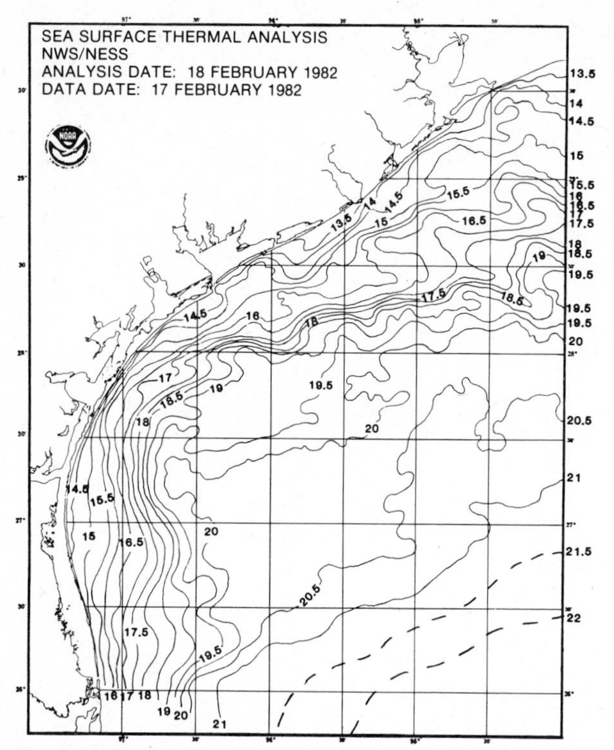

Figure 2.21. Sea-surface temperature based on satellite infrared imagery for 17 February 1982, South Texas Shelf. (Courtesy of R. Barazotto, NWS/NESS, Slidell, Louisiana.)

In the late summer the mean monthly data show little surface variance in temperature but a very strong cross-shelf thermal gradient at the bottom (Figure 2.20). The isotherms at the bottom are, in general, parallel to the local isobaths (Etter and Cochrane, 1975). The rapidity of the seasonal warming of the sea surface is shown in the satellite data from 6 April 1982 (Figures 2.23 to 2.24). Notice that the nearshore waters are 6 to 7°C warmer than they were in mid-February.

Etter and Cochrane (1975) also plotted the mean depth of the surface mixed layer for the January to February period (Figure 2.25). This layer thickens offshore to a maximum of 75 m near the shelf edge. Data from our moorings at the Flower Garden Banks from 1979 to 1981 and that from our conductivity, temperature, and depth (CTD) surveys at the shelf edge (Figure 2.26) agree with those of Etter and Cochrane (1975). The mean temperature recorded by our current meters at depths of less than 60 m (in the mixed layer) for February was 18.57°C, with a standard deviation of 0.6°C. The cycle of stratification for the shelf edge is also shown in Figure 2.26. Changes in density at the shelf edge are due almost entirely to changes in temperature.

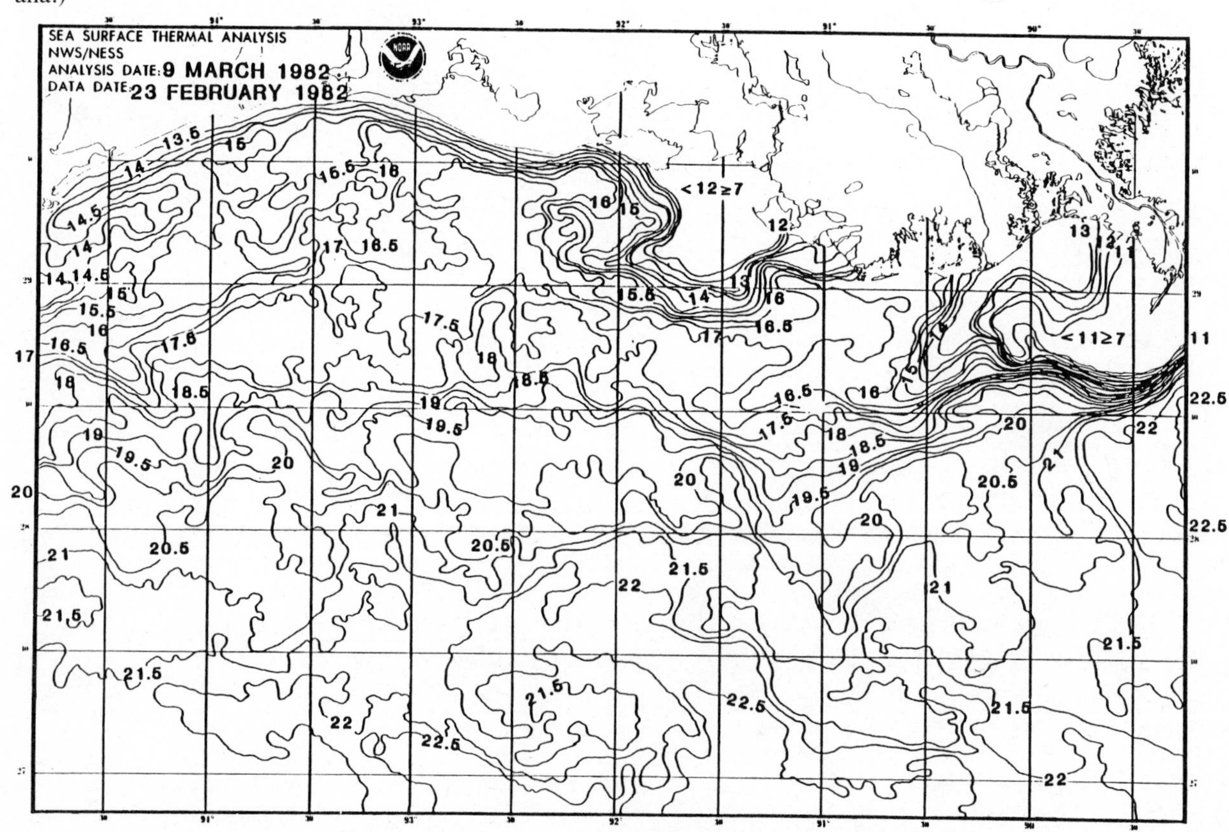

Figure 2.22. Sea-surface temperature based on satellite infrared imagery for 23 February 1982, Texas–Louisiana Shelf. (Courtesy of R. Barazotto, NWS/NESS, Slidell, Louisiana.)

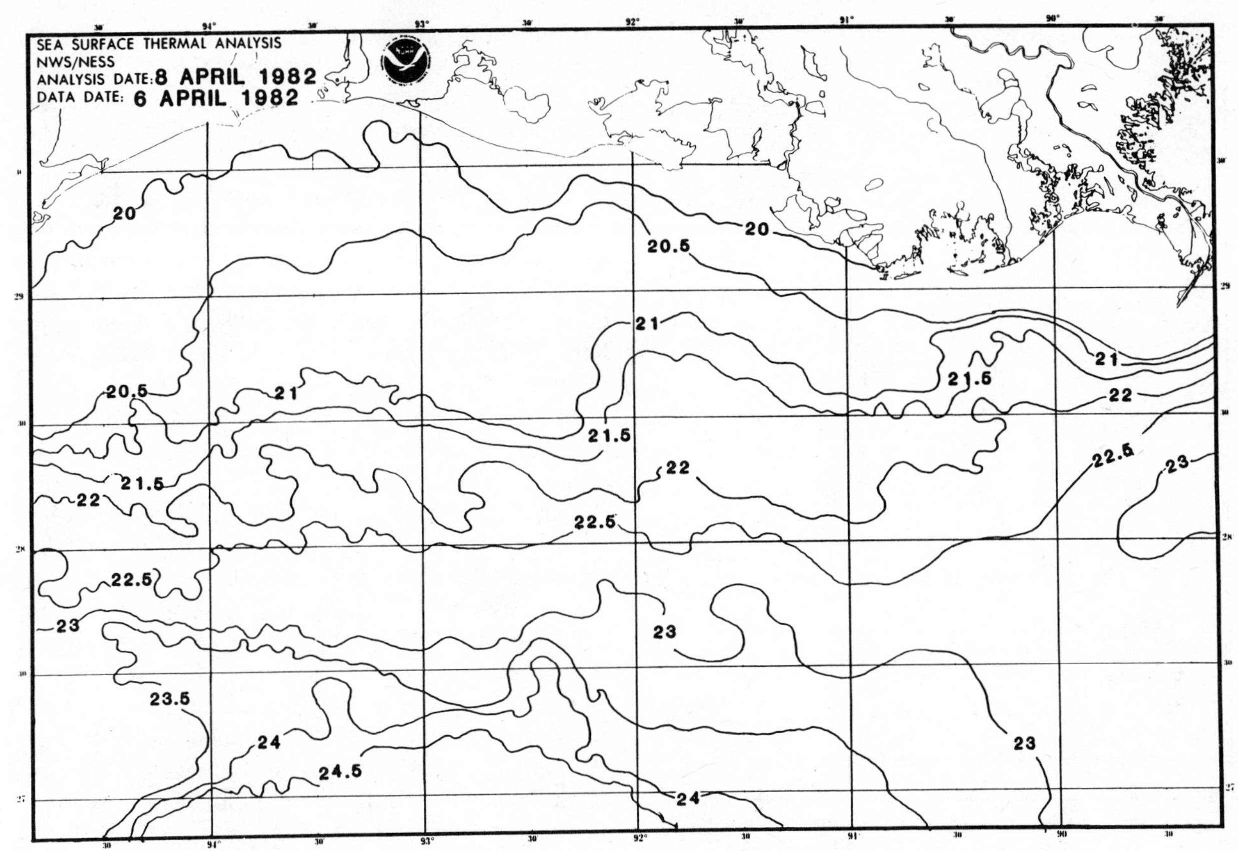

Figure 2.23. Sea-surface temperature based on satellite infrared imagery for 6 April 1982, Texas–Louisiana Shelf. (Courtesy of R. Barazotto, NWS/NESS, Slidell, Louisiana.)

Cochrane and Kelly (1982) proposed that the reason waters on the outer continental shelf stay so much warmer than nearshore waters is that the mixed layer on the outer shelf does not penetrate to the bottom. Nearshore, the winter storms do mix the water all the way to the bottom and each passing cold front extracts heat from a relatively thin layer of water. Offshore, however, the heat capacity of the water is such that the cold frontal passages actually remove more heat from the shelf edge waters than from the nearshore but the temperature of the relatively thick mixed layer remains relatively high.

Nowlin and Parker (1974) described the effect of cold air outbreaks on shelf waters in the Gulf of Mexico from data they gathered on two cruises in January 1966. The first cruise was just prior to a cold frontal passage and the second, about 15 days later, was just after the frontal passage, while the region was still under the influence of the cold air. They found that just offshore, in water of 10-to-20-m depth, the temperature decreased by about 5°C and salinities increased by approximately 1 ppt during the 2-week period. At the shelf edge 100 to 150 NM from shore, there was no change in salinity and the temperature decreased by only 1 to 2°C. Nowlin and Parker (1974) hypothesized that evaporation and heat loss to the cold, dry polar air and the nearshore waters greatly increased the density of that water, giving it the temperature and salinity characteristics of water found beneath the subtropical underwater of the deep Gulf. Data collected by researchers at Louisiana State University and Texas A&M University support this hypothesis (McGrail and Carnes, 1983; McGrail et al., 1982; Wiseman et al., 1982; Sahl et al., 1982).

Compare the sea-surface temperature maps of 6 April 1982 (Figures 2.23 and 2.24) with the map made from data obtained on 11 April 1982 (Figure 2.27). As shown in Figure 2.28, the 6 April data were taken at the onset of northerly winds at the beginning of a cold frontal passage and the 11 April map was obtained in the middle of a period of winds with a northerly component.

In the 11 April map the very nearshore waters have lost nearly 1°C in temperature but the offshore waters have actually become warmer in the area between 92 and 94°W, thus causing a sharp gradient near the coast. This gradient appears to be related

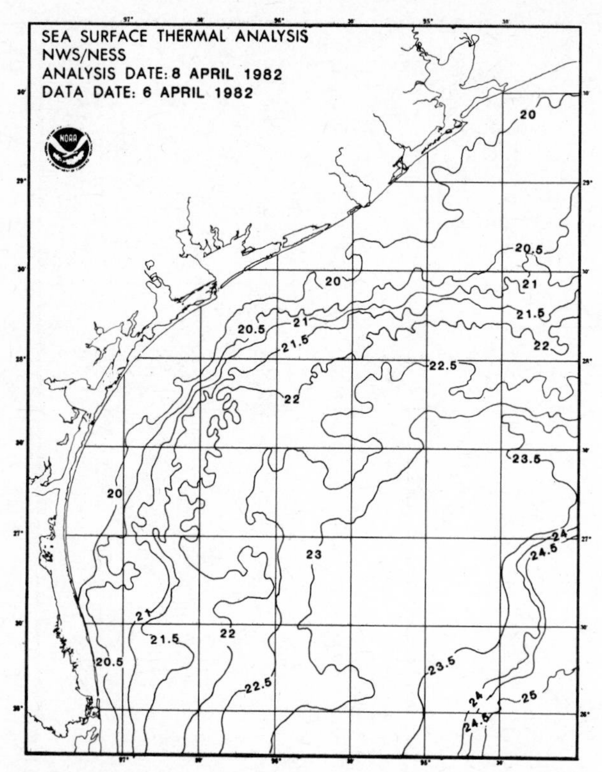

Figure 2.24. Sea-surface temperature based on satellite infrared imagery for 6 April 1982, South Texas Shelf. (Courtesy of R. Barazotto, NWS/NESS, Slidell, Louisiana.)

to competing events. Both the southerly and northerly winds during the period between 6 April and 11 April had strong alongshore components toward the east, which created a rapid downcast flow at the coast (Cochrane and Kelly, 1982). This flow appears to have produced a suction just west of the delta through the Mississippi Canyon area that drew warmer water in from offshore and moved it to the west. Notice that the kink in the 21° C isotherm located near 92°W in the 6 April 1982 map appears to have propagated northwesterly to near 93°15'W by the time of the 11 April data.

In March of 1982 we deployed three current meter moorings across the shelf break along the 95° W meridian (see Figure 2.13). This deployment included the time during which the April satellite data were obtained. Records of the wind at NDBO 42002 and NDBO 42008 were also obtained for the duration of the deployment. The progressive vector diagrams (PVD) for the winds at the two buoys and their vector average are shown in Figure 2.28. It is obvious that the winds with a component from the north are much stronger at buoy 42008, which is close to shore, than at buoy 42002. It is likely that this effect is due to the weakening of the cold front as winds passed over the relatively warm water of the shelf. This difference in strength of the north winds leads to a difference in direction of the wind stress for the

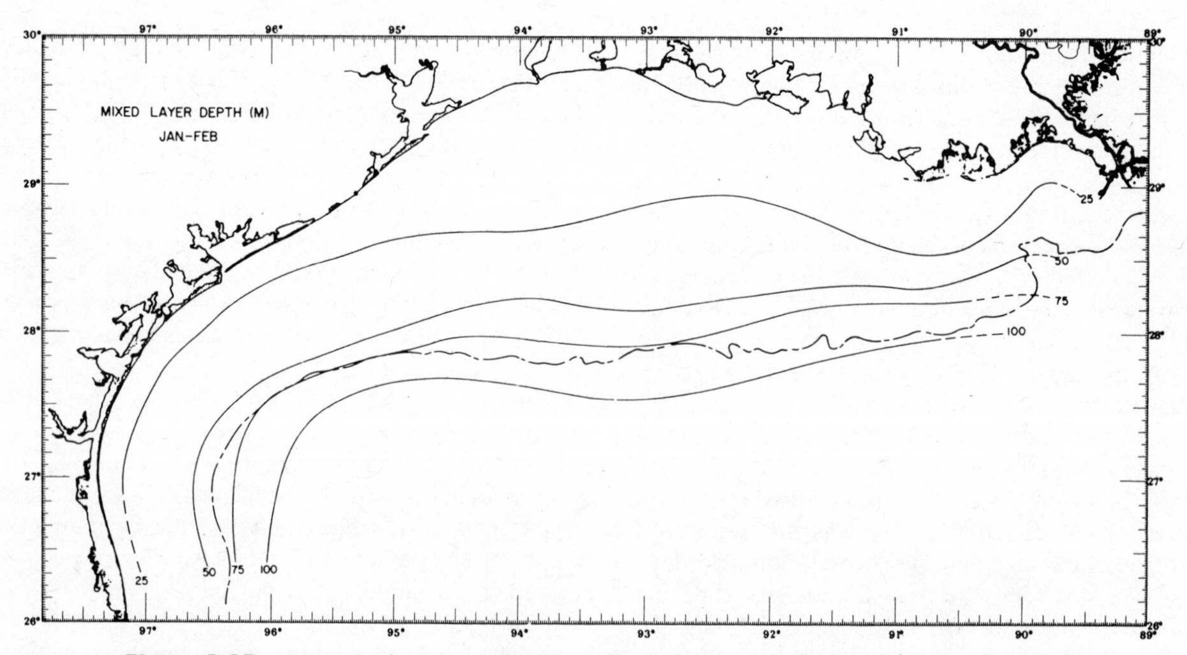

Figure 2.25. Map of mixed-layer depths (meters) for January–February. Dashed lines indicate lack of data and curvature of these lines suggests most probably tendency (from Etter and Cochrane, 1975).

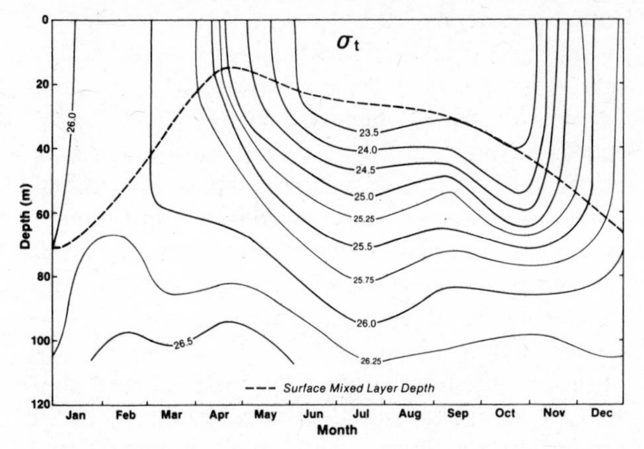

Figure 2.26. Seasonal progression of sigma-t versus depth near the Flower Garden Banks, constructed from CTD data collected by TAMU in April, September, and October 1979, October 1980, and March and July 1981; monthly averages of temperature time-series data recorded on TAMU current meter moorings near the banks; and from GUS III station W-6 (27°55′N, 94°36′W), cruises 13 to 24 of 1964 (Angelovic, 1976). The approximate depth of the surface mixed layer is indicated by the dashed line.

period. That near the coast is essentially to the west, whereas the wind stress at the site of buoy 42002 is directed toward the northwest. In both records, however, there are three major events during which the wind is out of the northeast. The first started on 22 March, the second, on 9 April, and the third, on 21 April (Figure 2.28).

The effect of these events on the currents may be observed in Figure 2.29. The strong wind event which started on 22 March reversed the surface flow at mooring 2 and turned it from easterly to westerly. At mooring 3 the flow from 43 to 89 m changed from north–northwest to west. The near-bottom flow on moorings 1 and 2 accelerated to the southwest and the deeper flow at mooring 3 was veered from southeast to west.

The event on 9 April appears to have been too short-lived to have made any significant imprint on the flow. The mid-depth current (60 m) at mooring 2 returned to easterly flow during the period of south and southeast winds following the 9 April event, as did the bottom flow at moorings 1 and 2. The current at the level of the two upper meters of mooring 3 was also swinging subtly to the northwest during this

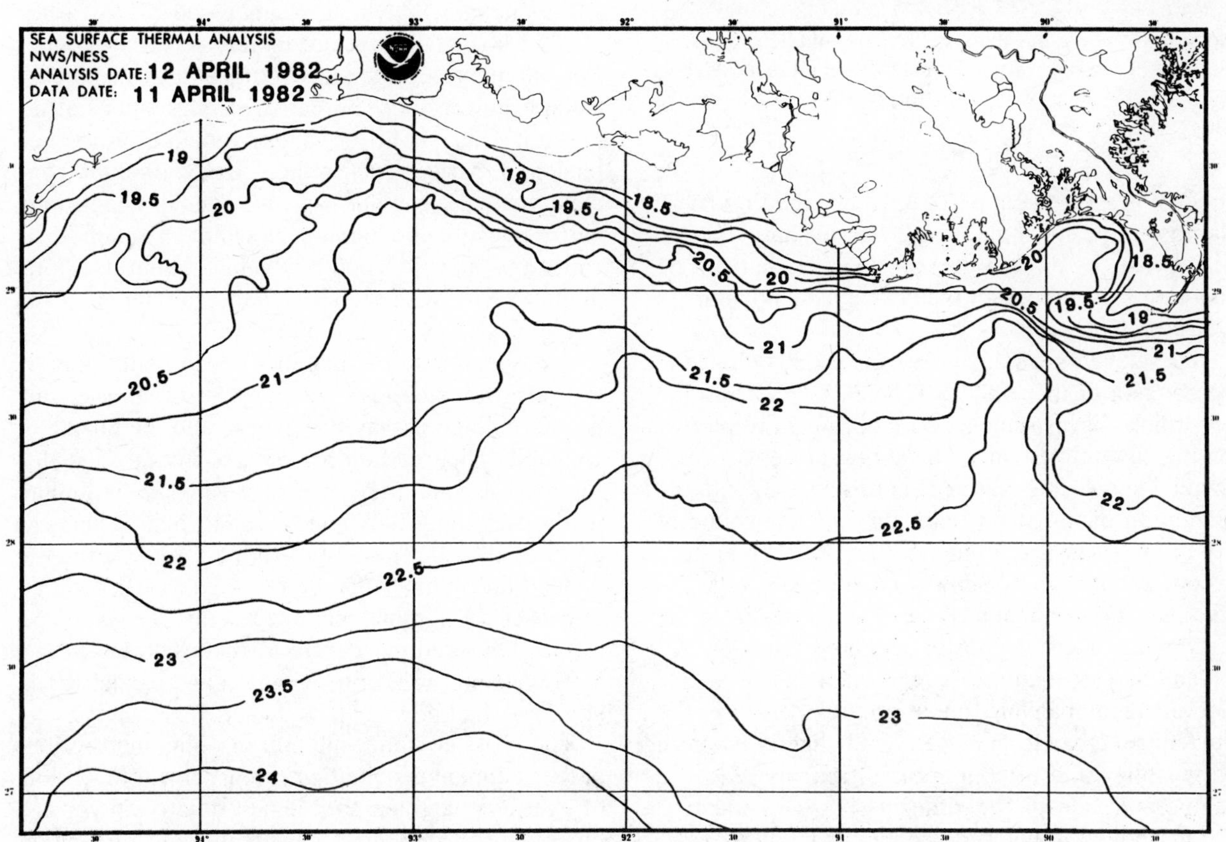

Figure 2.27. Sea-surface temperature based on satellite infrared imagery for 11 April 1982. (Courtesy of R. Barazotto, NWS/NESS, Slidell, Louisiana.)

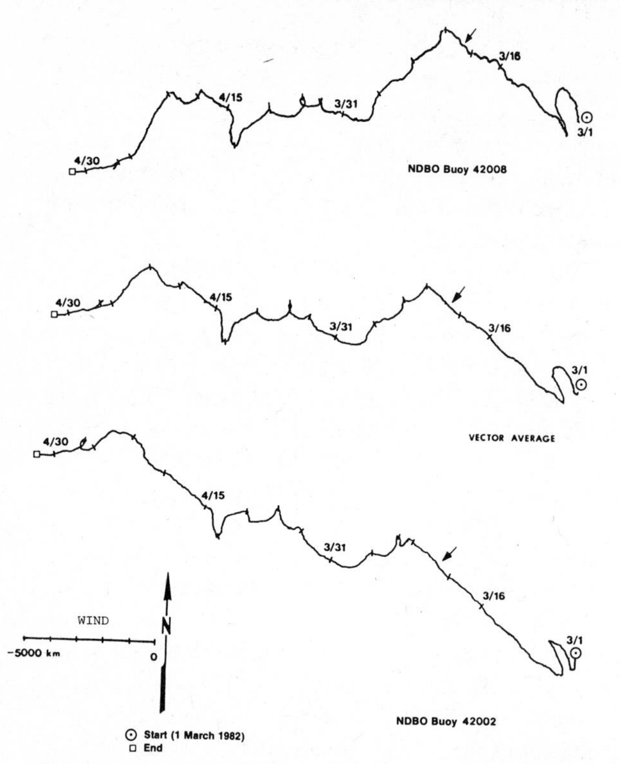

Figure 2.28. Progressive vector diagrams of wind from 1 March to 3 April 1982 at NDBO buoys 42008 and 42002 (locations shown in Figure 2.13). The middle diagram is the vector average of winds at 42008 and 42002. Arrows indicate the dates when current meters were deployed along the 95W transect.

time. When the event of 22 April hit, the flow at all depths at the site of mooring 2 veered sharply to the southwest, as did the current near the bottom on mooring 1. At mooring 3 the flow at all depths turned sharply offshore.

These events are shown in Figure 2.30, which the vector plot of the wind from NDBO 42002 and the recordings of instruments on mooring 2 are plotted on the same time scale. The temperature record of meter 1 shows little variance in this period, with the exception of a distinct warming about the time of the frontal passage in late April. The temperature record at meter 3 also shows only a general warming trend until the frontal passage (seen as a shift in the wind from out of the north) appeared in late April, when the temperature dropped nearly 3°C in six hours. Accompanying the drop in temperature, the current reversed from easterly and slightly onshore to westerly and offshore. Simultaneous with the temperature drop, the transmissivity on meter 2 (20 m above the bottom) dropped from 50 to 18%, which means that the water changed from rather clear to heavily laden with sediment.

Tide. The model of the tides in the Gulf of Mexico by Reid and Whitaker (1982) is in excellent agreement with the tidal currents we measured at the shelf edge near the Flower Garden Banks (Table 2.2). Our records do indicate phase and amplitude changes in the M_2 and K_1 tide, which implies that the baroclinic, or internal, mode is important to both constituents. Stronger K_1 internal tides are observed in the summer when stratification is at its maximum but the M_2 internal tide does not show an appreciable seasonal variation. From Figures 2.3 and 2.4, it is apparent that the tidal currents on the east–west segment of the Texas shelf are highly ellipitcal and increase in strength toward the shore. The tidal currents on the north–south segment appear to be greatly attenuated for the M_2 and K_1 constituents of the tide. In both areas the major axis of the tidal current ellipses are nearly perpendicular to the coast.

SEDIMENT DYNAMICS AND THE NEPHELOID LAYER

The term "nepheloid" is derived from the Greek word "nephele," which means cloud. Thus, a nepheloid layer is a layer of cloudy or turbid water. For sedimentologists it has come to mean sediment-laden water, usually, but not always, found near the bottom. There is no quantitative definition for nepheloidal, partly because there is a continuum from clear to opaque water as sediment is added and partly because the amount of sediment required to cause any specific amount of light attenuation (cloudiness) varies with the size, shape, and index of refraction of the sediment particles.

Measurements of suspended sediment concentrations are made in a variety of ways. In the simplest a volume of water is filtered and the filtrate is weighed. This method provides a measure usually reported as milligrams per liter (mg/L). In similar method an aliquot of water is run through a particle counter like the Coulter Counter. The sediment concentration may then be reported as the number of particles per cubic centimeter (cm^3). To reference a value reported per cm^3 to a value reported in the mg/L system the size distribution and particle density must be known.

The most common method of obtaining profiles of the sediment distribution is to use the attenuation of light over a given path length (relative transmittance) or the amount of light scattered (forward or back) over a given path. Instruments that use the first technique are called transmissometers; instru-

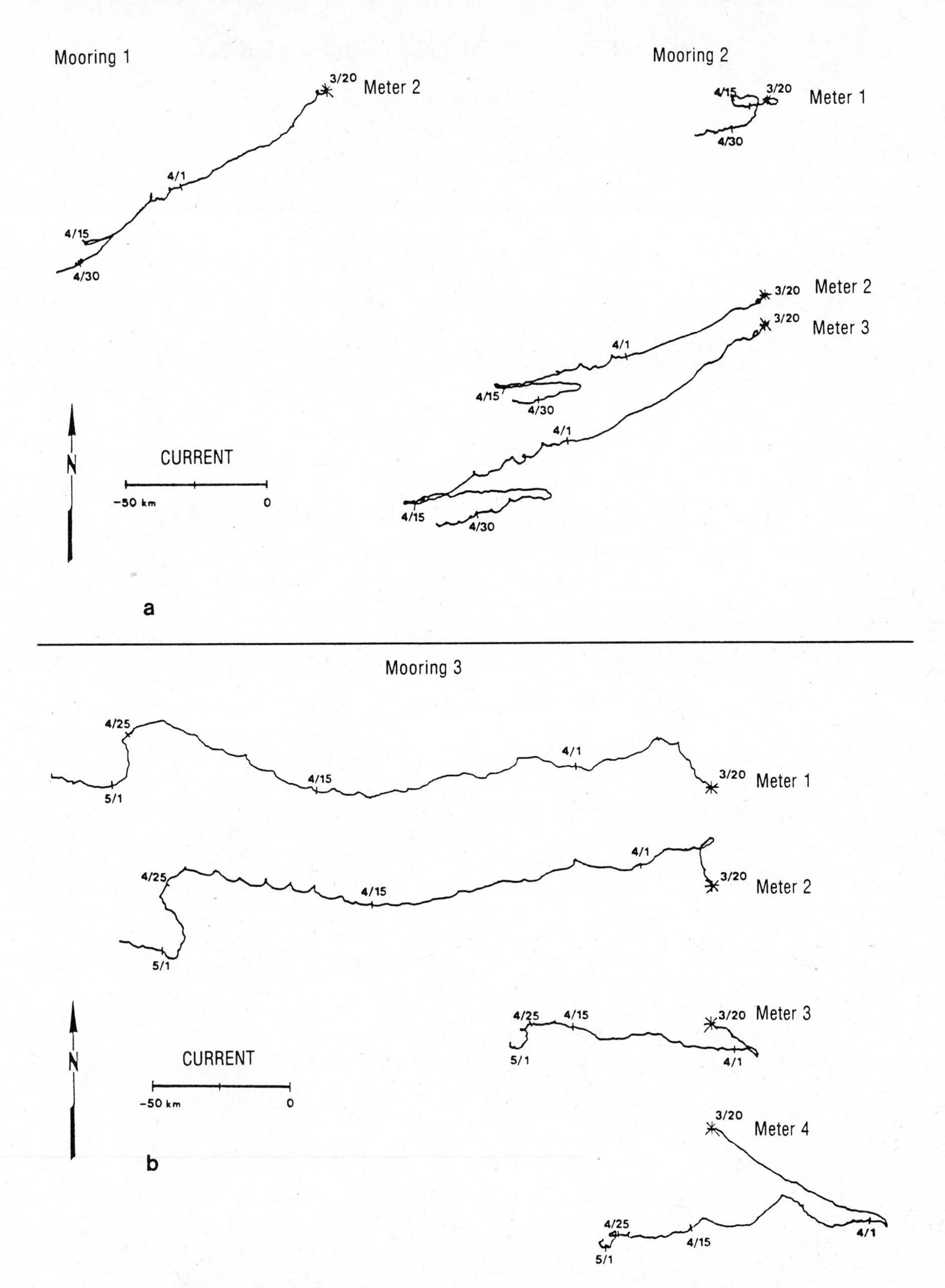

Figure 2.29. Progressive vector diagrams of currents measured at the 95°W array: (*a*) moorings 1 and 2 from 20 March to 30 April 1982; (*b*) mooring 3 from 20 March to 1 May 1982. (See Figure 2.13 for location.)

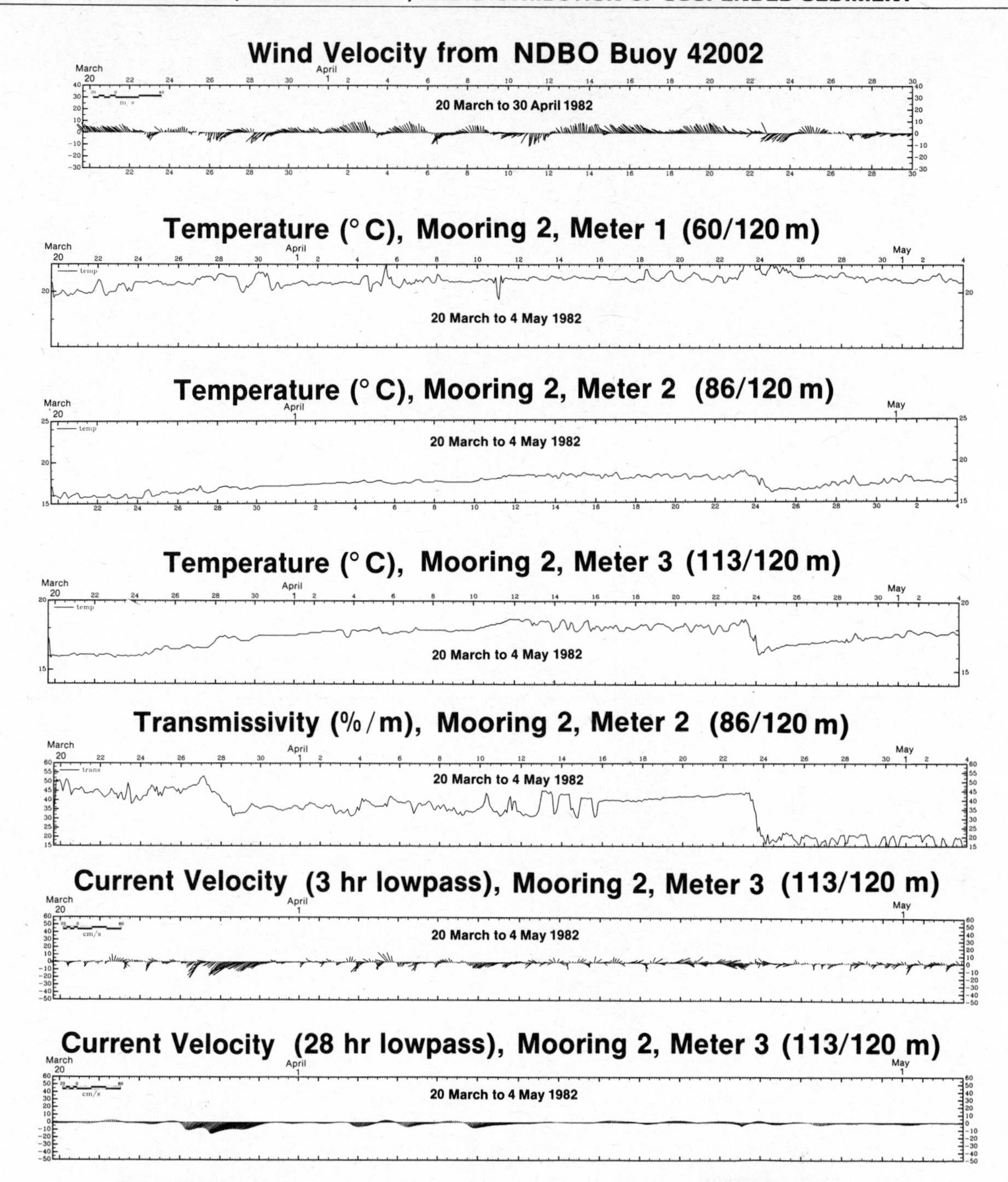

Figure 2.30. Time-series records of the wind from NDBO buoy 42002 and the water temperature, transmissivity, and current velocity from mooring 2 of the 95°W array between 20 March and 4 May 1982. Note the drop in temperature and transmissivity during the cold-front passage on 24 April 82.

ments using the second technique are commonly called nephelometers. Transmissometers yield values in percent transmittance per meter. The higher the value, the clearer the water. These values are referenced to mg/L by calibrations and by collecting water in a bottle at multiple points in the profile, filtering the samples from the bottles, and fitting the transmissivity values to the absolute mg/L values obtained by the filtration method.

Acoustic instruments which use acoustic reflectance from the particles over a known pathlength to obtain profiles have also been developed. These in-

TABLE 2.2. Amplitude and Phase of the Barotropic Tidal Currents Near the Flower Garden Banks for the M_2 and K_1 Constituents[a]

	u (East–West)		*v* (North–South)	
Methods[b]	Amplitude (cm/s)	Phase (°)	Amplitude (cm/s)	Phase[c] (°)
	M_2 *Tidal Constituent*			
Measured by vector averaging	0.6	99	1.4	354
Model, WFG	0.3	116	1.3	357
Model, EFG	0.3	132	1.1	354
	K_1 *Tidal Constituent*			
Measured by coherence method	0.7	335	1.1	200
Model, WFG	0.7	344	1.2	213
Model, EFG	0.6	349	1.1	209

[a]The measured values are those computed from current meters moored at mid-depth near the banks.

[b]Three techniques were used for obtaining values. The vector averaging and coherence methods were used to eliminate the effects of internal tides. The model values were obtained for the East Flower Garden (EFG) and West Flower Garden (WFG) regions from a numerical model of the barotropic tides in the Gulf of Mexico developed by Reid and Whitaker (in press).

[c]The phase is relative to the local tidal potential.

struments must be referenced in the same manner as light instruments.

Unfortunately, the calibration techniques for the profiling instruments possess so much scatter that most reports of suspended sediment distributions are in relative rather than absolute terms of concentration.

Nepheloid layers occur only when the turbulence of the water is high enough to offset the settling of the sedimentary particles under the influence of gravity. The larger the particles, the more intense the turbulence must be to maintain a suspension. Nepheloid layers are therefore usually composed of silt and clay particles because only the most energetic flows can maintain a sand suspension.

South Texas Shelf

The first comprehensive study of the nepheloid layer on the South Texas Continental Shelf was published by Shideler (1981), who used a transmissometer and Coulter Counter to study the distribution of suspended sediment on the shelf. Shideler and his colleagues made six cruises in 18 months, occupying the same stations on each cruise to examine spatial and temporal variances in the nepheloid layer. Shideler found that the nepheloid layer thickened offshore to a maximum of 35 m near the shelf break and that the concentration of suspended sediment in the nepheloid layer decreased from a maximum near shore to a minimum at the shelf break. He also found that the sediment in the nepheloid layer was dominated by inorganic detrital minerals. In addition to the offshore thickening of the nepheloid layer, Shideler found that it reached its maximum thickness in the central part of his study area, in the embayment between the drowned Rio Grande Delta and drowned Brazos–Colorado Delta. Further, he found that the nepheloid layer was thinner and of smaller areal extent in the fall than in the spring. He also shows a noticeable thickening of the nepheloid layer across the shelf toward the inlet at Matagorda Bay in his two March cruises (1976 and 1977) and two May cruises (1976 and 1977) but not in his November cruises (1975 and 1976).

Shideler concluded that the nepheloid layer is generated and maintained by resuspension of muddy seafloor sediment as a result of bottom turbulence. He then offered a number of suggestions regarding the agents that might cause such resuspension. He showed that surface gravity waves in calm weather could induce resuspension inshore of only about 10 m and even storm waves with 10-second periods are unlikely to be important in resuspension beyond 40-m depth. Beyond that his conceptual model is flawed by a lack of oceanographic data. He proposed that the mid-and outer-shelf resuspension is generated by turbulence along the front that migrates back and forth under the influence of the tide, wind, and intrusions of the Loop Current. However, as revealed in the data from the Reid and Whitaker (in press) model of the tides (Figures 2.3 and 2.4), tidal currents are on the order of 1.5 cm/s or less on that stretch of the shelf for the M_2 and K_1 constituents. Also, the Loop Current has never been reported west of the Yucatan Peninsula.

The flow presented by Cochrane and Kelly (1982) in Figures 2.11 and 2.12 does offer a suggestion for the trend of the nepheloid layer toward Matagorda Bay in the spring. During this period the coastal convergence zone is migrating upcoast and the turbid coastal waters carrying effluent from the Brazos and Colorado Rivers may be diverted offshore and to the south. In November, however, the flow appears to hug the coast.

The current meter records from the CSA site near Baker Bank show that near-bottom velocities at about the 75-m depth frequently exceed 20 cm/s (Figure 2.31), which is ample to keep silt and clay in suspension.

Holmes (1982) analyzed the mineralogy of the suspended sediment in the nepheloid layer and that of the substrate on the South Texas Continental Shelf. He reports that the dominant clay minerals in the bottom sediment are members of the smectite group and that these minerals make up 34 to 75% of the clay fraction; the mode is near 60%. In contrast, he found that the suspended sediment in the nepheloid layer is, in general, devoid of the smectite clays. He goes on to say that the near-bottom current velocity required to erode the mud on the mid- to outer shelf is in the order of 10 to 30 cm/s. He then says that no currents of this magnitude have been measured on the seafloor, even during a storm passage. On the basis of these two points—the absence of smectite in the nepheloid layer and the low supposed current velocities—Holmes (1982) concluded that the nepheloid layer is unrelated to the bottom sediment. He suggests that a thin veneer of sediment is altered biogenically to illite (another clay mineral) and made available for resuspension by annelid worms. Holmes appears to be wrong on both points.

A sample of the bottom current meter (69 m in 73 m of water) record from the Baker Bank site is shown in Figure 2.31. Note that the speed of the current almost always exceeded 10 cm/s and reached maxima of more than 35 cm/s. These peak velocities probably occurred during frontal passages. Also, in their report to Conoco, Inc., Continental Shelf Associates (1982) detailed the clay mineralogy of the bottom and suspended sediments and the sediment retained in sediment traps from the vicinity of their current meter mooring. They reported the smectite

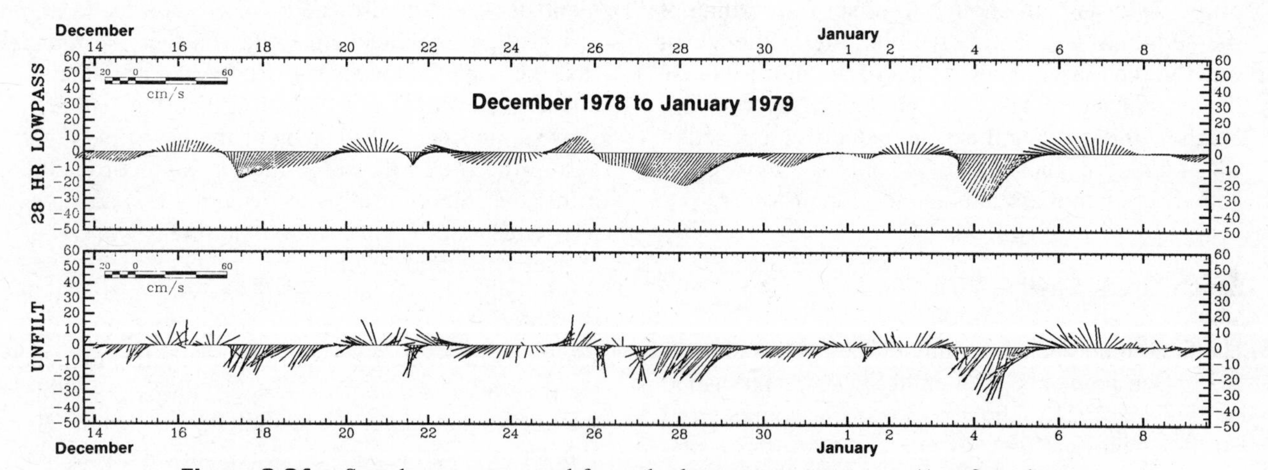

Figure 2.31. Sample current record from the bottom current meter (4 m from bottom in 73 m of water) of the CSA mooring near Baker Bank. The lower frame is the raw unfiltered data, whereas the upper frame shows the same record after filtering to remove currents with periods shorter than 28 hours. In this "stick plot" representation the length of each stick is the magnitude (cm/s) and the orientation is the direction (in map coordinates) of the current at each point in time.

group minerals under their alternate name, montmorillonite. Sediment from the traps averaged 43% montmorillonite; suspended sediment averaged 56.7% during one cruise and 53.2% during another; and the bottom sediment averaged 49.5% montmorillonite.

It is likely that Holmes used too small a sample of suspended sediment for X-ray analysis or suffered some other procedural error. Even without Continental Shelf Associates' evidence that the mineralogy of the suspended sediment and substrate is identical, it presses the credulity to suppose worms could change the mineralogy of the clays, then supply the altered sediment to the water column and not the substrate.

In its essentials, the hypothesis of Curray (1960) regarding the source of mud on the South Texas continental shelf if probably correct. However, we find no evidence to suggest that hurricanes or tidal currents are important agents of sediment transport beyond mid-shelf depths. The general picture is that surface gravity waves and swift alongshore currents keep water inside the 10 m isobath turbid with resuspended sediment and preclude significant deposition of silt and clay. When the shelf flow is toward the north (summer), fine sediment from the Rio Grande is swept to the north along the coast and in the bottom boundary layer. This must have been a much more important source prior to the diversion of the Rio Grande waters for agricultural purposes. When the shelf flow is to the south (fall and winter), sediment from the Colorado, Brazos, and perhaps even the Mississippi rivers is swept to the south in the bottom boundary layer. Depostion from these sources has produced the mud cover on the shelf between the two shelf-edge drowned deltas.

The nepheloid layer on the South Texas Shelf appears to be derived from both local resuspension of the fine sediment and advection in the bottom boundary layer from coastal sources.

East Texas–Louisiana Shelf

In the period from 1979 through 1982 we investigated sediment dynamics on the east–west segment of the Texas–Louisiana Shelf. Two approaches were used: (1) surveys in which profiles of salinity, temperature, transmissivity, and horizontal current velocity components were taken on closely spaced stations; and (2) moored instruments that provided time-series measurement of temperature, transmissivity, and current velocity. In addition, we used *in situ* flow visualization techniques to study the behavior of the bottom boundary layer. Results of these studies up to 1981 were summarized in McGrail and Carnes (1983).

Cross-shelf transects were run in March 1981, March 1982, and May 1982 with station spacings of approximately 10 NM on the shelf and 5 NM on the shelf break and upper slope. On the March 1982 cruise four shelf-perpendicular transects and one shore-parallel transect were run (Figure 2.32). Also, three current meter moorings were established at the shelf break along 95°W (Figure 2.13). The May 1982 transect was run from the shore to the current meter moorings along 95°W (Figure 2.33). The moorings were recovered at that time.

By combining and analyzing all of these data we are able to offer the following picture of shelf sediment dynamics.

In the winter cold-air outbreaks and their attendant wind events homogenize the nearshore waters and produce surface gravity waves capable of considerable bottom agitation to depths of perhaps 40 m. The data of Cochrane and Kelly (1982) indicate that nearshore currents may exceed 75 cm/s during these frontal passages. The waves and currents combine to fill the nearshore waters with high sediment loads. From Nowlin and Parker (1974) and Wiseman et al. (1982) it is known that the frontal passages also create cold saline waters in this region

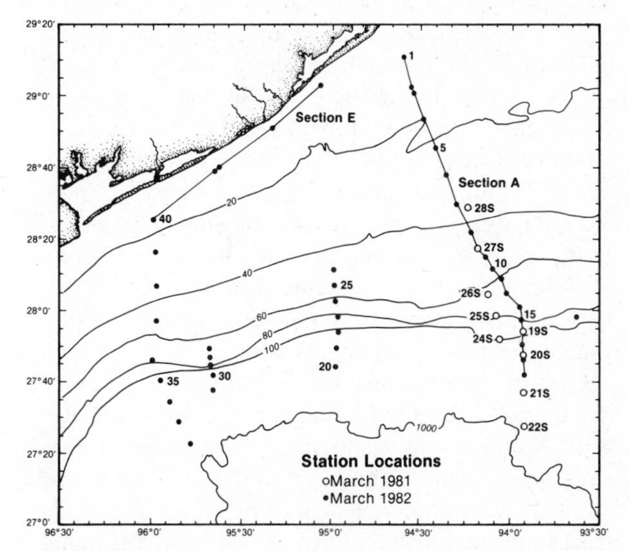

Figure 2.32. A map showing the locations of stations taken during March 1981 and March 1982 cruises over the Texas–Louisiana Shelf and slope. Sections A (across shelf) and E (along shore) are indicated by the connected stations.

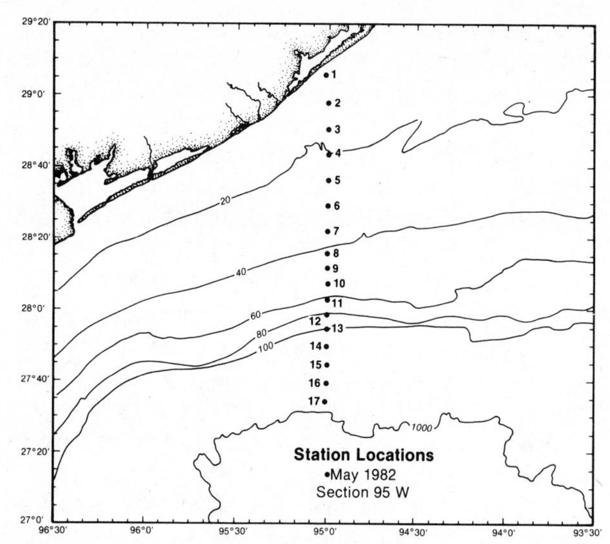

Figure 2.33. A map showing the location of stations on the cross-shelf transect along 95°W over the Texas–Louisiana Shelf and slope in May 1982.

which sink and run offshore along the bottom because of their excess density. Also, the bottom boundary layer of westward-flowing current in a rotating system possesses southwesterly flow (see Weatherly, 1975, for a discourse on Ekman bottom boundary layers). Sediment entrained in this bottom flow is carried to the outer shelf. The bottom mixed layer and nepheloid layer thicken in the offshore direction as the bottom waters flow off the sloping shelf and out over water of equal density on the outer shelf. The mean flow on the outer shelf at the location of the Flower Garden Banks, which is toward the east, even during the frontal passages, sets up a northeastward flow in the bottom boundary layer offshore and causes convergence in the bottom waters. The convergence appears to migrate back and forth over the shelf break. This migration shows up in the bottom current meter records as oscillation between onshore and offshore flow, with attendant decreases and increases in temperature and fluctuations in transmissivity levels.

Occasionally the flow on the outer shelf reverses and flows to the west, as it did in December 1980 at the Flower Garden site and during most of the 6-week period that the moorings at the 95°W site were out. It is not entirely clear whether these reversals represent a displacement of the eastward-flowing shelf-slope current seaward by unusually strong or persistent north winds or an actual reversal of the shelf-edge flow field.

During the March 1981 cruise the transect revealed a classic picture through the flow-field portrayed by Cochrane and Kelly (1982) (Figure 2.11 and 2.12). Inshore of the 70-m isobath the isotherms sloped up in the offshore direction and the flow was to the northwest at the surface, west at mid-depth, and southwest at the bottom. Seaward of the 70-m isobath the isotherms were steeply inclined down in the offshore direction with very strong (75 to 100 cm/s) flow to the east. Near-bottom flow in most outer stations had a northerly component to it. A plume of turbid water extended from the nepheloid layer on the shelf out over the slope along the 17°C isotherm (McGrail and Carnes, 1983).

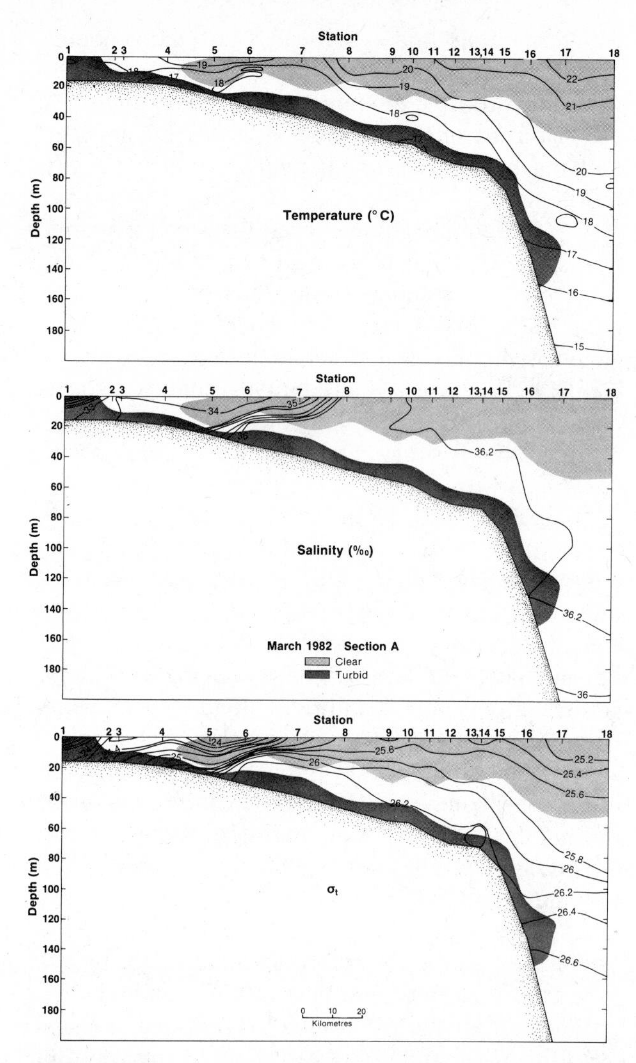

Figure 2.34. A cross-shelf vertical section of temperature, salinity, and sigma-*t* along section A, occupied in March 1982 (see Figure 2.32 for station locations). Regions of turbid and clear water are indicated by darker and lighter shading, respectively. Unshaded regions are of intermediate water transparency. The contour interval for salinity is 0.2 ppt down to 35 ppt and 1 ppt for salinities less than 35 ppt.

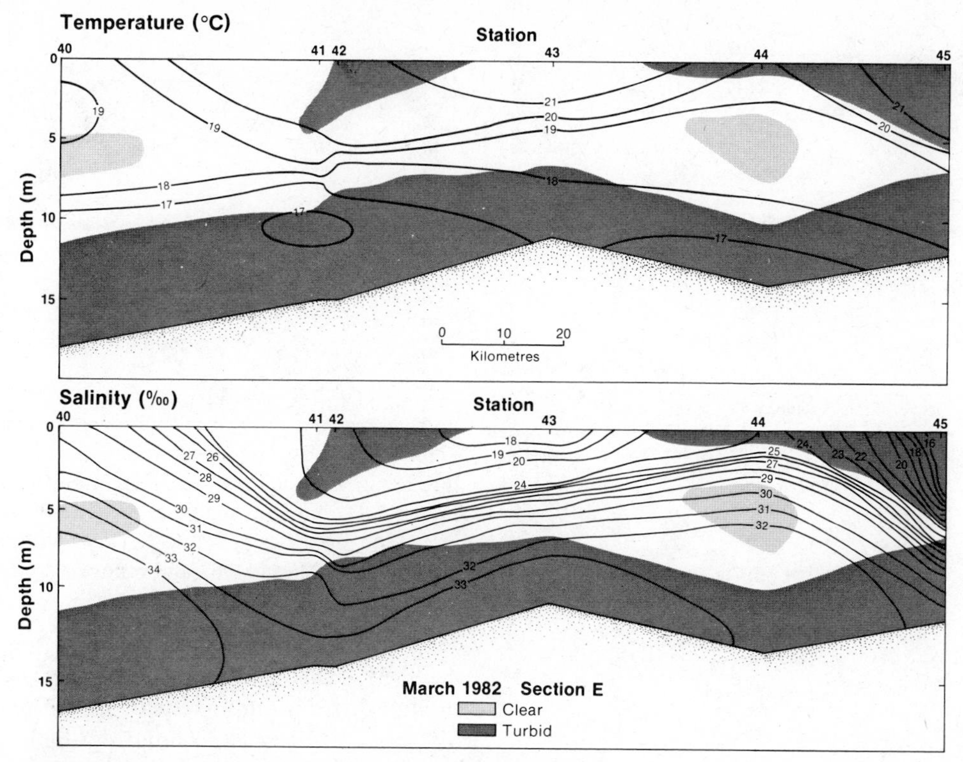

Figure 2.35. Along-shore vertical sections of temperature and salinity along transect *E* occupied in March 1982 (see Figure 2.32 for station locations). Regions of turbid and clear water are indicated by the darker and lighter shading, respectively. Unshaded regions are of intermediate water transparency. Regions marked clear in this figure are as turbid as regions marked turbid in transect *A* (Figure 2.34).

In March of 1982 the same transect revealed a somewhat different picture (Figure 2.34). The change in slope of isotherms occurred over the 40-m isobath, much closer to shore. The velocity profiles were rather chaotic, partly because of instrument malfunction and partly because the shelf seemed to be undergoing a transition, with strong southerly components to the nearshore flow. As shown in the plot of sigma-t (density minus 1 multiplied by 10^3), it is obvious that there is a strong nearshore front in the water column, with an inflection in the upper isopycnals at station 5. The slope of the isopycnals in the offshore region implies that flow was to the east, which is consistent with our profiling current meter records and with records from a current meter moored at the West Flower Garden Bank by Continental Shelf Associates for Union Oil Company.

Note the extensive wedge of low-salinity water extending nearly 90 km across the surface of the shelf waters. Also note that oceanic values of salinity extend within 60 km of the coast. This wedge is undoubtedly the leading edge of the February discharge maximum from the Mississippi River reported by Cochrane and Kelly (1982). The influence of this brackish water can also be seen in the lens of relatively dirty water that extended 50 km seaward in a low salinity wedge.

Inshore of the 10-m isobath the water was turbid from top to bottom. The nepheloid layer at the base of the water column up to 50 km offshore was heavily laden with suspended sediment. Transmissivity values as low as 5%/m for 660-nm light (red-light-emitting diode) were observed 3 m above the bottom in this region. This is, of course, the region in which the haline front intersects the bottom. Farther offshore the minimum transmissivity values were in the order of 20 to 30%/m, values that correspond to somewhat more than 1 mg/L. The nepheloid layer extended across the shelf in a well-mixed bottom layer, 10 to 15 m thick, and spilled over onto the slope.

Section E, run on the same cruise as section A in March of 1982, is a nearshore, coast-parallel section (Figure 2.35) which shows the alongshore variability in salinity, temperature, and suspended sediment concentrations. The transect cut across

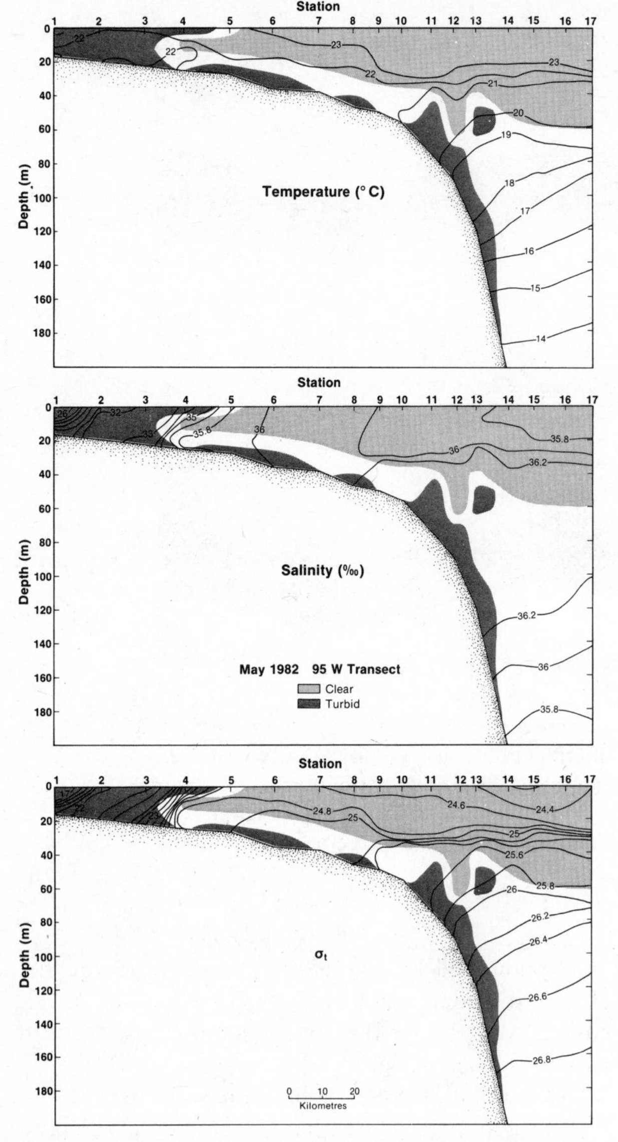

Figure 2.36. Cross-shelf vertical sections of temperature, salinity, and sigma-*t* along 95°W, occupied in May 1982 (See Figure 2.33 for station locations). Regions of turbid and clear water are indicated by darker and lighter shading, respectively. Unshaded regions are of intermediate transparency. The contour interval for salinity is 0.2 down to 35 ppt and 1 ppt for salinities less than 35 ppt.

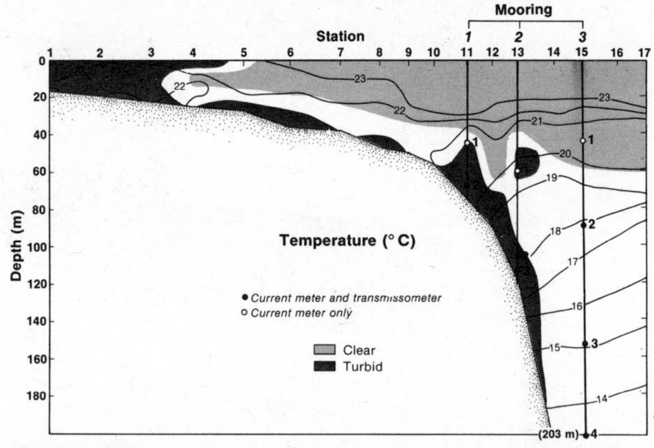

Figure 2.37. Cross-shelf vertical sections of temperature along 95°W across the Texas–Louisiana Shelf, as in Figure 2.36, but showing locations and depths of current meters on moorings 1, 2, and 3. These moorings were deployed on 19 March 1982 and recovered on 4 May 1982.

perturbations in the wedge of brackish, turbid water that serves to mix shelf and coastal waters. These eddies cause large-scale diffusion of sediment into offshore waters and onshore diffusion of salt into coastal waters. Figure 2.36 shows the distribution of mass properties and suspended sediment in May of 1982 when the current meters were recovered. The location of those current meters in relation to the 95°W transect is shown in Figure 2.37. From the salinity distribution it is clear that effluent from the Mississippi River is dominating the coastal zone. This wedge of brackish water extends out almost 50 km at the surface and 40 km at the bottom, making the front much steeper than it was in March.

The isopycnals, with few exceptions, all slope down toward the coast, which implies westward flow, as the current meter confirmed. Again the nearshore water, inside 15 m, was very turbid. The turbidity once again extended seaward in the brackish water at the surface out to approximately 50 km. A tongue of clear water penetrated within about 35 km of the coast between 8 and 13 m depth.

The cross-shelf distribution of suspended sediment in the nepheloid layer is substantially more complex than it was in section A. Over the middle shelf the nepheloid layer varies in thickness from about 5 to 15 m, with minimum values of transmissivity essentially the same as on section A. At the shelf break, however, the nepheloid layer wells up to more than 25 m in thickness, with minimum transmissivity values in the order of those in coastal water. In addition, there was an isolated core of turbid water off the shelf break centered at a depth of about 60 m.

From the records of the moored instrument it must be concluded that this turbid water was delivered to the shelf break to the east of 95°W during the episode of strong offshore flow or was actively resuspended by strong local currents at the time of the transect. The current meter records (Figure 2.30) strongly suggest that the former is the more likely explanation.

It appears that the sediment is kept in suspension over much of the inner shelf by swift currents and turbulence. It is then advected to the shelf edge, where episodic deposition may take place. Some of the sediment is clearly swept over the shelf edge to the slope and deep basin. It may be that the increase in tidal currents shown by Reid and Whitaker (in press) (Figures 2.3 and 2.4) help to keep sediment from accumulating to any great extent in the central area of this shelf segment. The surface gravity waves and strong wind-driven coastal currents keep the deposition of mud to a minimum in the nearshore.

In the summer months the supply of sediment to the outer shelf is more restricted because of the lack of major wind events and the diminished supply of sediment at the coast. Otherwise, the bottom boundary layer on the inner shelf should carry fine sediment to the southwest when it becomes entrained in the southeast flow.

Our current meter records from the shelf edge suggest that hurricanes are not the important agent of sediment distribution that Curray (1960) hypothesized. The reason is that hurricanes occur when the shelf is still stratified and do not directly force the bottom flow on the outer shelf. In fact, they appear to produce effects on the outer shelf similar to those of a frontal passage. Obviously a hurricane's influence in shallow water would be much more extreme. This is well illustrated by the current meter records of Hopkins and Schroeder (1981) from the

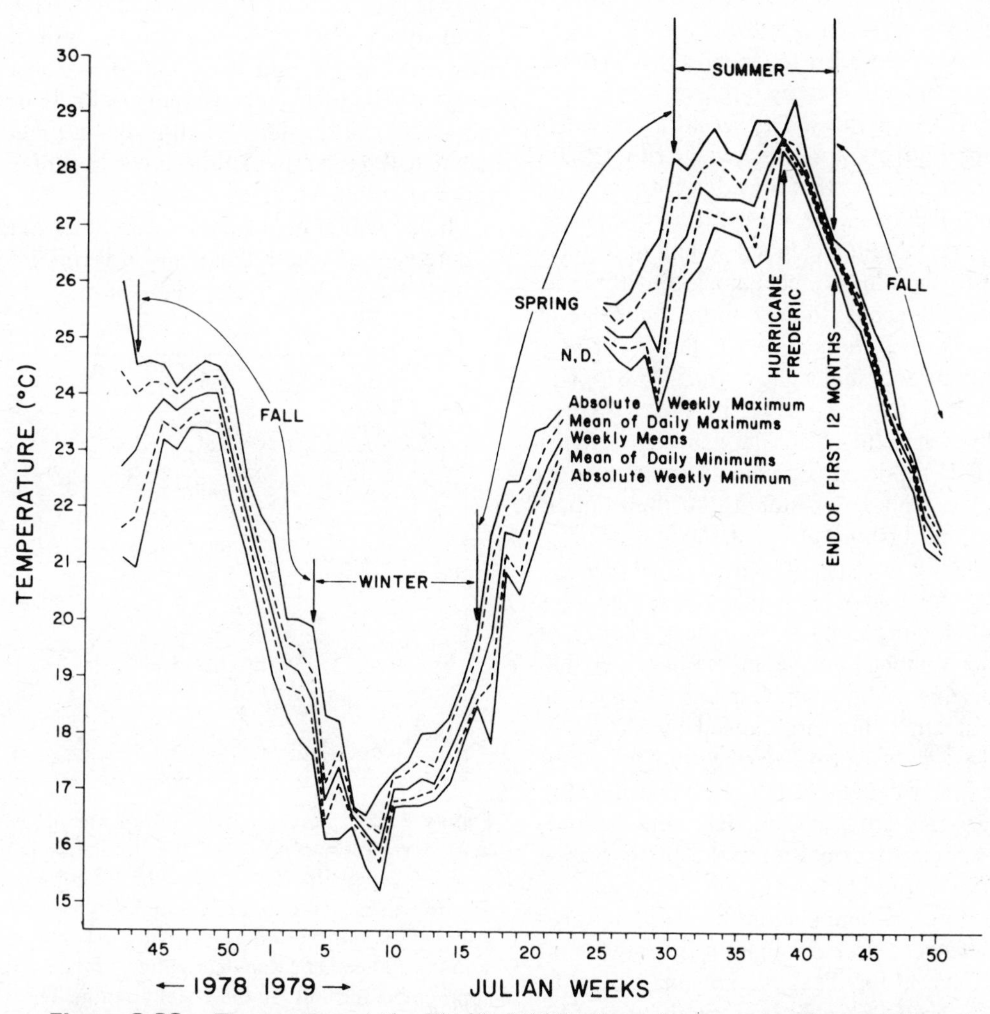

Figure 2.38. Temperature at the Florida Middle Ground versus time during 1978 and 1979 (Hopkins and Schroeder, 1981). The observation site is shown on Figure 2.1. Note that minimum temperatures in February and March range from above 15 to about 16.5°C.

Florida Middle Ground in 35 m of water during the passage of two major hurricanes, David and Frederic, in 1979. Still, the effects of the storm were short-lived.

In summary, active deposition on this segment of the shelf appears to be limited to the coastal waters of Louisiana and the outer shelf and upper slope of the whole region. It also appears that fine sediment is advected over long distances in the nepheloid layer before being deposited.

Circulation and Hydrography on the Florida Shelf

The Florida Shelf seems to undergo much the same seasonal temperature cycles as the Texas Shelf. The annual temperature record for the Florida Middle Ground is given by Hopkins and Schroeder (1981) (Figure 2.38; see Figure 2.1 for location). Currents at this site are characterized by large variances and low means (less than 2 cm/s). The West Florida Shelf receives comparatively little freshwater runoff. Because of that and the fact that the adjacent mainland is a carbonate platform, the West Florida Shelf receives no appreciable terrigenous sediment at all.

Niiler (1976) showed transects along the 26°N line of latitude from the coast to the continental slope in June and February. The salinity at the coast was 36.0 ppt in June and slightly less than 35.8 ppt in February. These values reflect the lack of fresh water contribution to the shelf. Mixing over the West Florida Shelf is reported by Niiler to penetrate to 100 to 150 m in winter. He stated that the temperature of the water inshore of about 100 m decreases through the winter because of its decreased heat capacity during the cold-air outbreaks. He also says that the shelf hydrography shows the presence of detached and semidetached segments of water that have their origins in the Loop Current. He goes on to demonstrate that they are caused by shear between the fast-flow Loop Current and the shelf waters. The shear causes instabilities that develop into meanders that grow and become detached eddies. These appear to propagate to the north against the prevailing southward flow at the shelf edge.

The absence of silt and clay in the sediment provides much clearer water throughout the water column than exists on the Texas–Louisiana coast, as shown by Steward (1981). His transmissivity profiles and hydrography from the Alabama Shelf, however, show that it is similar to the Texas–Louisiana Shelf, with a well-developed nepheloid layer.

CONCLUSIONS

With respect to the biotic communities of the Gulf it has been demonstrated that Caribbean populations can be carried north and east along the outer continental shelf off Mexico, Texas, and Louisiana in an anticyclonic circulation pattern. Similar communities could be delivered to the Florida Shelf by eddies that spin off the Loop Current and propagate to the north over the outer shelf.

The West Florida Shelf and Yucatan Shelf are similar in that both are floored by carbonate sediment (Figure 1.2) and neither suffers appreciable salinity variations.

The shelves of Mexico, Texas, Louisiana, Mississippi, and Alabama are all floored with terrigenous sediment that contains varying amounts of silt and clay (Figure 1.2). They also receive freshwater input that alters the nearshore salinity gradients. The presence of silt and clay, and other observations, suggests that nearshore waters are likely to be turbid top-to-bottom inside the 10-m isobath and that the shelf will have a nepheloid layer that may reach 35 m or more in thickness.

In the winter cold fronts create deep mixed layers. Shoreward of where these mixed layers intersect the

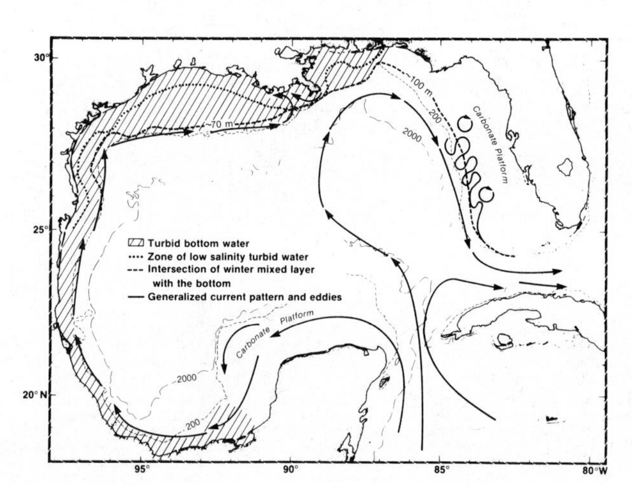

Figure 2.39. A map of the Gulf of Mexico which summarizes the primary near-surface circulation patterns that may be responsible for carrying biotic communities into the Gulf and from there anticyclonically around the western Gulf or to the Florida Shelf via the Loop Current and spin-off eddies. The diagonal pattern indicates regions of appreciable turbidity in the water column. The dotted line delineates seaward margin of near-shore, low-salinity turbid water. The dashed line near the outer continental shelf indicates depths at which the winter mixed layer intersects the bottom.

bottom, the temperature of the water may drop very low because of the limited heat capacity of the water. As each front comes through it extracts more heat so that coastal waters off Texas may drop to 10 or 12° C. Offshore of the intersection of the mixed layer with the bottom, water temperatures remain relatively high because of the large heat reservoir of the thick lens of water. At the shelf edge of Texas the water at a depth of about 50 m remains warmer than 17.5°C all year and above 18°C most of the year.

Figure 2.39 summarizes these characteristics of the shelf habitats.

3

A SUMMARY OF HARD-BOTTOM BIOTIC COMMUNITIES IN THE GULF OF MEXICO

The purpose of this chapter is to provide an overview of the biotic communities that occupy hard bottoms in the Gulf of Mexico as background for more detailed accounts to follow of these communities on the outer continental shelf off Texas and Louisiana. The inflow of warm tropical water from the Caribbean, which carries larvae from West Indian reef ecosystems, is of overwhelming importance in determining the nature of the Gulf of Mexico benthos. The resultant planktonic "ambience" of tropical forms provides a broad potential for adult community structure, which is selectively expressed at various locations on the continental shelf.

Characteristics of benthic communities depend primarily on the structural and sedimentological nature of the substratum, river outflow, and seasonal and regional variations in water temperature, salinity, and turbitidy (see Chapters 1 and 2). In the northern Gulf, which is in many ways a differentially stressed ecotone for tropical biota, the influence of coastal marine and climactic environments is particularly important in determining the degree of expression of tropical benthic communities. Thus assessment of marine biogeographical relationships in the Gulf and adjacent regions requires a view of the distribution of and general relationship between major hard-bottom communities from the shore at least to the edge of the continental shelf.

REGIONAL REEFS AND HARD BANKS

Rocky outcrops and hard banks are common on the continental shelf throughout the Gulf of Mexico, even on those parts covered by soft terrigenous sediments. The biota associated with them vary from decidedly tropical coral reefs in the southern Gulf to less diverse assemblages comparable to those of the coastal jetties in the north.

Emergent tropical coral reefs occur throughout the Caribbean, Bahamas, and offshore of the Florida Keys on the Atlantic side from the Dry Tortugas to Cape Florida (Miami). These reefs are most often dominated from low-tide level to 3 to 9 m, depending on conditions of wave energy and location, by the elkhorn coral *Acropora palmata*. Other important reef-building associates are the staghorn coral, *Acropora cervicornis;* fire coral, *Millepora complanata*; finger coral, *Porites porites*; brain corals, *Diploria* spp.; mountainous star coral, *Montastrea annularis*; and many more. Shallow-water alcyonarians, which include species of *Gorgonia, Pseudopterogorgia, Plexaura,* and *Eunicea*, are typically abundant. Submerged reef zones, down to 35 m or so in some areas of the Caribbean, are dominated typically by *Montastrea annularis, Diploria* spp., *Colpophyllia* spp., other faviid corals, and alcyonarians. The reefs support highly diverse communities of tropical algae, sessile and mobile reef invertebrates, and reef fishes.

Emergent coral reefs occur on the outer Yucatan shelf adjacent to islands and on the crests of submerged banks (Alacran reef, Cayo Arenas, Nuevo reef, Triangulos, and Arcas, Figure 3.1) (Logan, 1969). Similar emergent reefs are present nearshore off the city of Veracruz (Heilprin, 1890; Rannefeld, 1972) and off Cabo Rojo, near Tampico (Rigby and McIntyre, 1966). All have basically the same community structure and dominance patterns as emergent coral reefs in the Caribbean, the Florida Keys, and the Bahamas.

On the Atlantic coast of Florida elements of the tropical coral reef biota decrease in importance in a south-to-north gradient from Cape Florida (Miami) to Palm Beach. The two corals most responsible for reef building (*Acropora palmata* and *Montastrea annularis*) are rare north of Miami, and substantial tropical reef structures do not occur. Only a few isolated colonies of *Acropora palmata* are found north of Ft. Lauderdale, and the area is characterized by alcyonarian-dominated hardgrounds. Stony corals are present but do not build reefs (Jaap, personal communication).

From Palm Beach to Stuart (St. Lucie Inlet) heartier elements of the tropical reef biota occur,

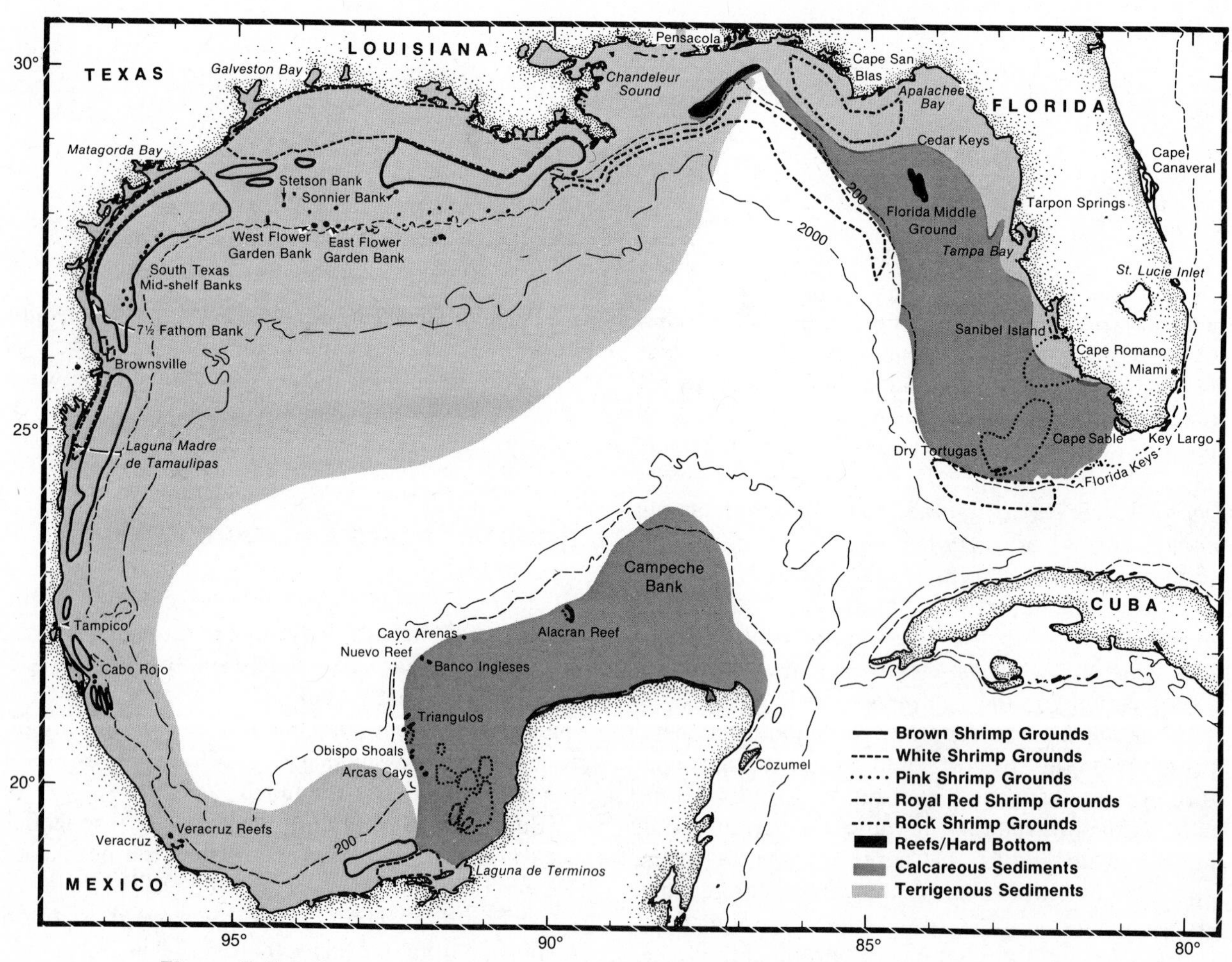

Figure 3.1. Zones and localities of major biogeographical significance in the Gulf of Mexico.

but there is a transition to low diversity *Oculina* (pretzel coral) bank communities which are the dominant coral communities northward to the Georgia border in 50-to-150-m depth (Avent et al., 1977; Reed, 1980). Fossil *Oculina* reefs occur on the continental shelf of the northwestern Gulf of Mexico, buried deep beneath recent sediment (e.g., the author has samples from an oil-well core that penetrated a 10-m-thick *Oculina* reef 30 m below the seafloor).

Several varieties of ahermatypic corals (*Astrangia, Phyllangia,* and *Oculina*), "coral heads," and a few alcyonarians, including *Leptogorgia virgulata* and *Leptogorgia setacea*, occur at Gray's Reef (17-m depth) 17.5 NM off Sapelo Island, Georgia (NOAA, 1980). Patches of coral heads (*Solenastrea hyades* and *Siderastrea siderea*), ahermatypic corals (*Oculina, Astrangia, Phyllangia,* and *Balanophyllia*), sea fans, algae, sponges, and a fair number of tropical Atlantic reef fishes occur in Onslow Bay off the South Carolina coast (Huntsman and MacIntyre, 1971). *Solenastrea* and *Siderastrea* are also components of the South Florida reef ecosystem, but they are among the heartiest, occurring in Florida Bay, where temperature, turbidity, and salinity conditions vary considerably. The Carolina coral patches experience winter water temperatures at least as low as 10°C.

The most northerly true tropical coral reefs in the Atlantic occur in Bermuda due to the northward path of the warm Gulf Stream current. Although parts of these reefs are quite shallow, they differ from other emergent reefs in that they lack *Acropora palmata* and *Acropora cervicornis* and are dominated by head corals (*Montastrea annularis, Diploria strigosa,* and *Porites astreoides*) and the typical Caribbean shallow-water alcyonarians.

Whereas numerous ledges and outcroppings on the west Florida shelf support associations of hearty corals and other tropical reef biota, shallow-water tropical coral reefs are nonexistent north of the Tortugas. The Florida Middle Ground communities, dominated by the fire coral *Millepora* and the scleractinians *Madracis* and *Dichocoenia*, exhibit moderate coral diversity and harbor an abundance of shallow-water alcyonarians, tropical invertebrates, and reef fishes. The Middle Ground is not, however, an active coral reef comparable to those found off the Florida Keys (Jaap, personal communication). The lack of true coral reefs on suitable substrata on the northwest Florida shelf is probably due to the occurrence of winter water temperatures (15 to 16°C) several degrees below that generally accepted as minimal for vigorous reef development (18°C).

Coral reefs do not occur nearshore in the western Gulf north of Cabo Rojo. Seven and One-Half Fathom Bank, a small sandstone prominence located 3.2 km offshore from central Padre Island, Texas, is largely covered by mats of tube-building polychaete worms (*Phyllochaetopterus*), sponges, ascidians, and hydroids. The bank, which crests at about 8.5 m depth from a surrounding bottom of 14 m, bears assemblages of mollusks and crustaceans of mixed warm-temperate and tropical affinities and a substantial number of species in common with the south Texas jetties and other very nearshore banks farther north, such as Heald Bank off Galveston (Tunnell and Chaney, 1970; Tunnell, 1973; Felder and Chaney, 1979). Seven and One-Half Fathom Bank is commonly subject to seasonal temperatures that range from 13 to 30°C, salinities of 28 to 36 ppt, and highly turbid to fairly clear water conditions (Tunnell, 1973). Being in an area of converging currents, the bank is bathed by coastal waters from the more tropical southwestern Gulf in summer and the northern Gulf in winter. Therefore, Felder and Chaney considered at least the decapod fauna of the bank and of the jetties on the nearby south Texas coast to be transitional between temperate and tropical.

The sediment-laden rock outcrops that form Sebree Bank about 15 NM offshore from south Padre Island harbor primarily the same types of coastal epifauna found on the other nearshore "live bottoms" in the northern Gulf and off the southern Atlantic states (?*Oculina, Phyllangia americana,* bryozoans, sponges, hydroids etc.). Some tropical fishes and invertebrates frequent this bank and the nearshore oil platforms and jetties (e.g., *Holacanthus ciliaris, H. bermudensis, Halichoeres, Chaetodon sedentarius, Pomacentrus, Hermodice carunculata,* and *Stenorynchus seticornis*) (Greg Boland, LGL Ecological Research Associates, personal communication). However, this is not unusual even as far north as the Carolinas, particularly in summer when water temperatures are high. These coastal banks exist in environmentally stressed and variable environments for hard-bottom biota, in which high turbidity and sedimentation, low winter temperatures, and, in places, substantial decreases in salinity occur. None of the banks, including Seven and One-Half Fathom Bank, bears biotic communities that resemble those of tropical Atlantic coral reefs.

In fact, coral reefs are absent in the northwestern

Gulf between the emergent reefs off Cabo Rojo and the submerged reefs atop the East and West Flower Garden Banks 100 NM southeast of Galveston, Texas. Here, coral reefs exist at the crests of the banks at depths of 15 to 52 m. The shallowest (15 to 36 m) are dominated by the star coral *Montastrea annularis,* brain corals *Diploria strigosa* and *Colpophyllia* spp., and the mustard hill coral *Porities astreoides*. About one-third of the species of Caribbean reef-building corals occur on the banks along with several hundred species of reef invertebrates and fishes. There are, however, no shallow-water alcyonarians and no corals of the genus *Acropora*. Although they are the northernmost true tropical coral reefs on the continental shelf of North America, the Flower Garden reefs exist in oceanic water with winter temperatures that do not typically drop below 18 to 19°C, barely above the accepted lower limit for reef development.

Siltstone–claystone banks closer to shore off Texas (Stetson, Claypile, and Sonnier Banks), although cresting at suitable depths, are occupied by communities weakly dominated by sponges and crusts of the fire coral *Millepora alcicornis*. Much bare rock is in evidence. These bank communities, although composed largely of Caribbean biota, including a fair number of tropical reef fishes, are of much lower diversity than the Flower Gardens and are not comparable to tropical coral reefs. The substantial variability of temperature, salinity, and turbidity of inner- and mid-shelf waters bathing these banks precludes active coral-reef development.

Deep-water reef-building communities dominated by coralline algae (e.g., rhodoliths, pavements, and crusts) exist at greater depths (30 to 85 m or more) on the Flower Gardens, other shelf-edge banks between the Flower Gardens and the Mississippi, outer shelf banks on the Yucatan Shelf, on deeper "level" shelf bottoms off southwestern Florida, seaward of the Key Largo reef tract, and throughout the West Indies (Bright and Rezak, 1977; Logan, 1969; Macdonald, 1982; MacIntyre, 1972). The coralline algae and sizeable attendant populations of stony corals *Agaricia, Helioseris,* and *Madracis* produce massive amounts of carbonate substratum in the northern Gulf, certainly much more than the few coral reefs are capable of. These deep-reef communities are Caribbean in their affinities and nearly as diverse biotically as the coral reefs. Tropical leafy algae are particularly conspicuous.

The same coastal-neritic hydrographic conditions that preclude coral-reef development on the shallower mid-shelf banks off north Texas and Louisiana apparently prevent the expression of the deepwater reef-building communities on mid-shelf banks that crest between 55 and 70 m depth off South Texas (Southern, Baker, Hospital, and others). These carbonate structures, the remains of relict Pleistocene reefs, currently support only minor encrusting populations of coralline algae. The most conspicuous fauna are large white antipatharians *Cirrhipathes* (=*Cirripathes*), smaller branching antipatharians *Antipathes*, deep-water alcyonarians, sponges, occasional plates of the coral *Agaricia*, and an assemblage of fishes typical of deeper parts of offshore banks throughout the northwestern Gulf (*Holanthias martinicensis, Chromis enchrysurus, Liopropoma eukrines, Serranus phoebe, Lutjanus campechanus,* and *Mycteroperca phenax*, to name a few). Exposed reef rock with a veneer of fine sediment is common and becomes the dominant condition below a depth of about 75 m where epifaunal diversity is minimal.

Carbonate outcrops and surrounding sand and shell bottom at a depth of 50 to 60 m on the western rim of the DeSoto submarine canyon 15 NM offshore near Pensacola, Flordia, bear low diversity communities of an apparently mixed tropical and temperate nature. Shipp and Hopkins (1978) described a "sand-shell-coralline algae slope" whereon they observed the sea pansy *Renilla,* crabs *Calappa,* rock shrimp *Sicyonia,* starfishes *Astropecten* and *Luidia*, and the sand dollar *Clypeaster*, all fairly representative of Shrimp Ground fauna. Rather more tropical forms included the starfishes *Narcissia, Linckia,* and *Goniaster* and the slipper lobster *Scyllarus*.

The hard, blocklike DeSoto Canyon substrate supports clusters of the stony coral, *Oculina*; alcyonarians, *Lophogorgia*; antipatharians; hydroids; crabs, *Mithrax, Stenocionops,* and *Stenorynchus*; a lobster, *Scyllarides*; hermit crabs, *Dardanus* and *Petrochirus*; gastropods, *Fasciolaria* and *Scaphella*; echinoids, *Arbacia, Diadema, Eucidaris,* and *Lytechinus*; the sea cucumber, *Isostichopus;* and large ophiuroids, including basket stars and *Ophioderma*. The fishes are basically Caribbean types. Many of the species are also found in the deeper zones of the shelf edge banks off north Texas and Louisiana and on the mid-shelf banks off south Texas.

The DeSoto Canyon hard-bottom community seems to be somewhat like the more tropical south Texas mid-shelf carbonate banks and possibly other poorly known mid-shelf banks off Louisiana. All exist in neritic waters that may fluctuate seasonally between temperate and tropical temperatures, vary widely in turbidity, and be subject periodically to substantial decreases in salinity caused by river outflow. None of these structures harbor effective reef-building populations at present and are probably best viewed as stressed habitats occupied by some hearty representatives of the Caribbean biota.

Moore and Bullis (1960) reported a sizable (1219 m long) deep-water (230 to 280 m) coral reef on the Gulf of Mexico continental slope 40 miles east of the Mississippi River. The reef is composed of "one or two species of ramose, branching, colonial (ahermatypic) corals," primarily *Lophelia prolifera*. *Lophelia* apparently forms similar structures in deep water in the Caribbean, north of the Bahamas, and in the North Atlantic (Moore and Bullis, 1960).

Man has added significant amounts of hard substratum to the continental shelf of the northwestern Gulf of Mexico by constructing offshore oil and gas platforms, primarily off Louisiana (more than 1100 structures), and Texas (about 120). Increasing numbers are expected off Texas and in Mexican waters as oil and gas production expands in the western Gulf. Fewer than five platforms have been built east of the Mississippi.

Shinn (1974), George and Thomas (1979), and Gallaway and Lewbel (1982) described biotic zonation on Texas–Louisiana platforms which can be classified as coastal (surrounding bottom depth 0 to 30 m), offshore (30 to 60 m), or bluewater (more than 60 m) on the basis of species composition, structure and zonation of the biotic communities associated with them.

Coastal platform biofouling communities off Louisiana are dominated by acorn barnacles above a depth of about 12 m (usually either *Balanus reticulatus* or *Balanus improvisus* off Louisiana) with biomass decreasing with depth. George and Thomas (1979) indicated thick concentrations of the filamentous green alga *Enteromorpha* on the barnacle fouling mat just below the sea surface, with interspersed hydroids, *Syncoryne*; xanthid crabs, *Neopanope texana;* amphipods, *Corophium;* and pycnogonids. Below a depth of 2.4 to 6 m sea anemones, *Aiptasia*, replace the algae and hydroids on the barnacles, and cryptic blennies, *Hypleurochilus*, become common. *Balanus improvisus* was rare below 7.5 m, leaving *Balanus reticulatus* the dominant. No living barnacles were found below 12.2 m, below which hydroids dominated down to the bottom (18 m). Small corals (probably *Astrangia*) were present among the hydroids. Gallaway and Lewbel (1982) found that the large Mediterranean barnacle *Megabalanus antillensis* (= *Balanus tintinnabulum*) is dominant on the coastal platforms off Texas rather than the smaller barnacles typical of the Louisiana platforms.

Dominant fishes on coastal platforms are the Sheepshead, *Archosargus probatocephalus;* Atlantic spadefish, *Chaetodipterus faber*; Moonfish, *Selene setapinnis*; Lookdown, *Selene vomer*; Gray triggerfish, *Balistes capriscus*; Bluefish, *Pomatomus saltatrix*; and Blue runner, *Caranx crysos;* they often occur in large mixed schools. A few reef fishes are common, such as the Belted sand fish, *Serranus subligarius*; Rock hind, *Epinephelus adscensionis;* Flamefish, *Apogon maculatus*; Sergeant major, *Abudefduf saxatilis;* Cocoa damselfish, *Pomacentrus variabilis*; Gray snapper, *Lutjanus griseus*; and groupers, *Epinephelus* and *Mycteroperca*.

Mid-shelf offshore platform assemblages differ from the coastal in that pelecypods rather than barnacles dominate the biofouling biomass, although barnacles are abundant. The tree oyster, *Isognomon bicolor;* oyster, *Hyotissa*, and the leafy jewel box, *Chama macerophylla,* are bivalves of particular importance. Significant populations of the sometimes dominant alcyonarian, *Telesto,* are considered indicative of offshore platforms (Gallaway and Lewbel, 1982). Red and green macroalgae are important in the upper 10 m.

Whereas Atlantic spadefishes dominate and most of the coastal species are common, the offshore platform fish assemblage differs from that of coastal platforms in having greater abundances of Gray and Red snappers, *Lutjanus griseus* and *L. campechanus*, and a richness of tropical species such as the Cocoa damselfish, *Pomacentrus variabilis*; Blue and French angelfishes, *Holacanthus bermudensis* and *Pomacanthus paru;* Sergeant major, *Abudefduf saxatilis*; Brown chromis, *Chromis multilineatus;* filefishes, Monacanthidae; surgeonfishes, Acanthuridae; Flamefish, *Apogon maculatus;* and Creolefish, *Paranthias furcifer*. Almaco jack, *Seriola rivoliana*, and Amberjack, *Seriola dumerili;* Bar jack, *Caranx ruber;* Rainbow runner, *Elagatis bipinnulata;* Great barracuda, *Sphyraena barracuda;* Cobia, *Rachycentron canadum;* Crevalle jack, *Caranx hippos*; and hammerhead shark, *Sphyrna*, are present (Gallaway and Lewbel, 1982).

On outer-shelf, bluewater platforms stalked barnacles (*Lepas)* and algal mats predominate at the surface and hydroids and pelecypods appear to dominate at greater depths. Invertebrates and leafy algae, characteristic of shelf-edge banks, are present

(Spiny lobster, *Panulirus argus*; urchins, *Diadema* and *Eucidaris*; algae, *Dictyota* and *Peyssonnelia*; and others). The Mediterranean barnacle, so important on Texas coastal platforms, although present, is not a dominant on the bluewater platforms (Gallaway and Lewbel, 1982).

The most striking difference between bluewater platforms and those nearer shore is in the fish populations. The large schools of Atlantic spadefish, Lookdowns, and Bluefish are absent from the bluewater platforms, seemingly replaced by numerous Creole-fish, Almaco jacks, and Blue runners. The Sheepshead is replaced by the Gray triggerfish and a number of tropical species. Abundant tropicals include schools of Creole wrasse, *Clepticus parrai*, and the more solitary varieties such as the Spanish hogfish, *Bodianus rufus*; damselfishes; angelfishes; surgeonfishes; Rock beauty, *Holacanthus tricolor*; Red spotted hawkfish, *Amblycirrhitus pinos*; Great barracuda; and hammerhead sharks.

The outer to inner shelf transition from tropical reef assemblages to assemblages apparently of more temperate nature in the northwestern Gulf of Mexico is of considerable ecological interest, particularly because it can be recognized on both natural hard bottoms and man-made structures. A better knowledge of the biotic assemblages that inhabit the "ephemeral" offshore platforms (they must ultimately be removed by the oil companies) and the spatial and temporal variations in assemblages would be of considerable use in determining the environmental factors that control the nature, distribution, and biogeographic relationships of hard-bottom benthos and reef fishes in the Gulf.

BIOGEOGRAPHIC RELATIONSHIPS

To place the hard-bottom assemblages in proper perspective it is necessary to consider the general distribution patterns of all major benthic marine communities within the Gulf of Mexico, including those of the estuaries, coast, and offshore soft bottoms. Biogeographically pertinent aspects of the distribution of these communities may be summarized as follows (See Figure 3.1):

1. Coastal

a. Temperate salt marshes predominate in the northern Gulf from Tarpon Springs, Florida, to at least Port Isabel, Texas (Humm, 1973).

b. Stressed populations of tropical Black mangroves co-occur with the salt marshes as far north as the Mississippi Sound (Humm, 1973; Hildebrand, 1957).

c. Well-developed tropical mangrove forests dominated by Red and Black mangroves replace salt marshes southward from Tampa Bay and Laguna de Tamiahua in the Gulf and Jupiter Inlet on the East Florida coast. They are not significant in the northern Gulf or north of Cape Canaveral (Humm, 1973; Yanez-Arancibia et al., 1980; Ayala-Castanares, 1981).

d. Temperate oyster reefs (*Crassostrea virginica)* are extensive in bays north of Cape Canaveral and in the northern Gulf from Cedar Keys, Florida, to Port Aransas, Texas. Reduced reefs occur in Gulf estuaries as far south as Laguna de Terminos, Campeche (Bahr and Lanier, 1981; Taylor et al., 1973; Yanez-Arancibia et al., 1980; Fotheringham, 1980; Oppenheimer and Gordon, 1972).

e. Tropical seagrass beds are abundant inshore throughout the Gulf except between Galveston and the Mississippi Delta, where they are limited by high turbidity and low salinities. Beds in the Mississippi Sound may be restrained reproductively by temperate conditions. Tropical seagrass beds are not important in the Atlantic north of Cape Canaveral (Humm, 1973; Hildebrand, 1957; Humm and Hildebrand, 1962; Yanez-Arancibia et al., 1980; Kim, 1964; Randall, 1965).

f. Jetties in the northern Gulf bear biota similar in many respects to those of jetties in the Carolinas. Tropical forms are more abundant on South Texas jetties (Port Aransas and Port Isabel). South Florida intertidal rocks support distinctly more tropical assemblages (Hedgpeth, 1954; Edwards, 1976; Fotheringham, 1980).

g. Leafy algae populations within the Gulf are basically tropical, with a diminishing number of tropical forms and an increasing admixture of temperate species on the coast proceeding north from South Florida (around Tampa) and South Texas (around Port Isabel). Many of the northern Gulf temperate species become inconspicuous inshore during the summer. Algae on the mid- and outer shelves throughout the Gulf are distinctly tropical (Humm, 1973; Kim, 1964; Bright et al., 1981; Earle, 1969, 1972; Edwards, 1976; Humm and Hildebrand, 1962; Sorenson, 1979).

2. Offshore "level bottoms"

a. The White Shrimp Ground (inner-shelf) and Brown Shrimp Ground (outer-shelf) assemblages are present throughout the Gulf where terrigenous sediments predominate. White Shrimp Grounds are best developed adjacent to hyposaline estuarine environments (Pensacola to Texas and off Laguna de Terminos). Offshore Brown Shrimp Grounds (less estuarine-dependent) are extensive from Pensacola to the Gulf of Campeche. Both are most extensive off Texas–Louisiana. Throughout their range the shrimp ground communities are fairly uniform and the geographic ranges of the species extend substantially northward toward the Carolinas and southward into the Caribbean. Tropical and temperate characteristics are apparent at the southern and northern extremities respectively, of the Brown Shrimp Grounds within the Gulf, at least for the fish assemblages (Kutkuhn, 1962; Chittenden and McEachran, 1976; Hildebrand, 1954, 1955; Defenbaugh, 1976; Hedgepeth, 1953, 1954; Chittenden and Moore, 1977; Flint and Griffin, 1979; Thistle and Lewis, 1979).

b. Pink Shrimp Grounds (communities of more tropical affinities) are restricted to carbonate sand bottoms on the Yucatan and West Florida shelves, being best developed just north of the Tortugas and off southern Campeche (Hildebrand, 1955; Kutkuhn, 1962; Sanchez-Gil et al., 1981; Chittenden and McEachran, 1976).

c. Epibenthic communities of tropical origin occupy the carbonate substrata of continental shelves of Yucatan and West Florida, extending as far north as the mouth of the DeSoto Canyon off Pensacola. They are held offshore north of Cape Romano, Florida by coastal terrigenous sands and increasingly temperate coastal conditions. Rock outcrops on the West Florida shelf are occupied by epibenthos derived from tropical reef communities out to depths of at least 90 m. The diversity of tropical species apparently decreases in a northerly direction (Sanchez-Gil et al., 1981; Hoese and Moore, 1977; Cervignon, 1966).

d. Limited populations of heartier tropical epibenthos and fishes occur as far north as the Carolinas on the South Atlantic outer continetal shelf (Hoese and Moore, 1977; Downey, 1973).

3. Reefs and Hard Banks

a. Emergent tropical coral reefs occur southward from Miami, the Dry Tortugas, and Cabo Rojo, Mexico. Tropical reef epibenthos diminish north of Miami and are replaced by more temperate submerged *Oculina* banks off St. Lucie Inlet. Certain hearty hermatypic corals occur as far north as the Carolinas. Tropical hard grounds (but not coral reefs) extend northward to the Florida Middle Ground, West Florida shelf. Submerged coral reefs and deep-water tropical reef assemblages are present on shelf-edge banks off North Texas and Louisiana. Similar deep-water tropical reef assemblages occur on the outer southwest Florida shelf south of Charlotte Harbor and in the Atlantic off Key Largo.

b. Depauperate tropical epibenthic communities occupy mid-shelf banks in the northwestern Gulf of Mexico and hard bottoms on the north rim of the DeSoto Canyon, where conditions tend to be seasonally variable, temperate, and neritic.

c. Environmentally stressed hard bottoms nearshore in the northwestern Gulf bear basically temperate epifauna with admixed tropical forms (mainly some tropical fishes and mobile invertebrates).

d. Epibenthos on oil platforms off Texas are tropical near the shelf edge, mixed tropical and coastal on the mid-shelf, and basically temperate with admixed tropical forms nearshore.

e. Throughout the northern Gulf some tropical reef fishes move inshore during summer to occupy coastal platforms and hard grounds.

By correlating biotic distribution patterns with patterns of physical conditions within the Gulf, it becomes readily apparent that substratum type is the primary regional factor that influences the nature of marine benthic communities offshore. The terrigenous soft bottom (Figure 3.1) supports two distinct shrimp-ground communities which have usually been considered warm temperate (Carolinian). These communities, however, reach their greatest degrees of abundance and diversity on the mid- and outer shelves of the northwestern and southern Gulf of Mexico, where, if the substratum were carbonate and bottom turbidity low, one might expect decidedly tropical epibenthos to occur. Carbonate substrata within the Gulf generally support elements of the tropical (Caribbean) epibenthos, although offshore tropical diversity apparently decreases gradually toward the north.

A tropical biotic presence throughout the Gulf is the natural result of the fact that virtually all shallow marine water entering the area comes from the Caribbean and carries a continual supply of tropical larvae. Substantial modification of this warm water mass occurs nearshore in the northern Gulf because of the river inflow and cold winter air temperatures and at the southern tip of Florida and the Gulf of Campeche because of the river inflow (cold air also occasionally affects the waters of Florida Bay and vicinity). Otherwise the immense epipelagic zone of the Gulf of Mexico is tropical; it carries tropical larvae prepared to occupy any appropriate substratum that exists in suitable conditions of temperature, salinity, and turbidity.

Inshore along the northern Gulf coast, winter water temperatures may dip below 10°C in the bays and shallow nearshore waters. The effects of winter cold diminish toward the south, and perpetual tropical to subtropical epipelagic conditions (year-round temperatures above 16 to 18°C) exist on the outermost shelf.

Nearshore, low winter temperatures exclude most, but not all, of the tropical species, which leads to a basically warm temperate coastal biota (marshes, oyster reefs, etc.) with a substantial tropical content (black mangroves, tropical grass beds, etc.) in the northern Gulf. Appearance of temperate species of algae in the winter (Earle, 1969) and tropical fishes during summer (Briggs, 1974) on coastal and nearshore hard bottoms and man-made structures is characteristic of the biologically transitional and environmentally variable northern Gulf coast. This condition extends down at least to Tarpon Springs in Florida and the Laguna Madre de Tamaulipas in Mexico. The transition to tropical is fairly complete at the Ten Thousand Islands, Florida, and Cabo Rojo, Mexico (Figure 3.1) (Briggs, 1974; Hedgpeth, 1953; Pulley, 1952; Earle, 1969).

To make proper sense of the biogeographic relationships offshore we must consider separately the biota of terrigenous soft bottoms versus those of calcareous substrata and hard banks. These two groups are neither similar in species composition nor interdependent ecologically.

Latitudinally wide-ranging fauna that occupy terrigenous sands and muds tend not to vary greatly throughout the Gulf but show affinities to both the Carolinian warm temperate biota and the tropical Caribbean biota (somewhat more for the former). Briggs (1974) suggested that the White and Brown Shrimp Ground biota may have evolved in the northern Gulf of Mexico and spread to the South Atlantic coast during a period when the Florida peninsula was submerged (last submergence was during the Pliocene, more than one million years ago). This might be modified and carried further by speculating as follows:

1. The center of distribution and evolution of these assemblages is the northwestern Gulf of Mexico.
2. The evolutionary process is quite active.
3. The pelagic dispersal capabilities of the organisms (having evolved in and living in at least seasonally tropical conditions) are such that the warm waters of the southern Florida peninsula are no barrier.
4. The observed similarities between the Gulf and South Atlantic communities result from contemporary dispersal and recruitment processes.
5. The reason that these species are absent or rare as adults off southern Florida is that the carbonate substratum in that area is an unsuitable habitat.

This line of reasoning makes it unnecessary to invoke an historical submergence of the Florida peninsula (with consequent continuous temperate marine conditions from the northern Gulf to the Carolinas) or its emergence as a barrier to dispersal to explain the disjunct distribution of the supposed warm-temperate Carolinian biota. In fact, the extremely high degree of affinity between the Gulf of Mexico north coastal and terrigenous offshore soft-bottom communities and those of Cape Hatteras to Cape Canaveral argues against historical isolations spanning a million years. Such similarities could not have been maintained that long without substantial transfer of genetic material in one direction or another. The patterns of the currents in the Gulf, the Straits of Florida, and the Gulf Stream, which move great volumes rapidly out of the Gulf and up the Atlantic coast, strongly favor transfer (by larval transport) from the Gulf to the South Atlantic. In addition, northern Gulf warm-temperate biota are more diverse and have a substantially greater degree of endemism (approximately 10%; Briggs, 1974) than those on the southern Atlantic coast, thus indicating a center of distribution in the Gulf.

The clearly tropical offshore biota are much easier to interpret. They extend as far north as substratum and water quality will permit. In the eastern Gulf this means at least to the mid-shelf Florida Middle Ground. Temperatures there may drop to slightly less than 16°C in winter and the fairly diverse tropical communities, though dominated by corals, do not form true coral reefs. In the western Gulf the Flower Garden coral reefs at the shelf edge rarely experience temperatures as low as 18°C. The mid-shelf banks, on which assemblages are biotically depauperate but are primarily of tropical affinity, experience winter temperatures somewhat less than 16°C. Thus even with appropriate substratum the mid-shelf bank tropical communities in the northern Gulf are inhibited by marginal winter temperatures (below 18°C), low salinity in the western Gulf, and periodic or chronic high turbidity.

The tendency for elements of the tropical biota to extend farther north offshore than they do on the coast has been generally recognized not only for the Gulf (Briggs, 1974; Earle, 1969; Chittenden and McEachran, 1976) but also for the South Atlantic coast northward to the Carolinas (Earle 1969; Downey, 1973; Huntsman and MacIntyre, 1971). Indeed, on the outer shelves tropical ecosystem components are, in places, virtually surrounded by supposed warm-temperate habitats and biota, particularly in the northwestern Gulf of Mexico. In view of this, and the coastal mixture of temperate and tropical species in the northern Gulf, sharp boundaries between the warm-temperate Carolinian, and tropical Caribbean biogeographic provinces are not appropriate. In fact, we may hypothesize that the entire Carolinian Province is largely a northerly expression of hearty species of southern affinity with a substantial boreal (cold-temperate) contingent.

SUMMARY

A universally valid latitudinal boundary between tropical and warm-temperate biota in the region cannot be prescribed. The impact of temperate climatic conditions on community structure is certainly evident on the coast and inner shelf of the northern Gulf of Mexico. Even there, however, the basically temperate estuarine and nearshore assemblages are accompanied by tropical ecosystem components in significant abundance. Progressing offshore, the benthic biota become increasingly tropical in nature so that the general impression is of a more or less crescent-shaped coastal and nearshore zone in the northern Gulf wherein an incomplete transition is made from tropical offshore to basically warm-temperature coastal biota. North–south, tropical–temperate transitional patterns offshore are more subtle and even less complete within the Gulf. Community types on the mid- and outer shelves may be more dependent on substratum type and bathymetry than latitude.

Whatever the regional biogeographic relationships may be, it is certain that biotic communities that occupy the hard banks of the outer continental shelf in the northwestern Gulf of Mexico are of tropical origin. Depending on their distance from shore, however, these communities are living barely within acceptable ecological tolerance levels for thriving tropical reef assemblages (Flower Gardens and other shelf-edge banks) or below these limits (Sonnier, Stetson, and the South Texas fishing banks). The studies described in the following chapters add significantly to our knowledge of the benthic communities on these banks and of certain ecological processes essential to regional reef construction and maintenance.

4

GEOLOGY OF THE FLOWER GARDEN BANKS

The Flower Garden Banks are similar in origin, general structure, and sediment distribution but differ in the details of structure, physiography, and sedimentology. This chapter focuses on the specific differences in the geology of the two banks.

Our work on the geology of the Flower Garden Banks began in 1961. As a result of research cruises to the banks during 1961, 1968, 1969, and 1970 a Ph.D. dissertation, published as a Texas A&M Sea Grant Publication, described the hydrology, biology, and geology of the West Flower Garden Bank (Edwards, 1971). Much of the descriptive material on the West Flower Garden Bank (WFG) presented here is based on the work conducted by Edwards, with details added from subsequent cruises to the bank. The geology of the East Flower Garden Bank (EFG) presented here is based on work performed under BLM-TAMRF Contracts AA550-CT6-18 and AA851-CTO-25 (Bright and Rezak, 1978a; Rezak, 1982a).

GENERAL DESCRIPTION

The Flower Garden Banks are located near the shelf edge, approximately 107 NM due south of Sabine Pass (Figure 4.1). The East Flower Garden Bank is at 27°54'32''N latitude and 93°36'W longitude in Lease Blocks A-366, A-367, A-374, A-375, A-388, and A-389 of the High Island Area (Figure 4.2). The bank is pear-shaped and covers an area of about 67 km^2. Slopes are steep on the east and south sides of the bank and gentle on the west and north sides (Figure 4.2). The shallowest depth on the bank is about 20 m in the northeastern part of Lease Block A-388. The surrounding water depths are about 100 m to the west and north and about 120 m to the east and south. An elongate depression in the north-central part of Lease Block A-389 has a depth of 136 m. Total relief on the bank is about 116 m.

The West Flower Garden Bank is 12 km west of the East Flower Garden at latitude 27°52'27''N, longitude 93°48'47''W (Figure 4.1) in Lease Blocks A-383-385, A-397-399, and A-410 of the High Island Area, South Addition, and Lease Block GB-134 of the Garden Banks Area. It is a much larger bank that covers about 137 km^2. It is oval-shaped and oriented in a northeast–southwest direction (Figure 4.3). The crest of the bank lies at a depth of approximately 20 m. Surrounding depths vary from 100 m to the north, to 150 m to the south. Total relief on the bank is approximately 130 m.

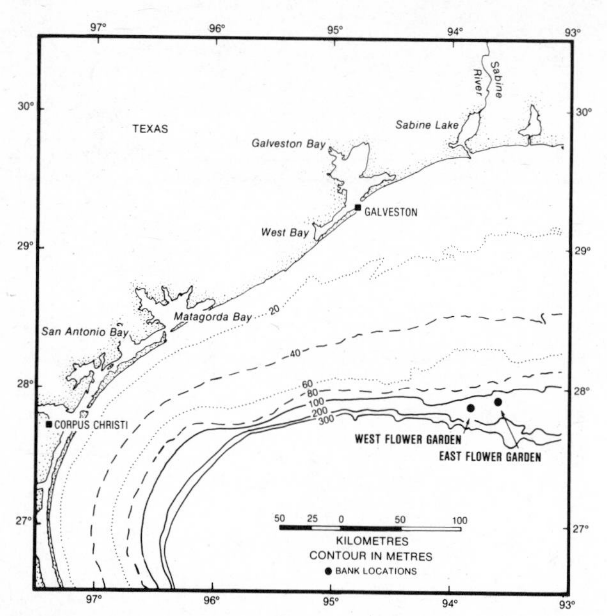

Figure 4.1. A location map of Flower Garden Banks, northwestern Gulf off Mexico.

PHYSIOGRAPHY AND STRUCTURE

Previous Research

In recent years it has been well documented that the East and West Flower Garden Banks are the surface expression of salt diapirs capped by living coral reefs at the crests (Nettleton, 1957; Edwards, 1971; Rezak, 1981, 1982a). Initial interest in the banks on the outer continental shelf was expressed in a paper on the Mississippi Delta (Trowbridge, 1930), and as early as 1937 Shepard suggested that some of these banks were due to salt diapirism. Carsey (1950) speculated that these topographic features could easily be underlain by salt plugs but warned that they might be igneous plugs rather than salt domes. Stetson (1953) stated that the banks were reefs that had kept pace with rising sea level or salt domes with a thin cap of calcareous organisms. Goedicke (1955) compared the banks to salt domes in the Persian Gulf region. In 1957 Nettleton published gravity data taken across the West Flower Garden Bank which clearly indicated a salt-dome origin for that bank. Levert and Ferguson (1969) published a short paper that reviewed previous studies and illustrated the living reef facies of both Flower Garden Banks. Edwards (1971), who presented the most comprehensive report on the West Flower Bank, used 3.5 kHz and air-gun seismic profiles for the first time to illustrate the salt dome subbottom structure of that bank.

East Flower Garden Structure

Figure 4.4 is a structure map prepared from a CDP Sparker survey conducted by Aquatronics in 1974. Reflectors are present on the records over a major portion of the bank and faults are much more easily delineated than on the 3.5-kHz profiles. The major fault trends are west–northwest to east–southeast, with minor trends toward the northeast and the southeast. North–south profiles show the faulting to be steplike, upslope, down-faulting typical of the tensional fractures produced during the domal uplift of sedimentary rocks over salt domes. The central graben (Figures 4.5, 4.6, 4.7, and 4.8) has developed partly because of tensional stresses developed during uplift and to removal of salt from the crest of the ridge by dissolution.

The shaded area on Figure 4.4 represents the trend and lateral extent of chaotic reflections, here interpreted to be salt. The reflections are mapped at 100 and 150-ms intervals (two-way travel time). Beyond the shaded area bedded sediments occur down to the limit of seismic penetration. The dashed line indicates the axis of the uplift, which corresponds nicely with the central graben. Reflectors dip more steeply to the south of the axis than to the north. This is most probably due to the location of the bank at the shelf break. The axis of the salt ridge lies under the shallowest portion of the bank and through the northwest corner of Lease Block A-389. The brine seep complex described below is a short distance to the southeast.

EAST FLOWER GARDEN BRINE SEEP COMPLEX

A brine seep complex at the East Flower Garden Bank has been described by Bright et al. (1980b) and Rezak and Bright (1981b). The primary brine system consists of interrelated components: numerous seeps (1), feeding a brine lake (2) which has an outflow into a canyon (3) that contains a mixing stream (4) that dilutes the brine to lower hypersaline concentrations (Figure 4.9).

The basin that contains the brine lake is 60 m from the edge of the bank, 4 m deep, oval-shaped, ap-

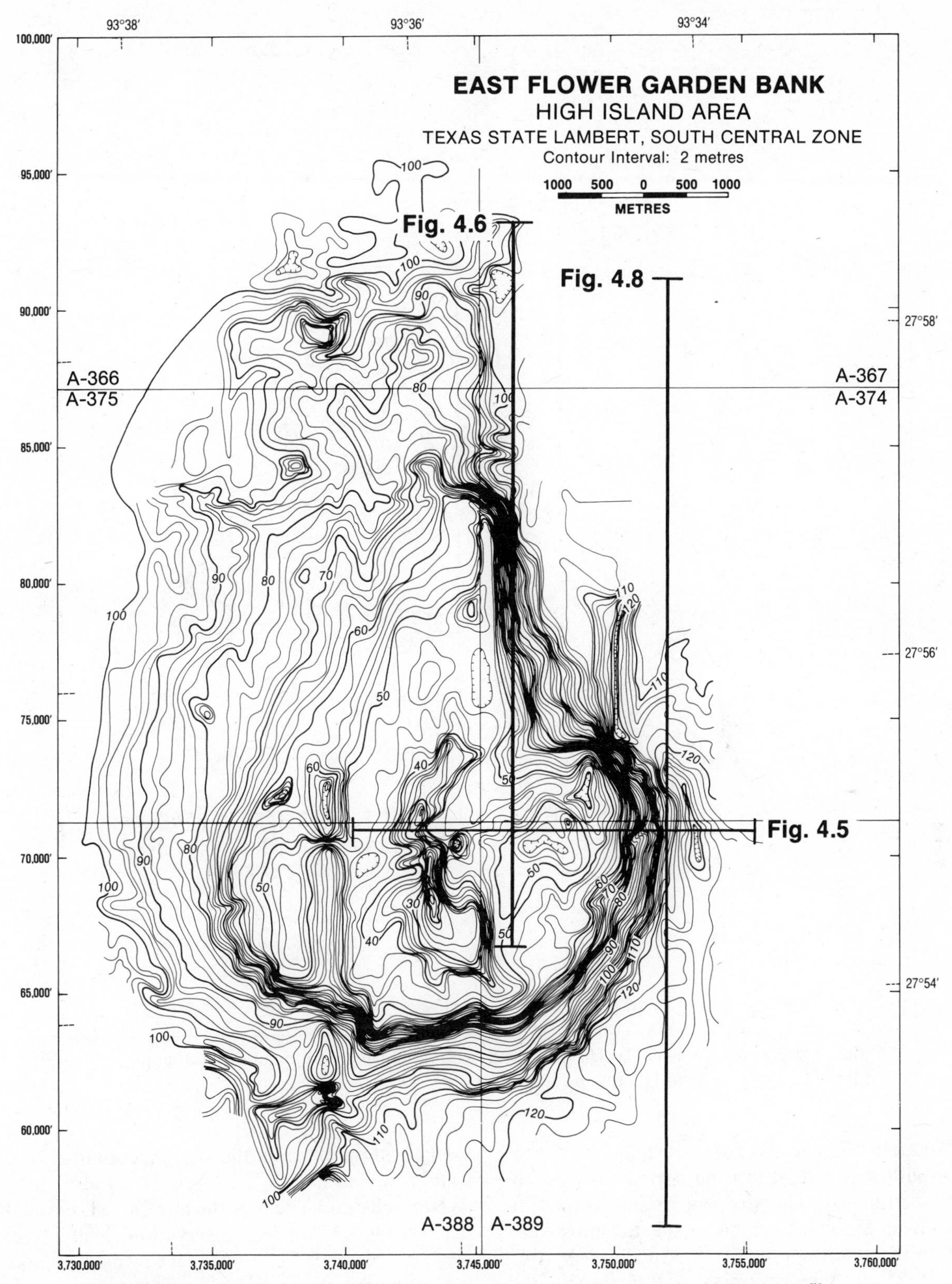

Figure 4.2. East Flower Garden bathymetry, showing the location of subbottom profiles in Figures 4.5, 4.6, and 4.8.

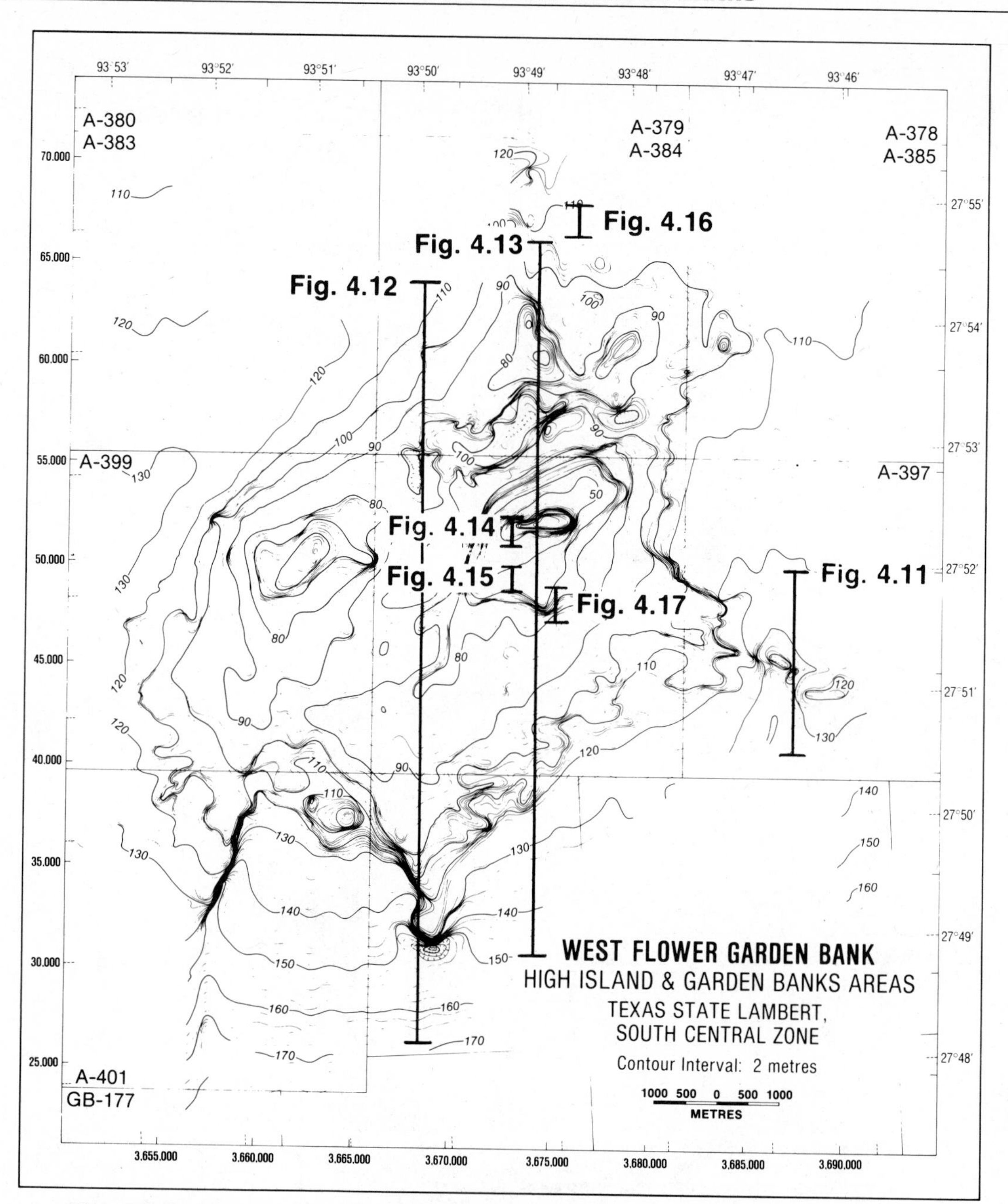

Figure 4.3. West Flower Garden Bank bathymetry, showing location of subbottom profiles and side-scan sonar records in Figures 4.11 to 4.17.

proximately 50 m across from west–northwest to east–southeast, and 30 m from north–northeast to south–southwest. The slope of the basin margin varies from 25° on the northwest side to almost vertical on the south–southeast. A brine lake approximately 25 cm deep occupies part of the slightly lower eastern and central basin floor at a depth of 71 m. The lake is irregular in shape, with a cuspate "shoreline." Numerous brine seeps occur in the sandy bottom between the shoreline and reef-rock wall of the basin.

Normally the surface of the brine lake is perfectly flat. During submersible observation small internal waves were created at the interface by movements of the research submersible. The waves were visible because of the presence of thin, white flakes (sulfuretum) floating at the interface. A canyon approximately 10 m wide at the bottom, 15 m wide at the top,

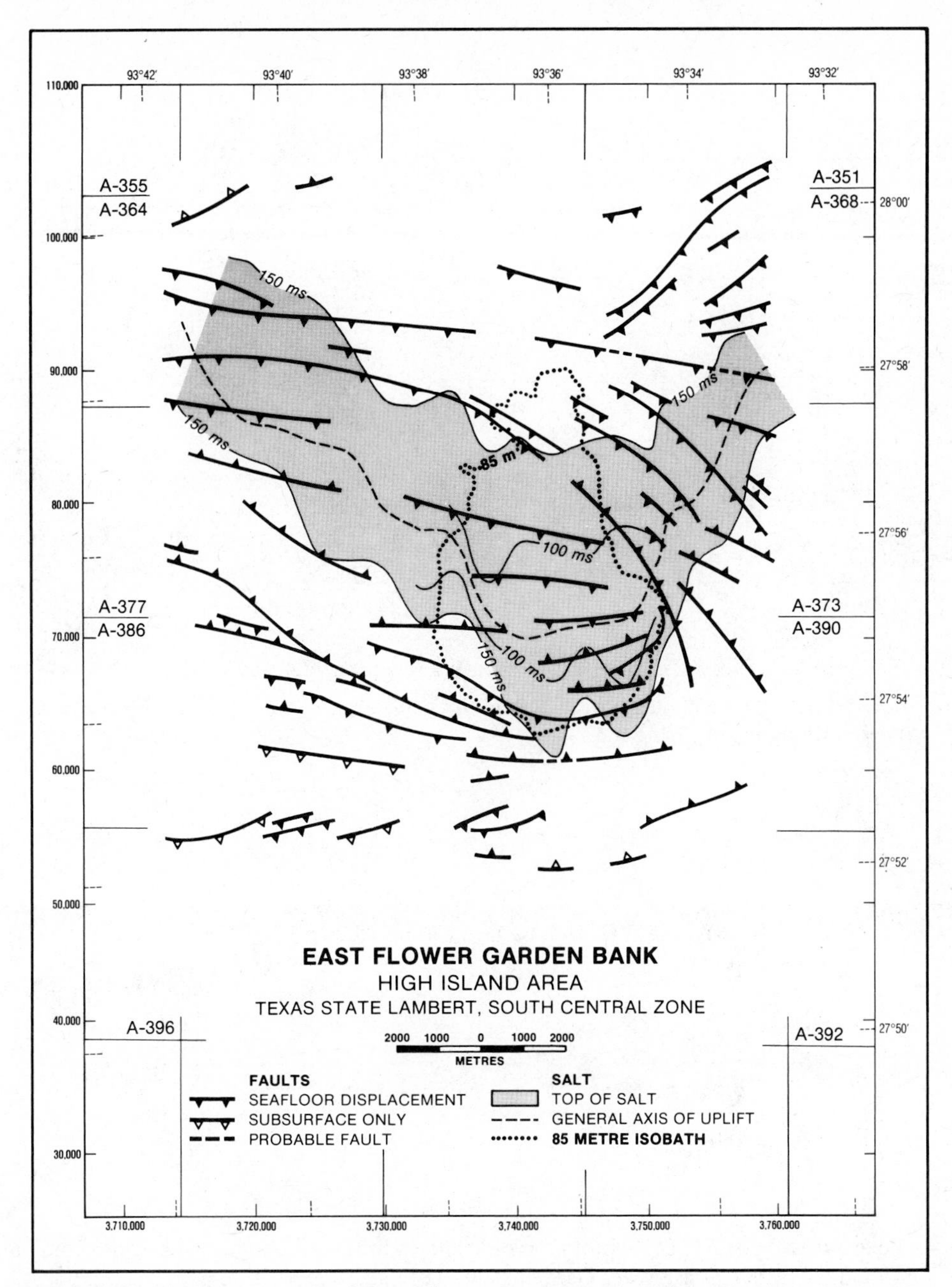

Figure 4.4. A structure map of East Flower Garden Bank region. The dotted line which delineates the 85 m isobath is given to indicate the shape of the bank proper.

and 60 m long extends from the east–southeast margin of the basin to the edge of the bank (79-m depth).

The residence time of brine in the basin is less than one day. The lake, 25 to 30 m long, 15 to 20 m wide, and generally less than 0.25 m deep, contains approximately 465 m^3 of brine. The overflow rate of brine from the lake into the canyon, calculated by current meter and direct measurement of outflow cross-sectional area, was determined to be 355 to 717 m^3/day. Residence time is therefore 0.65 to 1.3 days.

Both the depth of the brine lake and the overflow rate appeared unchanged during four years of observation (1976 to 1980). Seismic data indicate that the top of the salt may be within 30 m of the crest of the reef just to the northwest of the brine basin. The occurrence of another brine seep at 48 m on the bank suggests that the top of the salt may actually be slightly shallower than that. The brine results

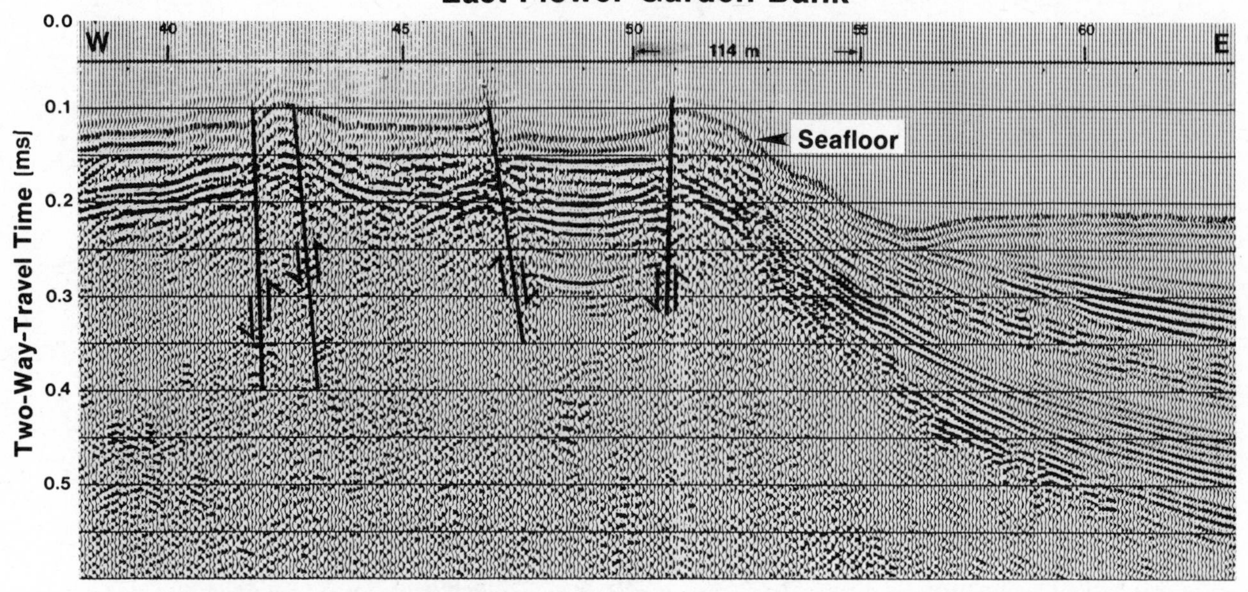

Figure 4.5. CDP sparker profile across the central graben at the East Flower Garden Bank, also shown in Figure 4.6.

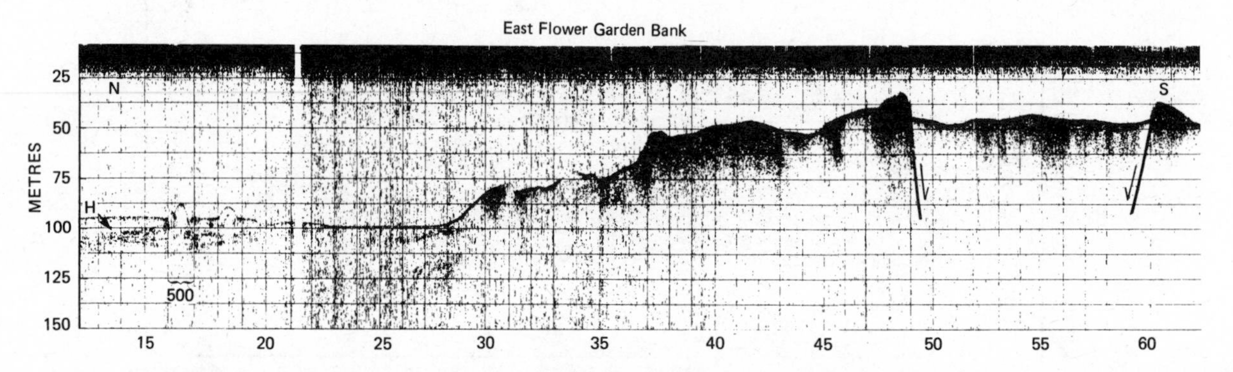

Figure 4.6. A 3.5 kHz subbottom profile shows the central graben (between shot points 49 and 60) at the East Flower Garden Bank. (See also Figure 4.5.)

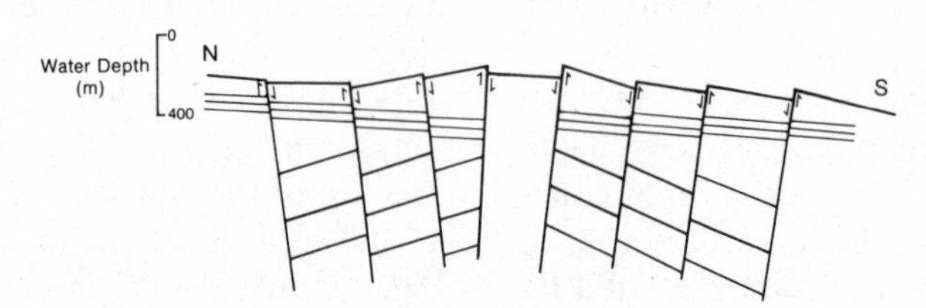

Figure 4.7. Diagrammatic representation of step faulting at the East Flower Garden Bank. (See also Figure 4.8.)

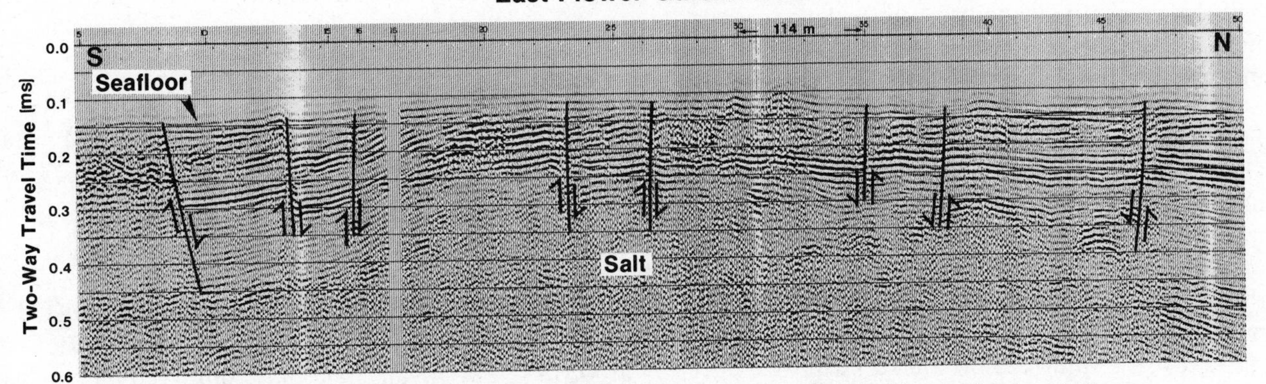

Figure 4.8. CDP sparker profile showing step faulting at the East Flower Garden Bank.

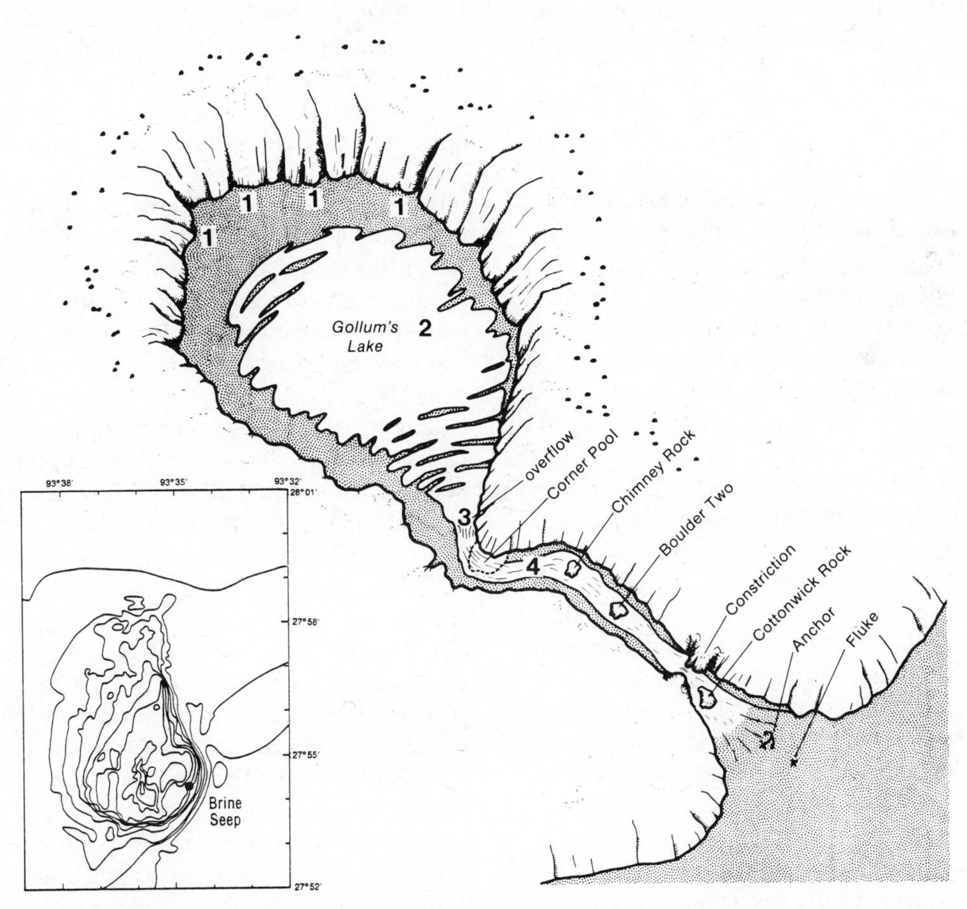

Figure 4.9. East Flower Garden Brine Complex: (1) brine seeps; (2) brine lake; (3) overflow; (4) canyon with mixing stream.

from dissolution of the salt by normal marine water that permeates the porous reef rock. Some of the numerous brine seeps that feed the lake are located several centimeters above the lake level around the shoreline. Dendritic rivulets of brine have been observed in the sands along the shoreline leading from the seeps to the lake.

Based on the measured outflow of the brine lake and the salinosity of the brine in the lake, calculations show that the amount of solid salt being removed from the crest of the salt diapir ranges from a minimum of 10,765 m^3/yr to 21,710 m^3/yr. Because other seeps are known to exist at the East Flower Garden Bank, this is a minimum range for the removal of solid salt. The removal of such large volumes of solids from beneath the crest of the bank must create sizable caverns under the cap rock and the overlying reef. Collapse of the crest of the reef into these caverns will depend on the strength of the cap rock and reef. If these structures are weak the collapse may be gradual, thus keeping pace with the removal of salt. If, however, the cap rock or reef is strong, the caverns may attain a considerable size before failure occurs and the collapse becomes catastrophic.

Chemical aspects of the East Flower Garden brine have been reported by Brooks et al. (1979) and by Bright et al. (1980b). The brine has a higher density than seawater, is anoxic, and contains exceptionally high levels of dissolved hydrocarbon gases (methane, ethane, and propane) and hydrogen sulfide. The density differential inhibits mixing of the lake brine with overlying seawater. Because of the lack of mixing, chemical characteristics of water above and below the interface differ drastically over a vertical distance of less than 2 cm (e.g., salinity: 36.7 vs 200 ppt).

Examination of the bathymetry at the East Flower Garden (Figure 4.2) reveals several shallow, closed depressions in the southern part of the bank. One large depression close to the northwest corner of Lease Block A-389 is a part of the central graben shown in Figures 4.5, 4.6, and 4.8. Evidence of faulting at the crests of salt domes is seen in Figures 4.5 (East Flower Garden) and 4.11 (West Flower Garden). The shallow and discontinuous nature of these closed depressions at the East Flower Garden indicates that graben formation has been active for only a very short time. It is expected that fault movements will continue as long as salt is being dissolved from the crest of the salt stock.

West Flower Garden Bank

Because the West Flower Garden is a much older salt dome, local relief on the bank proper is much greater than at the East Flower Garden. Although the structure of both banks is the result of normal faulting, there has been more movement along faults at the West Flower Garden. The chief cause of these structural differences is the greater amount of salt removal at the West Flower Garden Bank.

The structure of the West Flower Garden Bank is typical of a mature salt diapir in which crestal faulting has occurred. The trough running from the southeast corner of Lease Block A-399 to the southeast corner of Lease Block A-384 (Figure 4.3) is a crestal graben that exhibits far greater relief than the crestal graben at the East Flower Garden. Figure 4.10 is a structure/isopach map with faults delineated by Henry Berryhill (USGS; unpublished MS; on file at MMS, New Orleans) and those observed on our own 3.5-kHz records. Because of poor reflectivity in the area of the graben on our records, it is difficult to display the borders of the crestal graben on this map. The large east–west fault on the southeastern margin of the bank (Lease Block A-397) is illustrated in Figure 4.11). This fault displaces the seafloor and appears to be the result of recent movement. Figures 4.12 and 4.13 illustrate the abundance of normal faults on this bank. Edwards (1971, his Figure 20) illustrates a 1 $in.^3$ air-gun profile across the bank in which he shows the crestal graben and numerous normal faults to the southeast of the bank. The numerous bathymetric prominences on the bank represent horsts that stand above the surrounding grabens (Figure 4.3).

The living reef lies in the north-central portion of Lease Block A-398. It rises from depths of 40 to 50 m to a crest at about 20 m. Extending from near the base of the reef toward the northeast and the south is a broad terrace about 60 to 70 m deep (Figure 4.3). The surface of this terrace is characterized by large waves of sediment that consist primarily of the Coral Debris Facies sand and gravels of the *Gypsina-Lithothamnium* Facies (Figures 4.14 and 4.15). The gravel waves are oriented normal to the isobaths. Below these depths are numerous lineations (faults and outcrops of Tertiary bedrock covered by drowned reefs) (Figure 4.16) and patch reefs scattered to depths as great as 170 m (Figure 4.17). Most of the patch reefs above 90 m appear to have formed during the last rise of sea level.

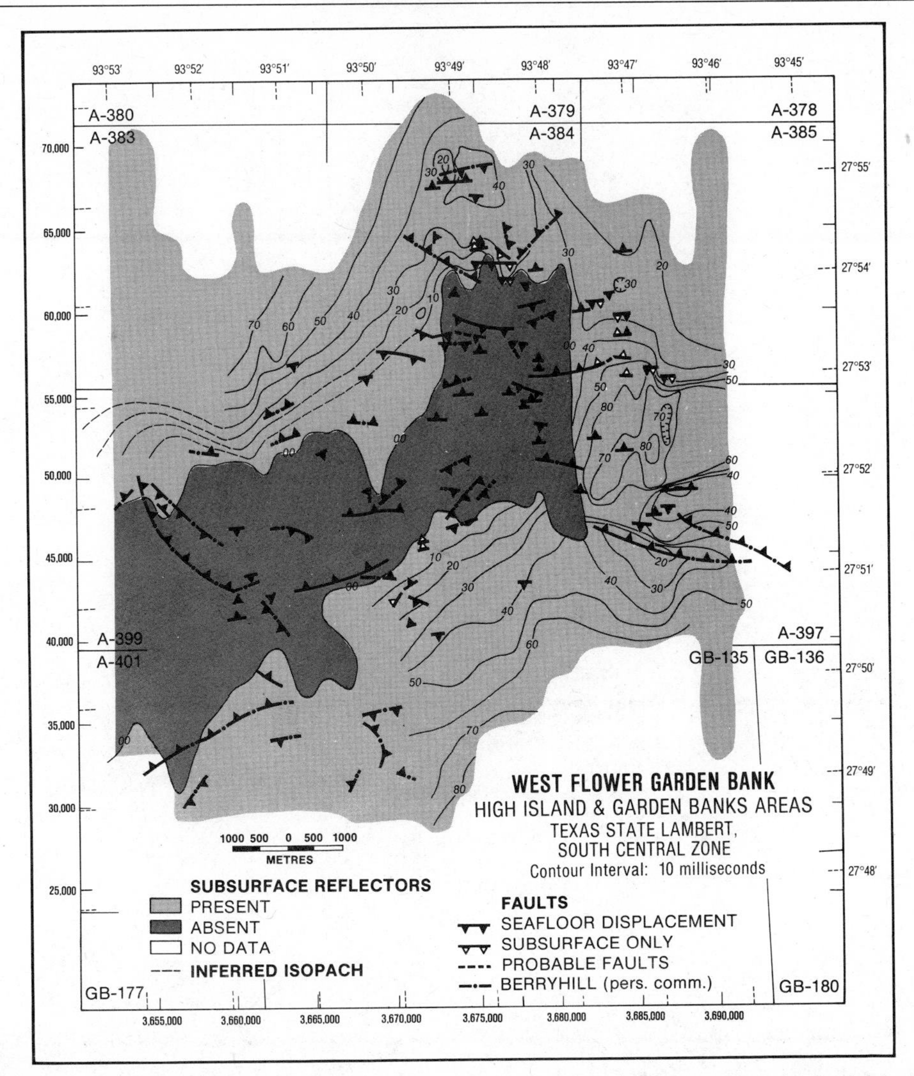

4.10. Structure/isopach map of West Flower Garden Bank. Contours are isopachs (sediment thickness) from the seafloor to horizon (*H*) shown in Figures 4.11 and 4.12.

SEDIMENTARY FACIES

General Description

Sediment facies at the West Flower Garden Bank were first described by Edwards (1971). Subsequent sampling augmented by direct observations of the bottom from the Texas A&M submersible DRV DIAPHUS confirm Edwards facies delineation and extend coverage to the East Flower Garden. The map of sediment distribution in the Flower Garden Banks area (Figure 4.18) is based on 43 samples from Edward's study of the West Flower Garden Bank, 35 samples taken at the East Flower Garden in 1979, 51 samples from selected areas away from the banks (these were selected from a large collection donated by Tenneco), and 10 additional samples taken in December 1979 to fill in the gaps in the off-bank areas.

The sediments of the two banks differ markedly in origin from the sediments of the open shelf surrounding the banks. Bank sediments are derived from the skeletons of organisms that have lived on the banks. The open-shelf sediments in the area of the banks, however, are sands and muds eroded from the North American continent and mechanically transported to the Gulf of Mexico by streams

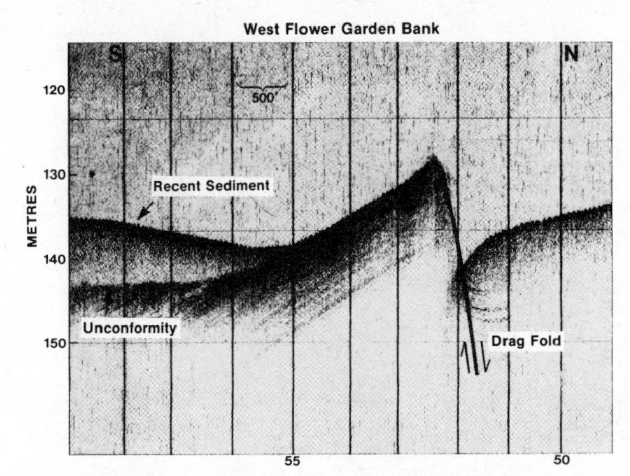

Figure 4.11. A 3.5-kHz north–south subbottom profile across the southeastern part of the West Flower Garden Bank (Lease Block A-397; indexed on Figure 4.3).

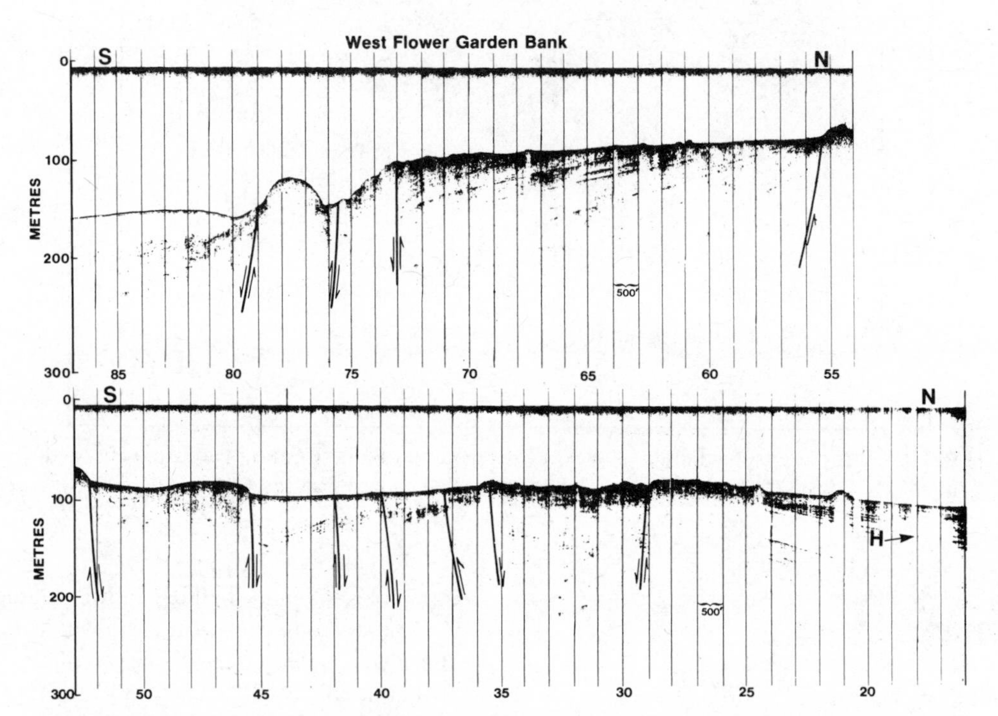

Figure 4.12. A 3.5-kHz north–south subbottom profile across the central part of the West Flower Garden Bank (location indexed in Figure 4.3). The right side of the upper profile joins with the left side of the lower profile: H = base of the isopach interval.

like the Mississippi, Trinity, Sabine, and Brazos Rivers. These sands and muds do not occur at depths shallower than 75 m at the Flower Garden Banks. The sediments above the 75 m level are coarse sands and gravels and the rocky, limestone structure built by corals and other reef-dwelling organisms. The loose sediments around the reef reflect the depth zonation of the biological communities that are present on the two banks. Table 4.1 illustrates the relation between the biological zones and sediment facies.

As shown in Table 4.1, the sediment facies are intimately related to the biological zonation and hydrological conditions at each bank. The sediment

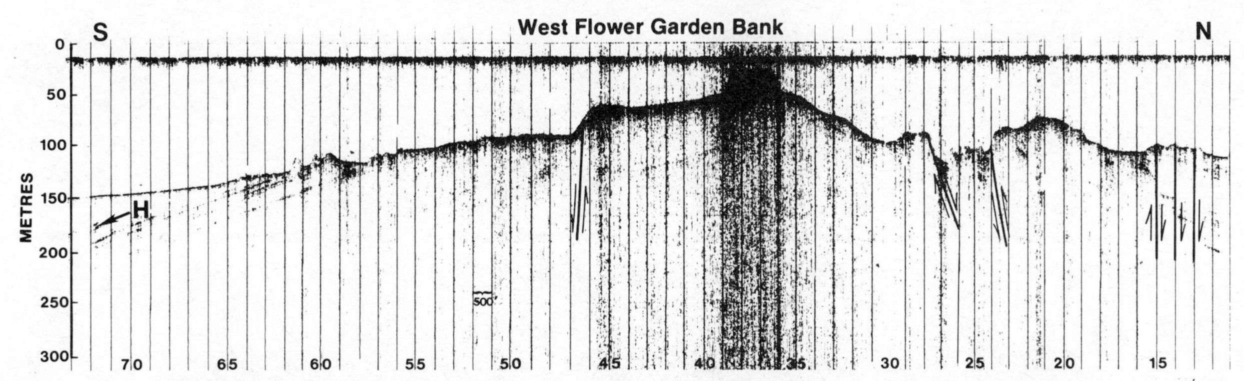

Figure 4.13. A 3.5-kHz north–south subbottom profile across the central part of the West Flower Garden Bank, including the living high diversity coral reef (location indexed in Figure 4.3). Note the lack of reflectors over most of the bank. *H* = base of isopach interval.

Figure 4.14. A side-scan sonar record showing surface features in a central portion of the West Flower Garden Bank, near the living high-diversity coral reef (Lease Block A-398; indexed in Figure 4.3). Water depth is 35 m on the reef, 40 m at sand waves, and 45 m at gravel waves. Light areas on the reef are sand.

boundaries, however, do not coincide with the biotic boundaries, due in part to the downslope movement of loose sediment by the force of gravity and in part to the use of soft-bodied organisms in delineating the biotic zonation. In Table 4.1, for example, the lower boundary of the Algal-Sponge biotic zone is based on the lower depth limit of *Neofibularia*, a colonial, soft-bodied sponge that also grows in the upper part of the *Amphistegina* Sand Facies.

By knowing the distribution of living, lime-secreting, skeletal organisms we can demonstrate that the direction of sediment transport on the banks is downslope. The Coral Debris Facies accumulates between coral heads of the Living Reef Facies and in a narrow band around the base of the living reef (Figure 4.14). The *Gypsina-Lithothamnium* Facies consists of coarse gravel and massive limestone that form *in situ* because of the growth of calcareous algae and encrusting calcareous protozoans. The *Amphistegina* Sand Facies, however, is composed of the recently dead skeletons of a small protozoan that lives and grows abundantly in the *Gypsina-Lithothamnium* Facies.The sand-sized skeletons of these protozoans are moved downslope by gravity to form the *Amphistegina* Sand. At depths less than 75 m the sediments of the banks are medium-to-coarse, calcareous sands and gravels (Figure 4.18). The living reef and the *Gypsina-Lithothamnium* Facies are natural sediment traps formed by their highly irregular surfaces. Any fine sediments, like those that occur below the 75 m depth, would be trapped in the irregular topography of these facies if they were

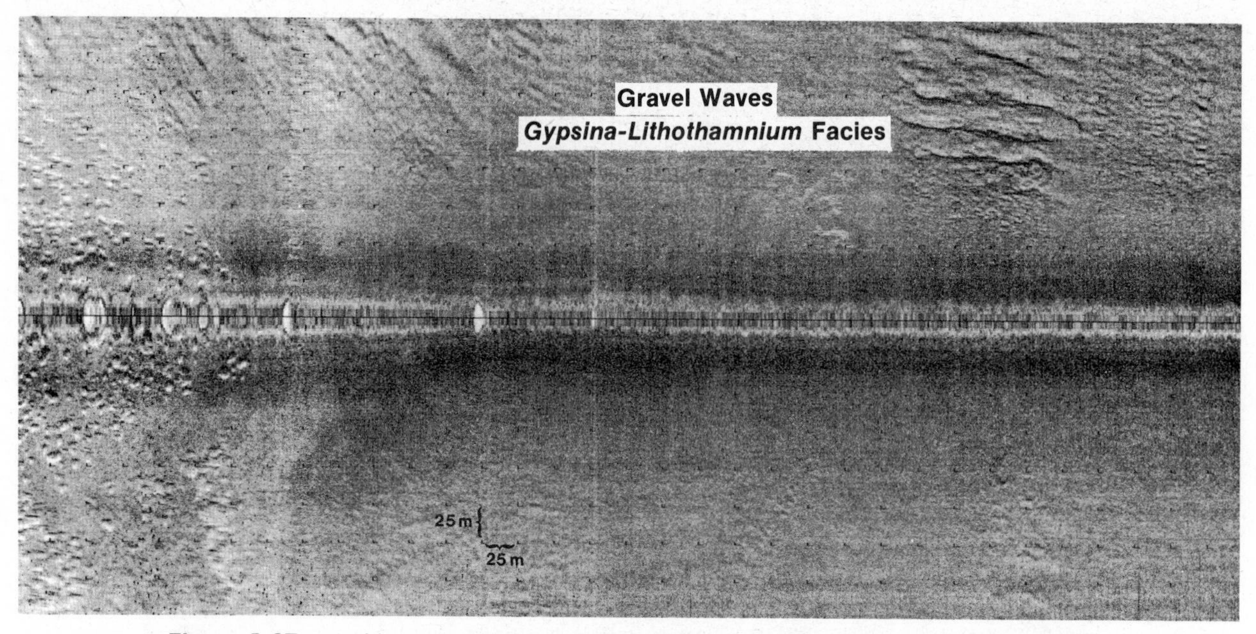

Figure 4.15. A side-scan sonar record showing surface features in a central part of the West Flower Garden Bank (just south of the record shown in Figure 4.13; indexed in Figure 4.3). Water depth is 58 m.

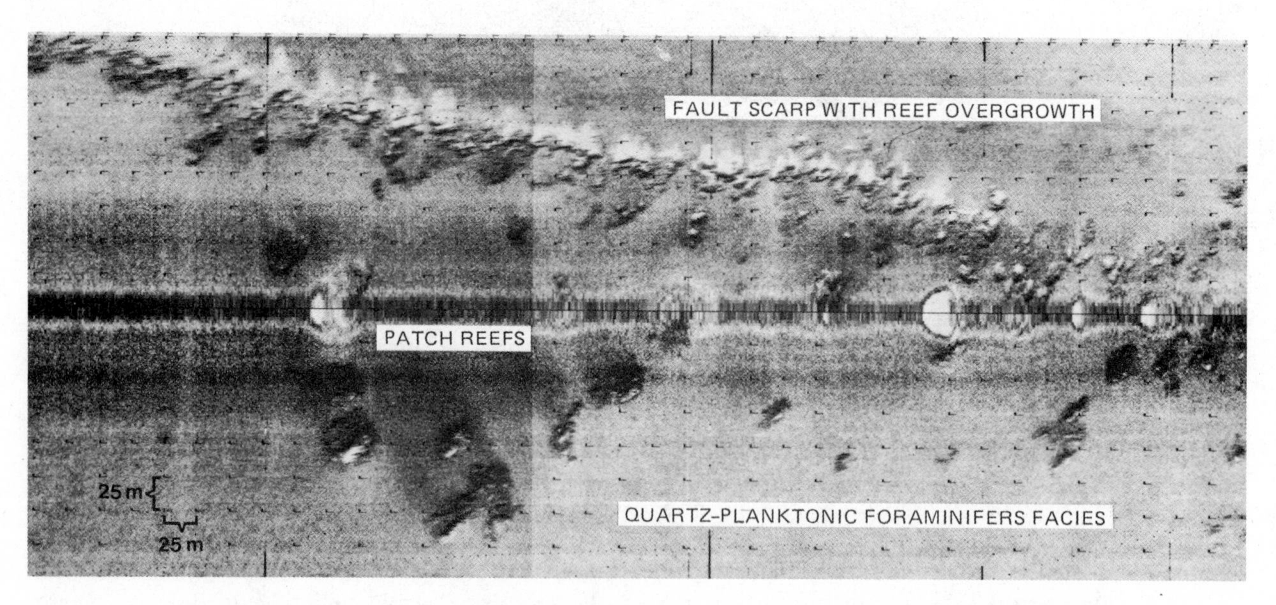

Figure 4.16. A side-scan sonar record showing a fault scarp with reef overgrowth in the northern part of the West Flower Garden Bank (Lease Block A-384; indexed in Figure 4.3). Water depth is about 105 m.

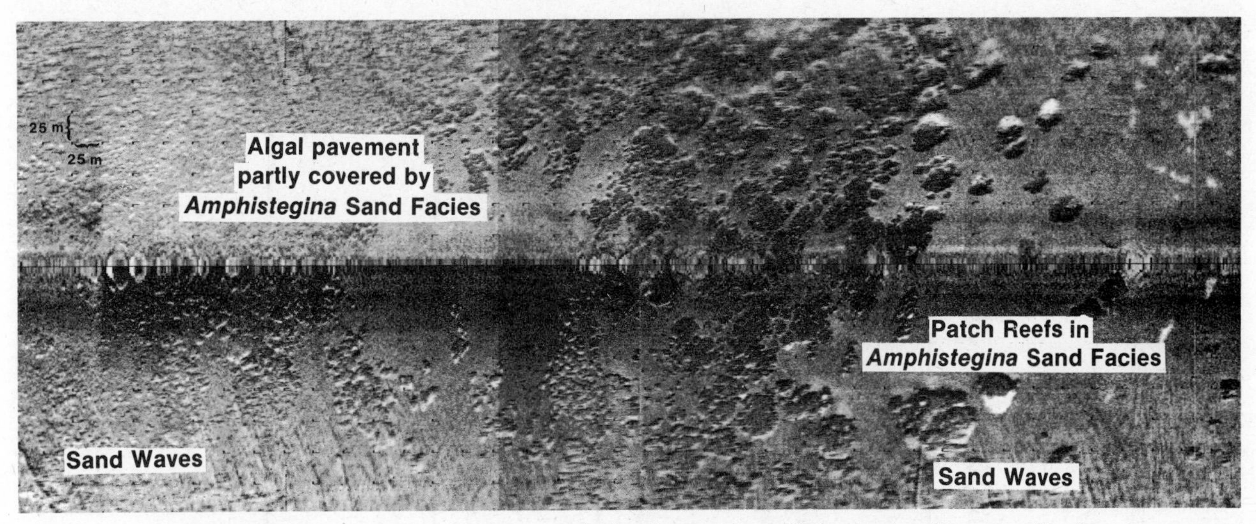

Figure 4.17. A side-scan sonar record showing patch reefs in the *Amphistegina* Sand Facies at the West Flower Garden Bank (Lease Block A-398; indexed in Figure 4.3). Depth is about 85 m.

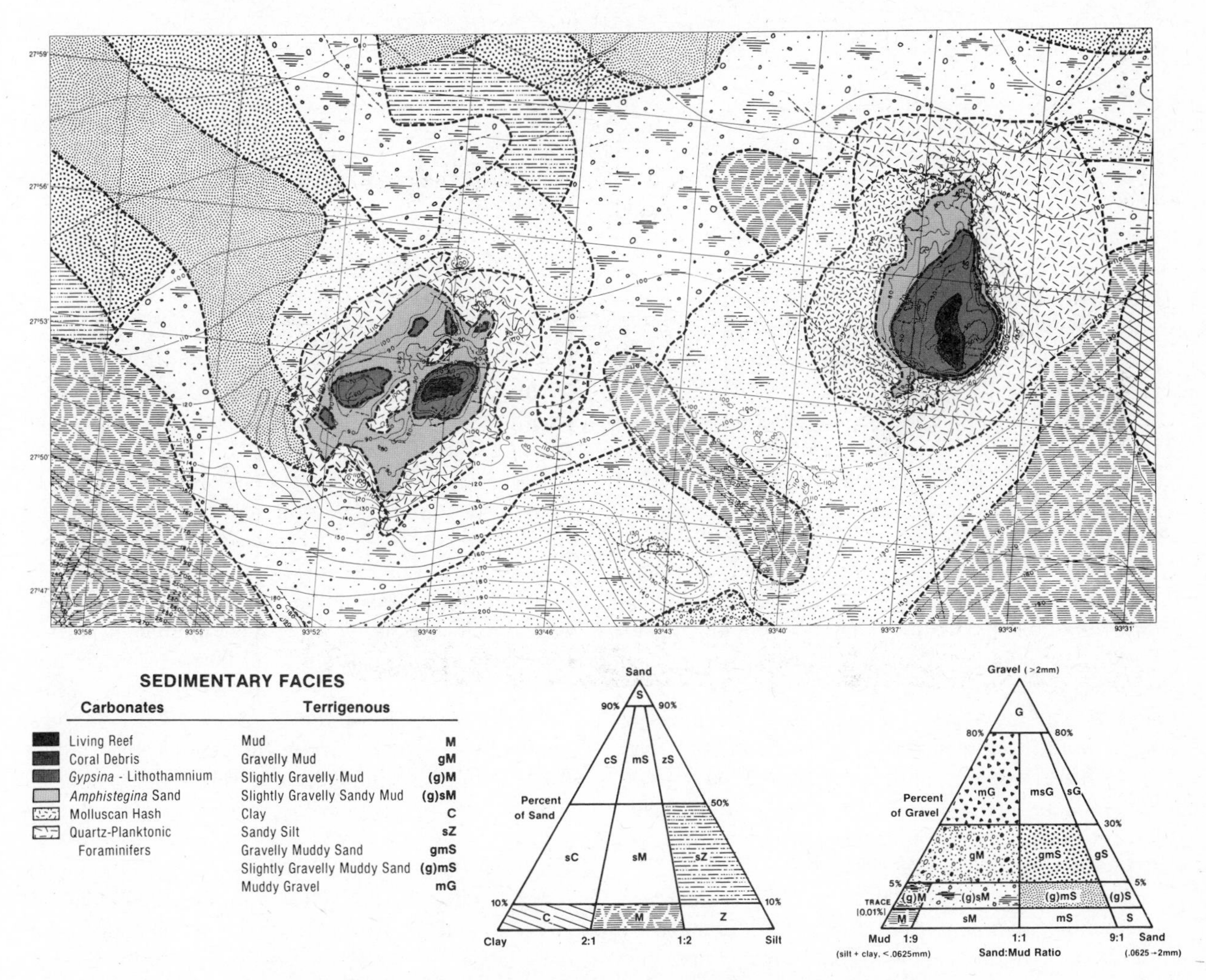

Figure 4.18. Sediment facies in the area of the Flower Garden Banks.

TABLE 4.1. Relationship Between Sediment Facies and Biological Zones at the East Flower Garden Bank

Sediment Facies	Depth (m)	Biological Zone	Depth (m)
Coral reef	15–50	*Diploria-Montastrea*	15–36
Living reef (massive limestone)	15–45	*Porites*	
		Madracis	28–46
Coral debris (coarse sand and gravel)	25–50	*Stephanocoenia*	36–52
Gypsina-Lithothamnium (coarse gravel and massive limestone)	50–75	Algal-sponge	46–88
Amphistegina sand (medium to coarse sand, muddy at depths greater than 85 m)	75–90	Transition	88–89
Quartz-planktonic forminifers (sandy mud)	90+	Nepheloid	89
Molluscan hash (muddy sand)	90+		

ever carried to the top of the reef by physical or biological processes. The surface irregularities would act as baffles that retard the velocity of the currents and cause deposition of fine sediment in nooks and crannies on the reef and in the *Gypsina-Lithothamnium* Facies.This process is analogous to a snow fence that traps snow away from highways and railroad tracks.

As indicated above, there is no land-derived mud in the bottom sediments above a depth of 75 m (Figure 4.19). This fact substantiates the conclusion based on studies of water and sediment dynamics that the nepheloid layer rarely rises to depths of 75 m (see below, Chapter 5). If the nepheloid layer were able to cover more of the bank, we would find land-derived muds mixed with coarser sediments at shallower depths. Moreover, the 80-m isobath is a major boundary in the biological zonation. That depth separates the turbid water fauna below from the clear water fauna and flora above. If the nepheloid layer were able to rise to shallower depths, the lower limits of the clear-water assemblages would also be raised by the same number of meters (Bright and Rezak, 1978b; Rezak and Bright 1981a, Vol. 3).

The boundary between the areas of carbonate and terrigenous sedimentation is an artifact caused by the use of two different classification systems: carbonate (genetic) and terrigenous (textural). A discussion of the problems created by the use of these two classification systems is presented in Chapter 1.

Description of Facies

Coral Debris Facies. As indicated in Table 4.1, the Coral Debris Facies ranges in depth from 25 to 50 m. It overlaps with the living reef, which has a depth range of 15 to 45 m. Where they overlap, the Coral Debris Facies consists of coarse sand and gravel lying between coral heads and in valleys or canyons leading down to the sediment apron around the reef (Figure 4.14). The narrow band of this facies around the living reefs at the East and West Flower Garden Banks attests to the minimal amount of transportation of coral debris. The dominant particle types are (1) coral fragments, 60 to 68%, (2) coralline algae, 16 to 30%; and (3) *Gypsina plana*, an encrusting foraminifer, 3 to 7%. Textural analysis of samples from this facies shows that 99.83% of the sample is coarser than coarse silt (0.062 mm).

Gypsina-Lithothamnium Facies. This facies ranges in depth from 50 to 75 m. The boundary between it and the Coral Debris Facies is transitional and its width is variable. Algal nodules and coral sand cover the bottom. The number of algal nodules

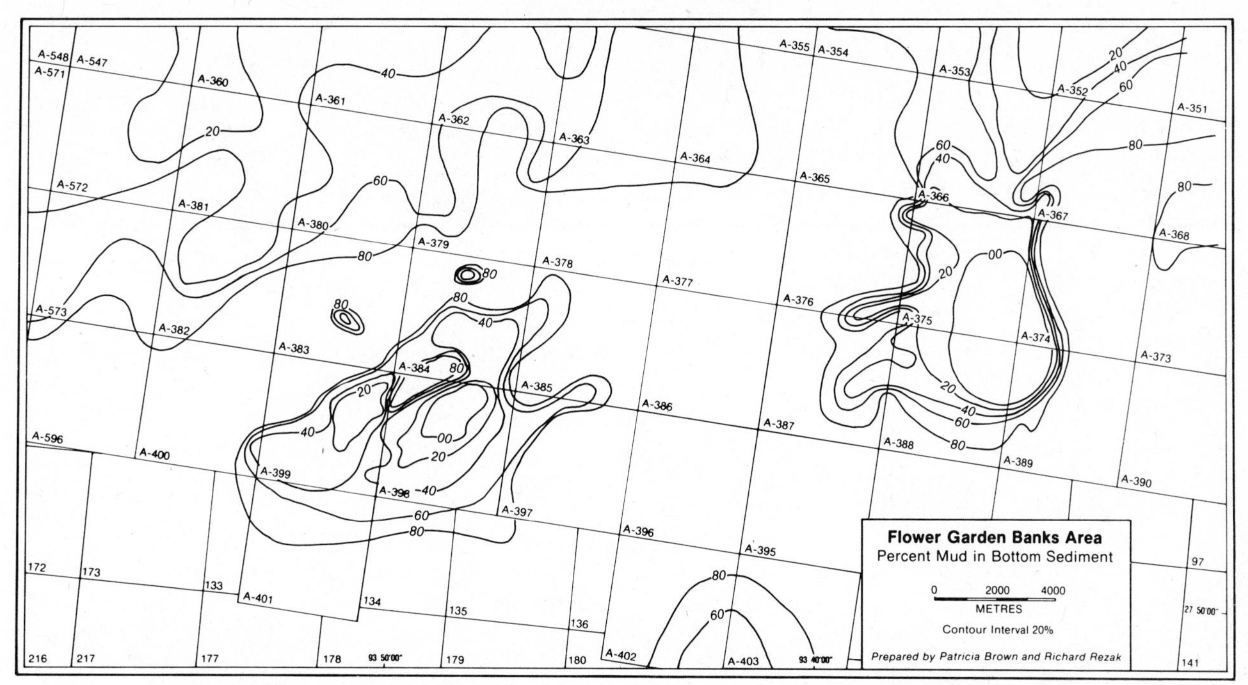

Figure 4.19. Percentage of mud (silt and clay) in bottom sediment in the area of the Flower Garden Banks.

increases with depth until the entire seafloor is covered. The sediment particles, which range from sand to cobble size, are 48.31% gravel and 49.9% sand. As much as 1.78% mud has been reported from this facies. However, the mud has been analyzed by electron microprobe and found to be carbonate; it is produced *in situ* by boring organisms within the algal nodules (G. Minnery, personal communication).

The major contributors to this facies are (1) coralline algae, 50 to 79%, (2) coral, up to 45%, (3) worm tubes, up to 21%, (4) *Gypsina plana*, up to 18%, (5) *Amphistegina* sp., up to 22%. The gravels are sediments that have been created by organisms with encrusting growth habits. Coralline algae, the major contributors to this facies, encrust skeletal grains (whole or fragmented molluscs or corals) or lithoclasts. The grains are rolled around by currents and grow by the addition of successive, more or less concentric layers of coralline algae to form coralline algae nodules similar to those described by Logan et al. (1969) from the Campeche Bank. The nodules attain diameters of 10 to 12 cm. A detailed description of the biota of the nodules is given in Chapter 6. One geologically important organism, the benthonic foraminifer *Amphistegina gibbosa*, grows profusely on the surfaces of the algal nodules. The skeletons of this foraminifer contribute to the coarse sand and granule fractions of the *Gypsina-Lithothamnium* Facies. In addition, the abandoned tests of *A. gibbosa* are transported downslope to accumulate in the *Amphistegina* Sand Facies. These grains are transported farther than any other skeletal grains at the Flower Garden Banks. The algal nodules, as displayed by the giant gravel waves on the side-scan sonar records of the West Flower Garden Bank (Figures 4.14 and 4.15), move in a direction parallel to the isobaths with little or no transport downslope.

Amphistegina Sand Facies.

The *Amphistegina* Sand ranges in depth from 75 to 90 m. Both upper and lower boundaries of this facies are transitional. The sediment varies in its texture from a clean sand in shallower depths to a sand that contains 34.29% mud at a depth of 94 m at the East Flower Garden Bank. At the West Flower Garden the *Amphistegina* Sand is 2.24% mud at 75 m and 18.9% mud at 90 m. *Amphistegina* tests form up to 32% of the sediment in this facies at both banks. Other major contributors include (1) coralline algae, 17 to 40%, (2) molluscs, 13 to 20%, and (3) lithoclasts, 6 to 16%. The lithoclasts are the result of bioerosion of drowned reefs within this facies.

Quartz-Planktonic Foraminifers Facies. This is the deepest facies in the area (> 90 m) and represents the most common sediment on the northwestern Gulf of Mexico outer continental shelf. However, because of its proximity to the banks, elements derived from the bioerosion of drowned reefs are also present. The facies consists of (1) quartz grains, 3.5 to 67%, and (2) planktonic foraminifers, 9 to 77%. Samples with low percentages of quartz grains have high percentages of planktonic foraminifers and vice versa. Added together, these components range from 53 to 83%. The silt-plus-clay fraction varies from 24 to 100%, with an average of 82%. The facies completely encircles the West Flower Garden Bank but occurs only on the north, east, and south sides of the East Flower Garden Bank. On the southeast margin of the East Flower Garden, where the steepest slopes occur, this facies is in direct contact with the *Gypsina-Lithothamnium* Facies.

Molluscan Hash Facies. This facies occurs on the western and southwestern margin of the East Flower Garden Bank. Edwards (1971) did not recognize the facies on the West Flower Garden Bank because of the lack of samples far enough to the south and southwest of the bank. It is composed of 15 to 54% sand-sized mollusk fragments and 0 to 34.5% quartz grains. The silt-plus-clay fraction ranges from 5 to 26%, with an average of 22%. It is easily distinguished from the Quartz-Planktonic Foraminifers Facies by its low content of planktonic foraminifers (0.5 to 13.3%) and its low mud content. Also, the much lower percentage of molluscs in the Quartz-Planktonic Foraminifers Facies ranges from 1.0 to 14.6%.

Patch Reefs. Both living and dead (drowned) patch reefs are abundant below a depth of 50 m (Figures 4.16 and 4.17). Living coralline algae cover substantial areas on these patch reefs down to depths of 85 to 90 m. Below 90-m water depth the patch reefs are dead because of reduced light penetration and the increased deposition of fine-grained clastics by the nepheloid layer. At the East Flower Garden the dead reefs have been observed to a depth of 120 m and on the West Flower Garden to a depth of about 150 m. Rezak (1977) illustrated rhombohedral crystals from a 91-m drowned reef and identified them as freshwater cement. Neither stable isotope nor electron microprobe analyses were run on this sample, however, Edwards (1971), assuming a static

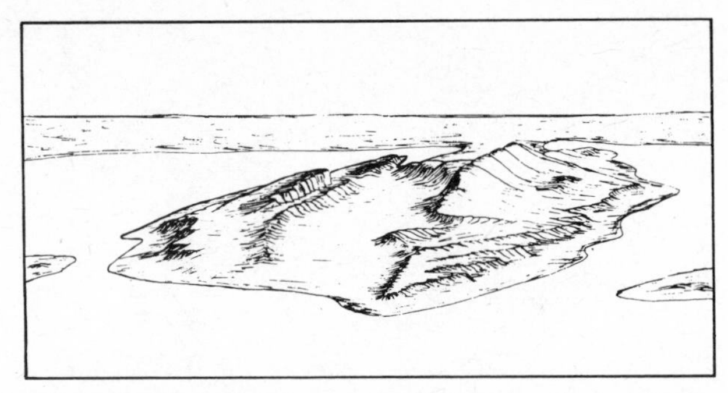

Figure 4.20. Sketch of the West Flower Garden Bank as it may have looked during the 121-to-134-m stillstand of the Gulf of Mexico (Edwards, 1971).

level for the crest of the West Flower Garden reef since Late Wisconsin time, illustrated the probable configuration of the West Flower Garden Bank during the 121-to-134-m sea level stillstand (see Figure 4.20).

Recent work by Stafford (1982) and Minnery (1984) on the diagenesis of reef rock at both Flower Garden Banks has included petrographic, stable isotope, and electron microprobe analyses. Blocky sparry mosaics in what appear to be spherulitic aragonite cements have been referred to as low-Mg calcite alteration products produced in the subaerial environment (Rezak, 1977). Microprobe analyses of these mosaics (Stafford, 1982) indicate that they are aragonite. They have a stable isotope composition typical of marine limestones. The significance of the blocky sparry mosaics is not known at this time.

SUMMARY

The East and West Flower Garden Banks (Figures 4.1 to 4.3) are bathymetric prominences caused by salt diapirs. Bedrock outcrops on the seafloor at the crest of these prominences, caused by fracturing of the rocks overlying the salt diapir (Figure 4.10), served as substrates for the initial growth of reef-building organisms. Because the conditions of water depth, temperature, salinity, and water clarity were favorable, a complex of reef communities (Figure 4.14) drastically changed the nature of the bottom sediments in the area of the banks.

The normal sediments on the open shelf surrounding the banks are sands and muds that have been eroded from the North American continent and mechanically transported to the Gulf of Mexico by streams like the Mississippi, Trinity, Sabine, and Brazos rivers. The banks rise from surrounding shelf depths of 100 to 180 m to crests as shallow as 20 m. The normal shelf sands and muds do not occur on the banks at depths shallower than 75 to 80 m. The sediments above the 75-m level are coarse, skeletal sands and gravels and rocky, limestone structures built by corals and other lime-secreting, reef dwelling organisms. The loose sediments around the reef reflect the depth zonation of the biological communities that are present on the two banks. The distribution of sediment types, however, does not coincide with the boundaries between biotic communities mainly because of the downslope movement of sediment by the force of gravity.

The downslope movement of sediment produced on the banks, the absence of land-derived muds in the bottom sediments above a depth of 75 m, and the major biological boundary at approximately 80 m substantiate the conclusion derived from water and sediment dynamics that the currents flow around the banks rather than up and over them. The nepheloid layer rarely rises to depths of 75 m.

The presence of a series of brine seeps and a brine lake, which has been documented, indicates the removal by dissolution of prodigious amounts of solid salt from the crest of the salt diapir. The removal of large volumes of solid salt from the shallow subbottom beneath the crests of the reefs creates hazardous seafloor instability in those areas.

5

FLOW, BOUNDARY LAYERS, AND SUSPENDED SEDIMENT AT THE FLOWER GARDEN BANKS

There are two reasons for using the East and West Flower Garden Banks as model systems with which to compare the other banks of the northwestern Gulf of Mexico. First, these two banks have the most complete sequence of depth zonations of biotic communities and sedimentary facies known to exist on the Texas–Louisiana Shelf. Second, as the site of a multifaceted monitoring program from January 1979 through July 1981, they have been the most intensively studied.

It is the living coral reefs at the crests of the East and West Flower Garden Banks that set them apart from otherwise similar banks in the region. By determining why these banks possess the reef and why they are zoned as they are, we are well on the way to understanding the conditions that govern the zonation on all the banks in the northwestern Gulf of Mexico. The biotic zones, sedimentary facies, and depth relationship between the two at the Flower Garden Banks are listed in Table 4.1.

In this chapter the East and West Flower Garden Banks are placed in their physical context; that is, the two banks are discussed in terms of the temperature, salinity, flow, and suspended sediment of the waters that bathe them and the relation of those variables to the banks' zonation.

GENERAL SETTING

The Flower Garden Banks, which lie between 93°30′W and 94°W just south of 28°N, rise above a mud substrate at the shelf edge at the widest point of the continental shelf in the northwestern Gulf of Mexico (Figure 1.5). They lie in the eastward flowing shelf-slope current (see Chapter 2) and rise in water that does not mix all the way to the bottom in winter. It should be remembered that they are high-relief banks that extend 80 to 100 m above the surrounding seafloor.

The fact that the banks are at the outermost edge of the widest point on the Gulf of Mexico Continental Shelf has two important implications. The first is that the volume of water on the shelf decreases both east and west of the longitude of the banks. Therefore continuity demands that alongshelf flow in either direction must diverge and slow in the vicinity of the banks or additional water must be drawn in from offshore to accommodate this increase in volume. The distribution of surface isotherms in Figures 2.22 and 2.27 implies that additional water was drawn in from offshore in February and April of 1982. In both instances warmer offshore waters appear to have been drawn up into the arch of the coastline. The second implication of the location of the Flower Garden Banks is that they lie at a maximum distance from the turbid surface water of the coast and should therefore receive maximum solar irradiance for shelf-edge banks in the northwestern Gulf of Mexico.

The high relief of the banks, combined with the fact that even in deep winter the waters do not mix to the bottom, precludes the possibility of resuspended sediment from the adjacent seafloor rising to the top of the banks.

METHODS

In January 1979 time series measurements of currents and temperature were established at moorings to the northeast of the East Flower Garden Bank (mooring 1), at the crest of the reef (EMCM), and at the southwest margin of the bank (mooring 2) (see Figure 5.1 for locations). In September 1979 a third mooring was added just northwest of the West Flower Garden Bank (Figure 5.1). Later, in April 1980, the third mooring was moved northeast of the West Flower Garden Bank and a fourth mooring was established northwest of the bank (Figure 5.1). Statistics regarding the position of moorings, depth of the instruments, sampling intervals, and duration of the recordings are given in Table 5.1.

In addition to the time series measurements, two other types of datum were collected. Profiles of temperature, salinity, transmissivity, and horizontal current velocities were obtained on seasonal cruises. At first, this was accomplished by separate lowerings of a salinity, temperature, and depth (STD) sensor, a transmissometer, and a profiling current meter. In 1980 we developed an instrument that contained a conductivity, temperature, depth (CTD) probe, a 25-cm pathlength transmissometer (XMS), and an electromagnetic profiling current meter in a single instrument package. Data from the instrument package are fed up a 7-conductor armored cable to a Z-80-based microcomputer deck unit which is also interfaced with a Loran C receiver for continuous acquisition of position data. The system was named the Profiling Hardwired Instrumented System for Hydrography (PHISH). On all cruises after October 1980 the PHISH was used for the acquisition of station data. Seasonal sampling cruises were made in April, July and October of 1979, October 1980, March 1981, and July 1981. Additional stations were taken near the West Flower Garden Bank in March 1982.

For the third type of data dye emission obser-

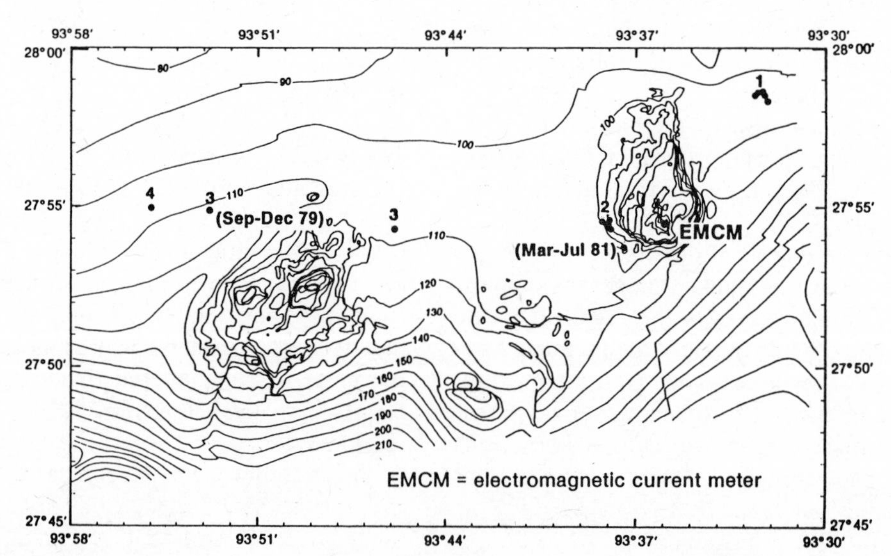

Figure 5.1. Mooring positions for all six current meter array deployments from January 1979 to July 1981. The two cases in which a mooring position was significantly changed are indicated by the deployment dates in parentheses.

TABLE 5.1. Current Meter and Associated Time Series Data Inventory, East and West Flower Garden Banks, All Deployments from January 1979 to July 1981[a]

Instrument Disk File Code	Month and Year of Recovery	Location	Depth (m) Meter/Bottom	Date and Time (GMT) Beginning		End	
		Savonius Rotor Current Meter					
Mor1Met1	Apr 79	27°58.63′N, 93°32.42′W	56/96	1/16/79	10:34	3/26/79	23:52
Mor2Met1	Apr 79	27°54.65′N, 93°38.02′W	60/100	1/17/79	21:16	4/10/79	19:22
Mor2Met3	Apr 79	27°54.65′N, 93°38.02′W	96/100	1/17/79	21:52	3/25/79	22:46
Mor1Met1	Sep 79	27°58.38′N, 93°32.19′W	60/100	7/12/79	18:05	9/4/79	19:11
Mor1Met2	Sep 79	27°58.38′N, 93°32.19′W	94/100	7/12/79	18:21	9/5/79	11:15
Mor1Met3	Sep 79	27°58.38′N, 93°32.19′W	96/100	7/12/79	18:38	9/1/79	21:26
Mor2Met1	Sep 79	27°54.57′N, 93°38.23′W	60/100	7/11/79	22:05	9/3/79	14:41
Mor2Met2	Sep 79	27°54.57′N, 93°38.23′W	94/100	7/12/79	02:09	9/4/79	12:21
Mor2Met3	Sep 79	27°54.57′N, 93°38.23′W	96/100	7/11/79	21:41	9/4/79	07:47
Mor1Met1	Dec 79	27°58.55′N, 93°32.32′W	53/99	9/9/79	15:01	12/10/79	04:19
Mor1Met2	Dec 79	27°58.55′N, 93°32.32′W	95/99	9/9/79	15:10	11/25/79	07:10
Mor2Met1	Dec 79	27°54.60′N, 93°38.23′W	53/99	9/11/79	16:19	11/28/79	14:19
Mor2Met2	Dec 79	27°54.60′N, 93°38.23′W	95/99	9/11/79	16:20	12/13/79	05:44
Mor3Met1	Dec 79	27°54.93′N, 93°52.79′W	61/107	9/11/79	18:15	12/9/79	22:39
Mor1Met1	Sep 80	27°58.56′N, 93°32.61′W	53/99	4/22/80	23:20	8/12/80	20:20
Mor1Met2	Sep 80	27°58.56′N, 93°32.61′W	95/99	4/23/80	18:00	9/7/80	04:20
Mor2Met1	Sep 80	27°54.43′N, 93°38.00′W	49/95	4/23/80	20:00	10/5/80	17:20
Mor2Met2	Sep 80	27°54.43′N, 93°38.00′W	90/95	4/23/80	20:00	8/20/80	07:40
Mor3Met1	Sep 80	27°54.35′N, 93°45.90′W	58/104	4/23/80	16:00	9/7/80	04:20
Mor3Met2	Sep 80	27°54.35′N, 93°45.90′W	100/104	4/23/80	16:00	8/20/80	06:00
Mor1Met1	Jan 81	27°58.63′N, 93°32.52′W	54/96	10/13/80	19:40	1/19/81	21:20
Mor1Met2	Jan 81	27°58.63′N, 93°32.52′W	80/96	10/13/80	17:40	1/23/81	18:00
Mor2Met1	Feb 81	27°54.39′N, 93°37.95′W	32/99	10/25/80	09:35	2/11/81	15:15
Mor2Met2	Feb 81	27°54.39′N, 93°37.95′W	57/99	10/25/80	08:15	2/9/81	16:35
Mor2Met3	Feb 81	27°54.39′N, 93°37.95′W	83/99	10/25/80	09:15	2/10/81	03:55
Mor2Met4	Feb 81	27°54.39′N, 93°37.95′W	95/99	10/25/80	09:55	2/10/81	11:35
Mor3Met1	Feb 81	27°54.34′N, 93°45.89′W	52/101	10/20/80	17:50	2/9/81	17:11
Mor3Met2	Feb 81	27°54.34′N, 93°45.89′W	63/101	10/20/80	21:10	2/11/81	20:10
Mor3Met3	Feb 81	27°54.34′N, 93°45.89′W	90/101	10/20/80	21:10	2/10/81	06:50
Mor3Met4	Feb 81	27°54.34′N, 93°45.89′W	97/101	10/20/80	21:30	12/5/80	05:50
Mor1Met1	Jul 81	27°58.58′N, 93°32.53′W	47/97	3/6/81	00:40	7/16/81	00:00
Mor1Met2	Jul 81	27°58.58′N, 93°32.53′W	58/97	3/6/81	00:00	7/11/81	10:00
Mor1Met3	Jul 81	27°58.58′N, 93°32.53′W	85/97	3/6/81	00:20	7/16/81	00:00
Mor1Met4	Jul 81	27°58.58′N, 93°32.53′W	91/97	3/6/81	00:20	7/16/81	00:00
Mor2Met1	Jul 81	27°53.79′N, 93°37.47′W	50/103	3/5/81	22:20	6/3/81	21:40
Mor2Met2	Jul 81	27°53.79′N, 93°37.47′W	72/103	3/5/81	22:20	7/15/81	16:00
Mor2Met3	Jul 81	27°53.79′N, 93°37.47′W	85/103	3/5/81	22:20	7/15/81	16:40
Mor2Met4	Jul 81	27°53.79′N, 93°37.47′W	97/103	3/5/81	22:20	4/14/81	03:20
Mor3Met1	Jul 81	27°54.38′N, 93°45.90′W	53/103	3/5/81	17:40	7/16/81	05:40
Mor3Met2	Jul 81	27°54.38′N, 93°45.90′W	64/103	3/5/81	17:20	7/15/81	16:20
Moe3Met3	Jul 81	27°54.38′N, 93°45.90′W	91/103	3/5/81	17:20	7/15/81	15:00
Mor3Met4	Jul 81	27°54.38′N, 93°45.90′W	97/103	3/5/81	17:20	7/15/81	15:00
Mor4Met1	Jul 81	27°55.01′N, 93°55.01′W	47/97	3/5/81	22:20	7/15/81	15:00
Mor4Met2	Jul 81	27°55.01′N, 93°55.01′W	58/97	3/5/81	16:20	7/15/81	22:40
Mor4Met3	Jul 81	27°55.01′N, 93°55.01′W	85/97	3/5/81	16:20	7/15/81	13:40
Mor4Met4	Jul 81	27°55.01′N, 93°55.01′W	91/97	3/5/81	16:00	7/15/81	13:20

(continued)

TABLE 5.1. (Continued)

Instrument Disk File Code	Month and Year of Recovery	Location	Depth (m) Meter/Bottom	Date and Time (GMT)			
				Beginning		End	
Electromagnetic Current Meter							
The first four deployments were made on top of the East Flower Garden Bank. The last three deployments were on mooring 2, near the southwest edge of the East Flower Garden Bank.							
EMCM.EMA	Apr 79	27°54.65′N, 93°35.92′W	28/30	2/10/79	18:00	4/16/79	05:30
EMCM.EMA	Jul 79	27°54.65′N, 93°35.92′W	28/30	4/29/79	18:00	7/07/79	05:20
EMCM.EMA	Sep 79	27°54.65′N, 93°35.92′W	28/30	7/14/79	23:30	8/29/79	14:20
EMCM.EMA	Dec 79	27°54.65′N, 93°35.92′W	28/30	9/25/79	00:00	10/16/79	13:50
MOR2.EMCM	Sep 80	27°54.43′N, 93°38.00′W	15/95	4/17/80	15:40	5/08/80	00:40
MOR2.EMCM	Feb 81	27°54.39′N, 93°37.95′W	20/99	10/24/80	22:25	1/11/81	19:54
MOR2.EMCM	Jul 81	27°53.79′N, 93°37.47′W	38/103	3/05/81	16:53	4/06/81	13:13
Temperature Probe							
Deployed on mooring line adjacent to the Marsh–McBirney electromagnetic current meter. The first three deployments were on top of the East Flower Garden Bank. The last deployment was on mooring 2, near the southwest edge of the East Flower Garden Bank.							
CONTMP.EMA	Apr 79	27°54.65′N, 93°35.92′W	27/30	2/10/79	16:44	4/24/79	09:10
CONTP.EMA.	Jul 79	27°54.65′N, 93°35.92′W	27/30	4/29/79	13:40	7/10/79	18:46
CONTMP.EMA	Sep 79	27°54.65′N, 93°35.92′W	27/30	7/16/79	14:10	8/30/79	23:22
CONTMP.EMA	Sep 80	27°54.43′N, 93°38.00′W	14/95	4/23/80	20:00	6/8/80	04:00

[a]Mooring locations are shown in Figure 5.1.

vations were made in the bottom boundary layer on and adjacent to the Flower Garden Banks. These observations were carried out *in situ* from the DR/V DIAPHUS, a two-man submersible. The design of the dye emitters used in the October 1980 experiments and shown in Figure 5.2 consisted of a plexiglass canister open at the top, with a small screen-covered orifice in the bottom; an electromagnet; a timing circuit; a rubber diaphragm; and a battery pack. The electromagnet consisted of a carriage bolt wound with a thin copper wire, screwed into the bottom of the plexiglass canister so that the head of the bolt was just below the top of the canister. A metal disk was attached to the middle of the rubber diaphragm to position it above the head of the carriage bolt when the diaphragm was stretched over the top of the canister. The leads from the electromagnet were attached to a timing circuit which could be set to pulse at intervals of 4 to 10 seconds. Powdered fluorescent dye was placed in the emitter so that when the emitter was immersed in water a concentrated dye filled the canister. When the timing circuit pulsed the powered electromagnet attracted the metal disk on the diaphragm. This caused the diaphragm to depress, thus expelling a puff of dye through the orifice in the bottom of the canister.

Four of these devices were placed on a length of aircraft wire, as shown in Figure 5.3. Below each emitter a fiberglass rod was fitted on the wire to permit it to rotate freely. The rod, a fishing rod blank, was marked every 5 cm with brightly colored reflecting tape and terminated with a vane. Before the array was deployed the emitters were turned on and synchronized so that all pulsed at the same time. A buoy was attached to the upper end of the aircraft wire for flotation and a weight of about 80 lb to the bottom to serve as an anchor. An acoustic release was attached to the buoy at the top of the array, and it, in turn, was attached to a bright yellow line 10 m longer than the water depth at the dive site. At the other end of the bright yellow line was still another buoy.

At the beginning of an experiment the whole array was put over the side of the tending vessel; then the submersible was launched. It maneuvered over to the surface float and followed the yellow line down to the dye emitters. The acoustic release was then triggered to separate the emitter array from any surface motion that might have been carried by the yel-

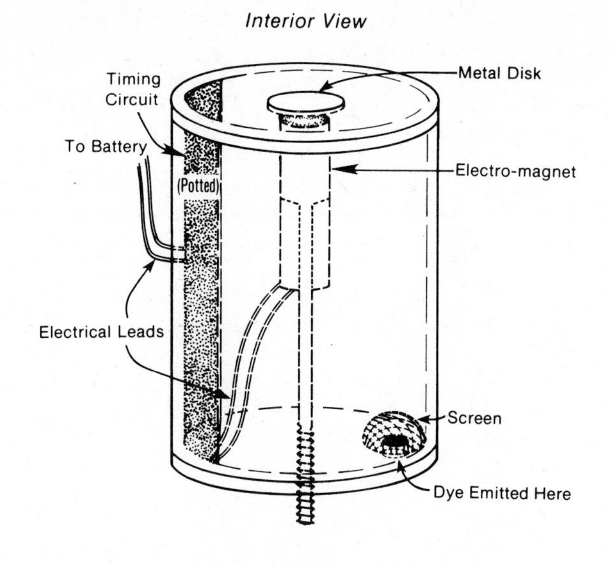

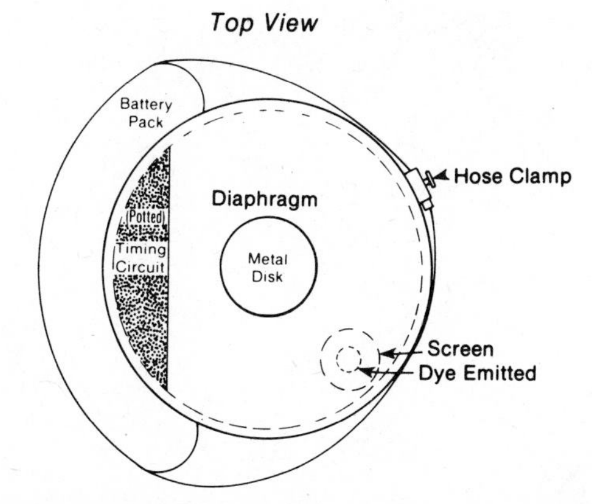

Figure 5.2. Diagrammatic representation of dye emitter apparatus used in October 1980 boundary layer experiments at the Flower Garden Banks.

low line. Once on the bottom the submersible moved to a position perpendicular to the flow and in front of the emitter array. The dye emission was then recorded on super-8-mm color film and color videotape cassettes. While these observations were taking place PHISH profiles were taken from the surface ship.

The recorded data were used to calculate velocity profiles, Reynolds stresses, and diffusivity. They were also used, qualitatively, to examine various flow phenomena in the bottom boundary layer (BBL).

TEMPERATURE AND SALINITY

The annual cycles of temperature and salinity (*T–S*) at the Flower Garden Banks are best illustrated by reexamining the mean annual cycle of density (Figure 5.4) and comparing it with composite *T–S* profiles for early fall, late winter, and midsummer (Figure 5.5). The location of the stations used in the composites is shown in Figures 5.6 to 5.9.

In winter, repeated cold-air outbreaks cool the surface waters and mix them down to approximately 70 m by January. Minimum temperatures in this layer, recorded by an instrument on the reef at the East Flower Garden (27-m depth), are in the order of 19.5°C. The thickness of the mixed layer decreases through the spring as the number and intensity of the polar cold-air outbreaks decline. Rapid warming begins in April and by June there is a strong thermocline with a surface mixed layer of minimal thickness. As surface heating continues through the summer, this layer thickens and the temperture increases from about 28°C in June to about 30°C in August. In late September or early October the polar fronts again push their way out to the region of the Flower Garden Banks. Wind mixing and destabilization of the density structure by surface cooling once again cause the surface mixed layer to thicken and decrease in temperature.

Angelovic (1976) reports that the minimum vertical temperature gradient in the vicinity of the Flower Garden Banks occurs in late February. At that time the surface mixed layer is 30 to 50 m thick, with temperatures in the range of 19 to 20°C. Temperatures at the bottom (125 m) are 17 to 18°C. The progression of the temperature cycle at 27 m for the period from deep winter to spring is shown in Figure 5.10. From February through late March the mean temperature remains near 20°C with very modest variance. About 22 March the temperature rises above 20°C and begins to climb toward 20.5°C. At that same time there is an onset of rather large 1- and 2-day oscillations in temperature. This is at the very time the base of the mixed layer rises above 30 m so that the top of the Flower Garden Banks is in the thermocline. The large excursions in temperature are the result of vertical oscillations of the thermocline isotherms past the fixed measurement point and are caused by inertial currents, shelf waves, and tides.

The composite profiles from October 1980 (Figure 5.5*a,b*) show a mean mixed layer depth of 35 m. Those stations that possessed significantly thicker mixed surface layers were located just downstream of the East Flower Garden Bank and the thickening appears to be an orographic effect. The temperature in the surface mixed layer is on the order of 27°C.

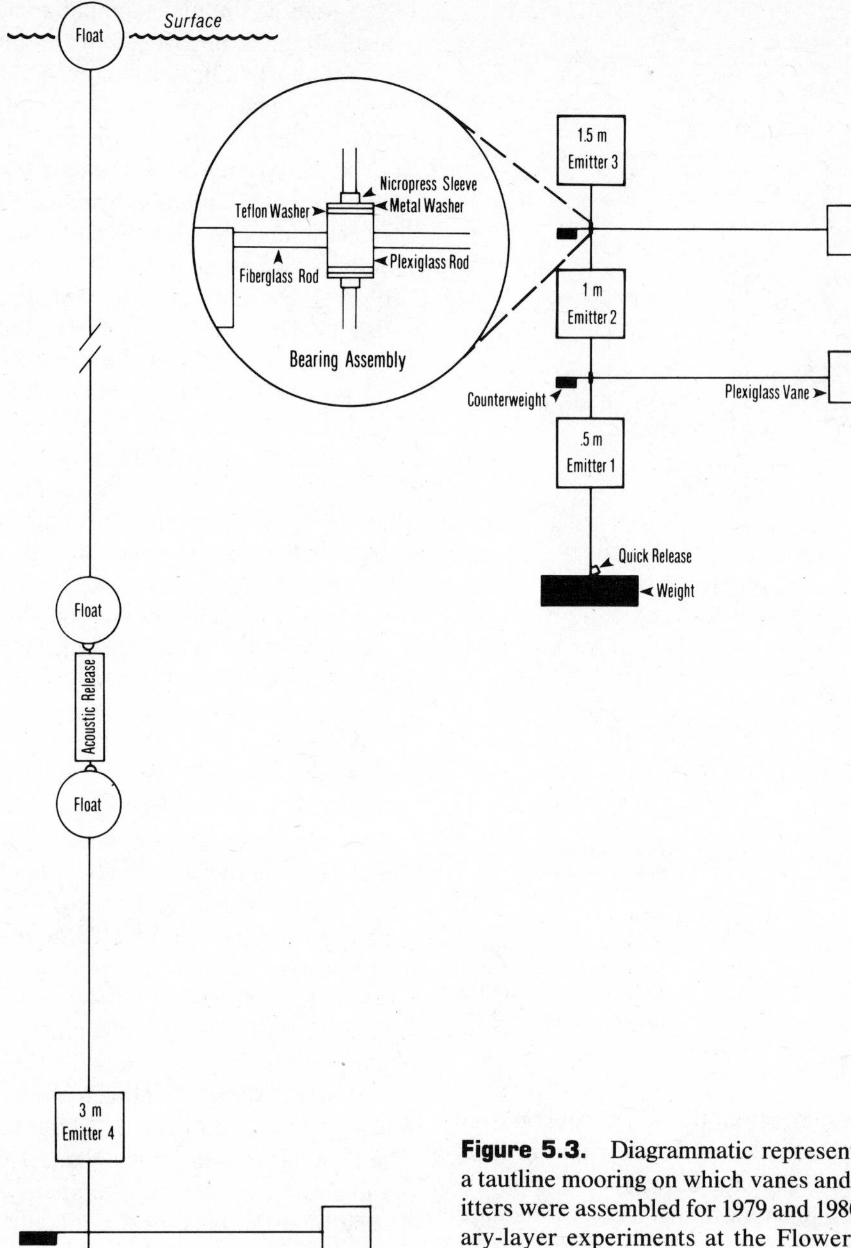

Figure 5.3. Diagrammatic representation of a tautline mooring on which vanes and dye emitters were assembled for 1979 and 1980 boundary-layer experiments at the Flower Garden Banks.

Beneath the mixed layer was a strong thermocline. The spread of the envelope around the profiles is also due to orographic effects.

The March 1981 profiles illustrate the minimum thermal gradient reported by Angelovic (1976). The variations in salinity during both cruises were caused by a slight mismatch in the time response of the temperature and conductivity sensors. This was cured before the July 1981 cruise by attaching a pump to the conductivity sensor to prevent it from lagging the temperature sensor.

The July profiles reveal a warm (30°C), thin (13 m) surface mixed layer with a strong thermocline beneath it. At the time of the survey water of 25°C or warmer extended to a depth of 50 m.

In October and March profiles are essentially isohaline with salinities slightly over 36 ppt. The July profiles, however, display a slight depression in the salinity in the 13-m thick surface mixed layer to about 35.5 ppt. This is consistent with the annual pattern of surface salinities shown in Figure 2.17. However, major year-to-year variations in midsum-

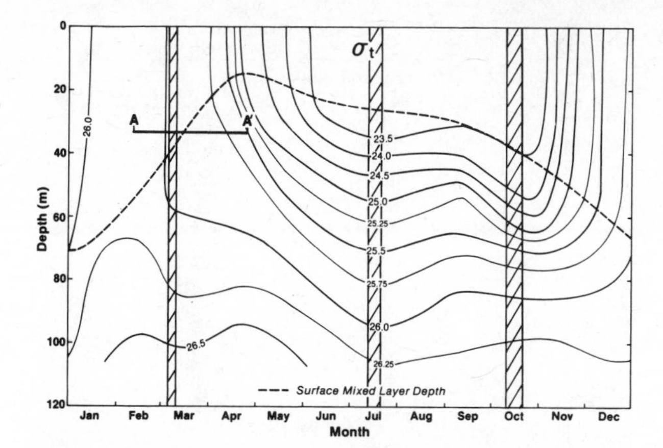

Figure 5.4. Seasonal progression of sigma-*t* versus depth near the Flower Garden Banks, as in Figure 2.26. *A* to *A′* represents the period of the temperature recording shown in Figure 5.10. The shaded portions indicate periods of survey data shown in Figure 5.5.

mer surface salinities occur at the shelf edge, as shown by the profiles taken at the southwest margin of the East Flower Garden Bank in July 1979 (Figure 5.11). Notice that even though the surface salinity in that profile was near 31.5 ppt, the salinity at 15 m was more than 35.5 ppt; or, to state it more emphatically, the water of relatively low salinity did not reach the level of the reef crest at the Flower Garden Banks.

Bottom temperatures at the shelf edge near the Flower Garden Banks undergo a peculiar double yearly cycle of minimums (17°C) and maximums (20°C). The minimums occur in spring (March–April) and midsummer (July–August), whereas the maximums occur in early summer (May–June) and midwinter (November–January) (Angelovic, 1976).

FLOW FIELD AROUND THE BANKS

Monthly Mean Current Vectors

Vector means were computed, on a calendar-month basis, for each current meter record available in the January 1979 to July 1981 interval. The monthly mean vectors were then segregated into three groups by depth. The first group (Figure 5.12*a*) represents instruments positioned near the middle of the water column (32 to 64 m). The second group (Figure 5.12*b*) represents flow at levels between 11 and 18 m above the bottom. The third group (Figure 5.12*c*) was derived from instruments deployed 8 m or less from the bottom. Bottom depths for all moorings ranged from 95 to 107 m. The monthly mean vectors are also shown as current roses in Figure 5.13. The time sense of the vectors is lost in this presentation, but the orientation of the vectors is more easily perceived.

With the exception of two periods, the mean monthly flow at mid-depth was directed toward the east, often at speeds exceeding 15 cm/s (Figure 5.12*a*). More modest speeds (on the order of 5 cm/s) were recorded in late winter (December–January) and midsummer (June). The two periods in which the monthly vector averages possessed a westerly component were July of 1979 and December of 1980.

During July 1979 Claudette, a major tropical storm, crossed over the Flower Garden Banks, then stalled over the coast. This storm produced sustained, high-velocity, southerly winds as far out as NDBO buoy 42002 (Figure 2.13) for a period of five days. Also, just prior to Claudette, a major hurricane (Bob) passed approximately 200 km to the southeast of the banks. Bob seems to have induced the reversals in the summer of 1979 and set up weak westerly flow along the shelf break; then Claudette produced intense northwesterly flow from 24 July through the end of the month (Figure 5.14). During December 1980 mid-depth currents had an average velocity vector toward the northwest. The reason for this anomalous flow pattern appears to be the cumulative effect of a long series of cold fronts that show up in the wind field as strong winds with northerly components (Figure 5.14). The reversal of the mean flow for this period should not be pictured as a simple phenomenon. The progressive vector diagrams for the Flower Garden deployments from 25 October through 9 February are shown in Figure 5.15. The vectors for December are marked as weeks 5 through 9.

With these two notable exceptions, flow at mid-depth in the vicinity of the Flower Garden Banks conforms to the model described in Chapter 2; that is, the flow at the shelf edge off the Texas–Louisiana segment is persistently to the east. The salinity of this water implies that it is oceanic rather than coastal in origin. From the midsummer profiles (Figures 5.5 and 5.11) it is apparent that coastal waters can become entrained at the surface.

Near-bottom currents (Figures 5.12*b,c* and 5.13*b, c*) tend to flow toward the east or southeast. In general, the stronger the flow, the more southerly it tended to be. The overall average for records in the 11 to 18 m above the bottom group was toward the southeast. Speeds for this level clustered between

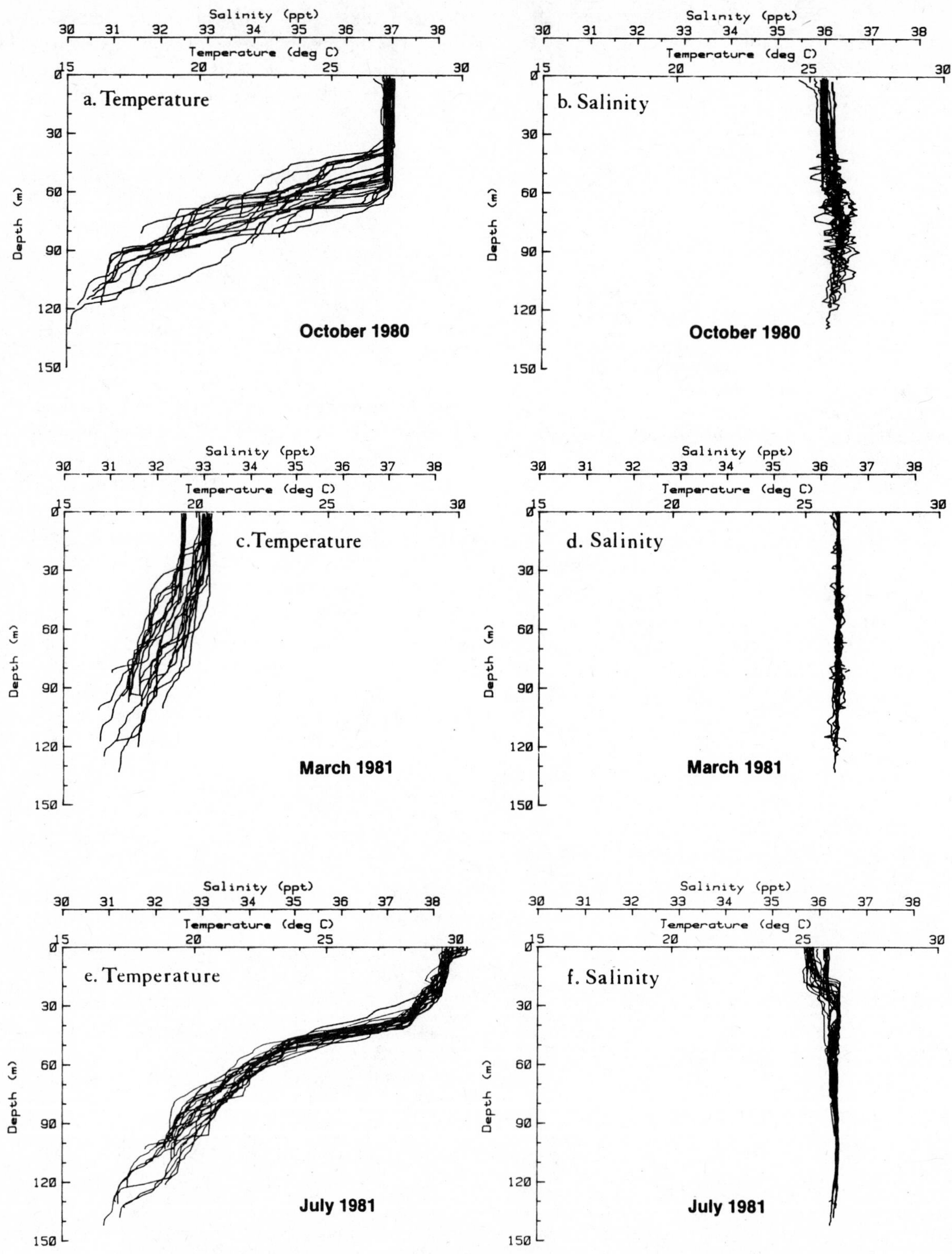

Figure 5.5. Composite plots of temperature and salinity profiles taken near the East and West Flower Garden Banks on three cruises.

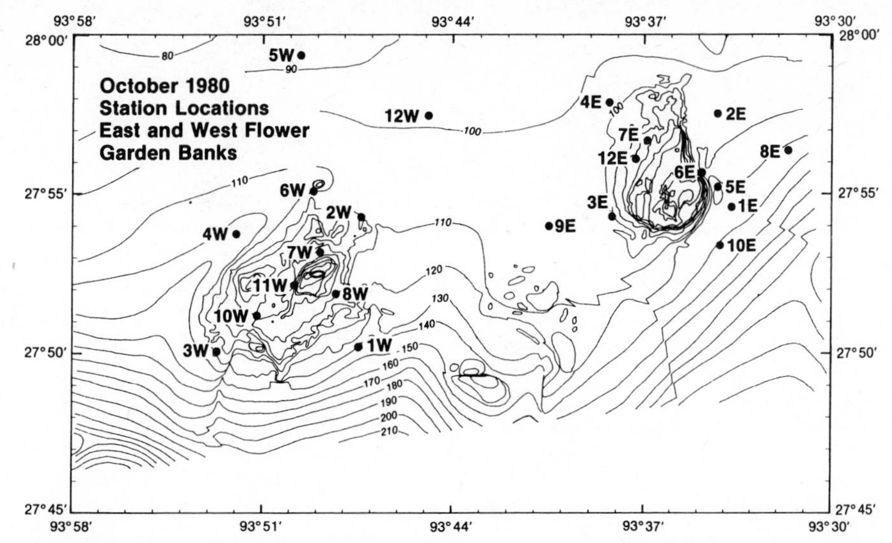

Figure 5.6. Hydrographic station location map for the BALTIC SEAL cruise, 10–20 October 1980.

1 and 5 cm/s. The large magnitude vectors pointing toward the south–southeast in Figure 5.13*c* are from mooring 2. These vectors are the result of strong topographic steering. If records for mooring 2 are removed for the very near bottom group the overall average is toward the east–southeast at 1.6 cm/s. The monthly mean vectors are only rarely toward the north. It did happen in March and January at 11 to 18 m above bottom and March, April, and November in the very near bottom flow. It is interesting to note that moorings 1 and 3 do not indicate near-bottom flow in the same directions at these times.

If we use the estimate of Ekman boundary layer thickness given by Weatherly et al. (1980),

$$\delta = k\frac{U^*}{f}$$

where k = von Karman's constant, approximately 0.4
U^* = the square root of the shear stress at the wall divided by the density of the fluid
f = the Coriolis parameter

and use the values we obtained for U^* in the boundary-layer dye studies, thickness (δ) is roughly 18 m. Therefore both near-bottom velocity groups should

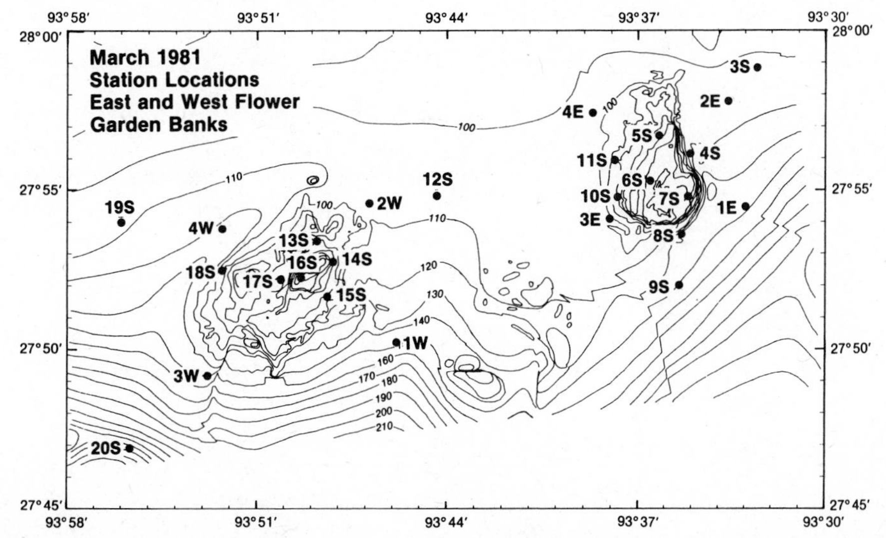

Figure 5.7. Hydrographic station location map for the GYRE cruise, 3–8 March 1981.

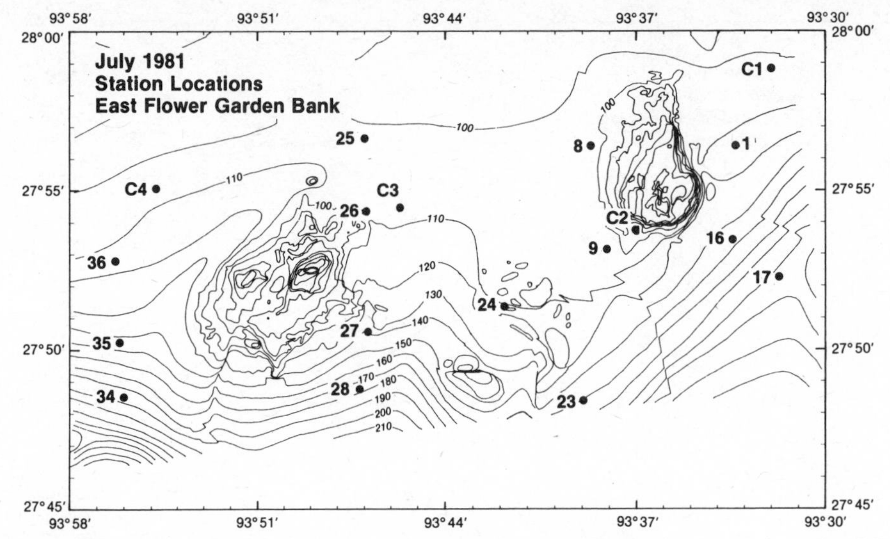

Figure 5.8. Hydrographic station location map for the NORTH SEAL cruise, 13–17 July 1981. (For complete East Flower Garden stations see Figure 5.9.)

have been in the Ekman bottom boundary layer. Eastward flowing geostrophic currents, however, should produce a boundary layer flow that is rotated counter-clockwise (looking down) to the northeast. The monthly mean vectors for the bottom flow are not representative of this rotation. There is some possibility that even moorings 1 and 3 were within the region of flow deformed by the banks and that centrifugal accelerations may have contributed to the clockwise rotation of the near-bottom vectors. It is also possible that during the winter the cold, dense water, which is formed on the mid- and inner shelf during cold frontal passages (Nowlin and Parker, 1974), tends to produce an offshore deflection of the bottom waters. It is more likely, however, that the rotation is due to baroclinic shearing. The idea that the near-bottom flow does move offshore is certainly consistent with the transects shown in Chapter 2 (Figures 2.34, 2.36, and 2.37). The transmissivity measurements on those transects show offshelf transport of fine sediment clearly in every case.

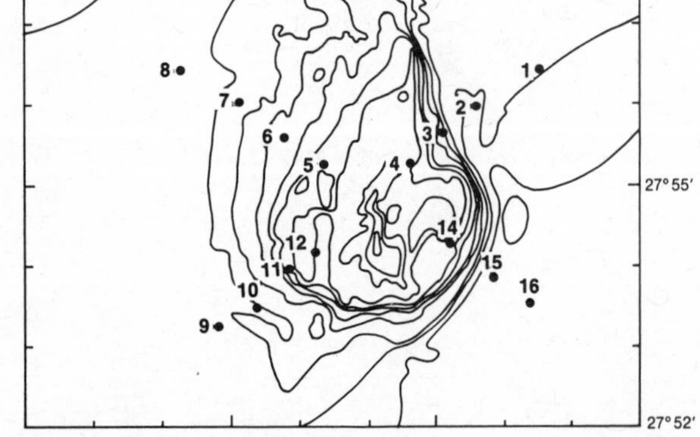

Figure 5.9. Hydrographic stations at the East Flower Garden Bank for the North SEAL cruise, 13–17 July 1981.

Variations on the Mean Currents

Each velocity (v) recording in a current meter record consists of the mean $(\bar{v})$ where the mean is taken over a period on the order of a month, for example—and a fluctuating component (v') that is,

$$v = \bar{v} + v'$$

It is not at all unusual to find that the fluctuating parts of the velocity have magnitudes larger than that of the mean flow. These variations about the mean velocity consist of periodic or nearly periodic oscillations and random white noise. A great deal of information regarding the nature of the flow at any given location can be extracted from a current meter record by examining these fluctuating components.

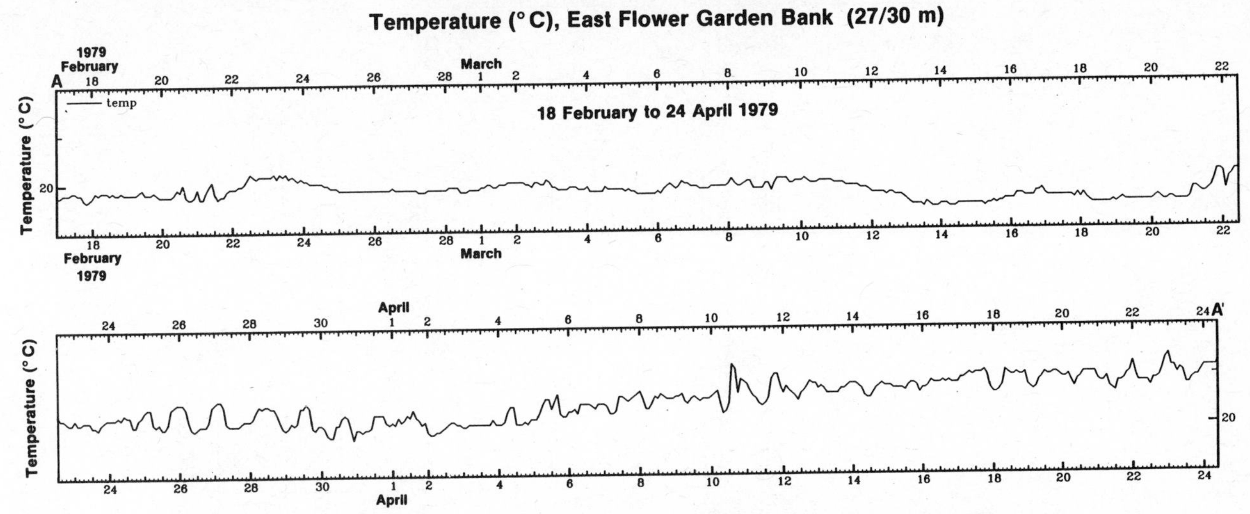

Figure 5.10. The temperature on the East Flower Garden reef at 27 m represents the transition from deep winter conditions to those of spring. The base of the mixed layer apparently reached this level on 22 March. Thereafter the instrument was in the thermocline. Compare this with the mean annual conditions in Figure 5.4.

The most common method of studying the oscillations in current meter records is by spectral analysis. Usually the record is resolved into two orthogonal vector components, u (east–west) and v (north–south). The mean subtracted from each series and the amplitude of the variance (cm^2/s^2) within frequency bands is calculated for frequencies that range between half the sampling frequency and the inverse of the record length. Typically, these analyses are accomplished by Fourier transforms of the series. A detailed description of the techniques used on the data collected at the Flower Garden Banks is given in McGrail et al (1982a).

In addition to being able to resolve how much variance is concentrated at a given frequency, it is also possible to determine the coherence and phase relationship between two time series over the same range of frequencies as above. In the following example, series y and series x have a coherency squared of 1 at the frequency shown because they contain the same sinusoidal oscillation. The X series,

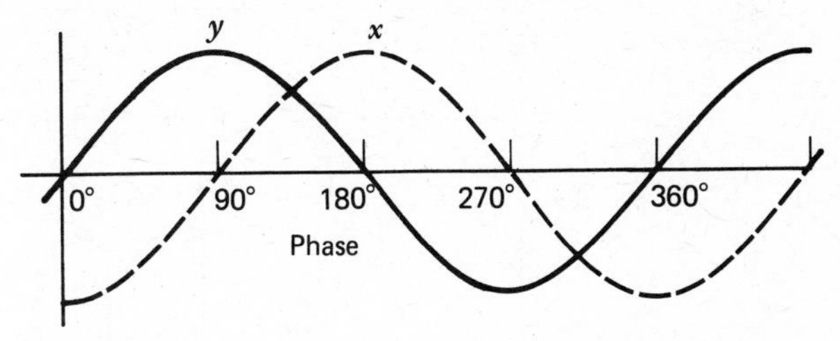

however, has a phase lag of 90° in relation to the y series. This situation might exist between the v velocity components from two current meters for a tidal oscillation if the two instruments were separated by one-fourth of the tidal wavelength in the direction of propagation.

It is possible therefore to determine (in a statistical sense) whether two time series are oscillating coherently at a common frequency. This is a great aid in determining whether two series are responding to the same forcing mechanism or whether two series are causally related (one forcing the other) at some

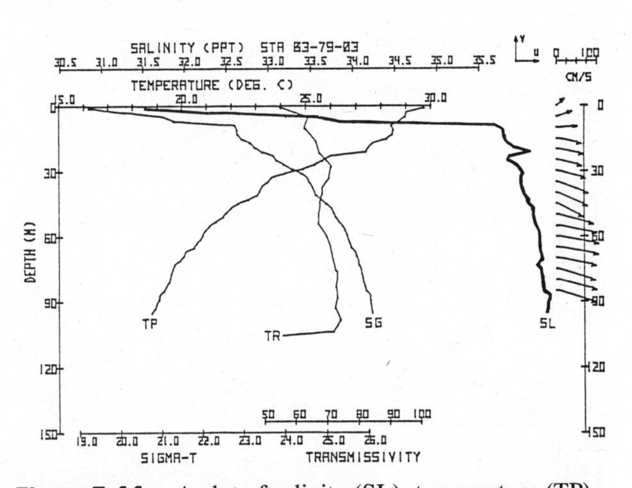

Figure 5.11. A plot of salinity (SL), temperature (TP), transmissivity (TR), and sigma-t (SG) from the East Flower Garden (July 1979). Note the surface salinity minimum near 31.5 ppt. Normal salinities (> 35.5 ppt) occur at depths of 15 m and greater.

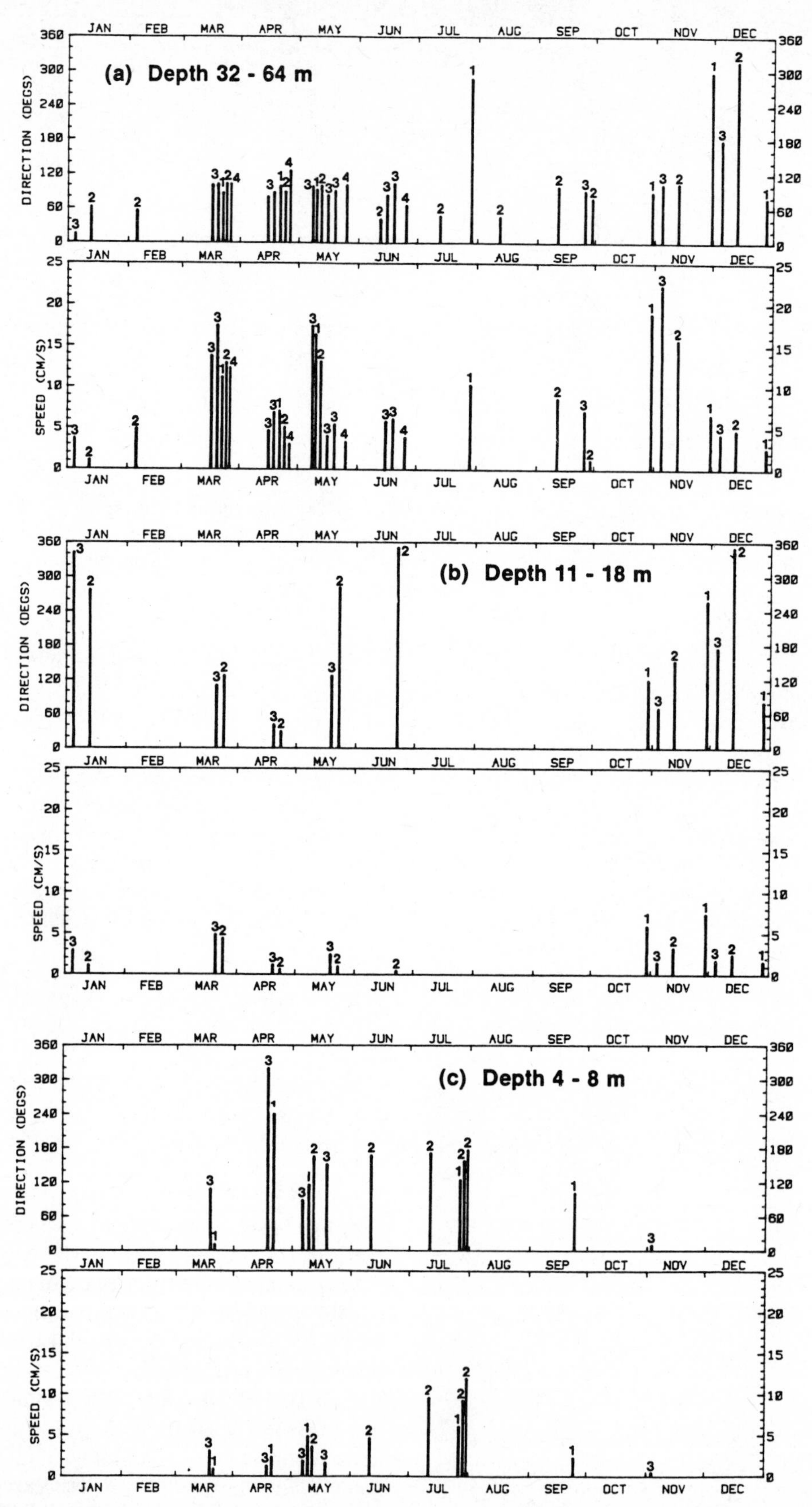

Figure 5.12. One-month averages of speed and direction for current meters in three depth ranges. The mooring number is printed above each average.

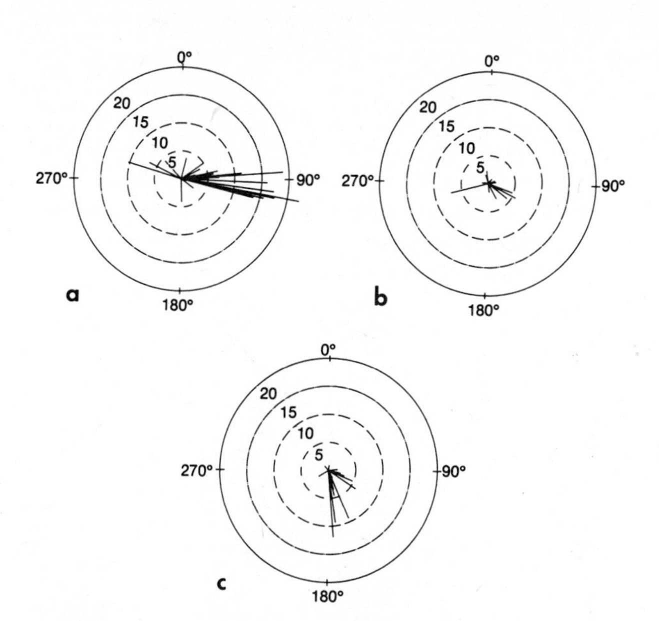

Figure 5.13. Monthly mean current vectors for all deployments and moorings. Speed labels are in cm/s. The data set is divided into three depth ranges: *(a)* 32 to 64 m depth (all measurements but one were taken in the depth interval from 47 to 64 m; *(b)* 11 to 18 m from bottom; *(c)* 4 to 8 m from bottom.

frequency, although it is not possible to distinguish between the two processes.

Finally, a technique called rotary spectral analysis is used to determine whether an oscillation in the velocity record of a time series is rotational, and if it is whether the sense of rotation is clockwise or counterclockwise. This is helpful in determining the phenomena that cause the oscillation. Inertial oscillations, for example, can rotate only in a clockwise sense (in the northern hemisphere).

Spectral analyses were performed on all time series records obtained from the moorings at the Flower Garden Banks. These records include velocity, temperature, and transmissivity series. Time series records of wind velocity from the nearby NDBO buoy (42002) were also subjected to spectral analyses.

Examples of these spectra are shown in Figures 5.16 through 5.24 to help illustrate the behavior of the flow at the Flower Garden Banks. Figures 5.16 and 5.17 are the spectra for mooring 1, mid-depth and near-bottom meters, respectively, during the summer. Figures 5.18 through 5.20 are spectra for the mid-depth and two near-bottom current meter records from mooring 3 (see Figure 5.1 for location) during late spring through midsummer. Spectra for the winter period at mooring 3 appear in Figures 5.21 through 5.23. Figure 5.24 represents the spectra of velocity fluctuations at 32 m, in the winter, at mooring 2.

Velocity Spectra. The contribution of random velocity fluctuation to the autospectra of u and v may be found in the slope of spectral density $G(F)$ with respect to frequency F on the log–log plots (Figures 5.16 to 5.24). The relationship is

$$G(F) = AF^{-a}$$

where the slope parameter a is a positive number.

It can be observed from these spectra that in general the value of A decreases with depth, which means that the amplitude of random fluctuations over all frequencies is smaller near bottom than in the upper portion of the water column. There are probably two reasons for this: (1) almost all driving mechanisms for the fluctuations are input through the surface and (2) random fluctuations are attenuated by friction near the bottom.

Notice that in almost all the spectra, but particularly in those from instruments deployed near the bottom, there is a change in the slope of $G(F)$ near the one cycle per day (cpd) frequency. Above 1 cpd the slope parameter a is relatively large but below 1 cpd the slope flattens out. This suggests that long-period random fluctuations are suppressed, particularly near the bottom. The one spectral set that does not exhibit a noticeable break in slope at 1 cpd is that for mooring 2, meter 1 (Figure 5.24). This record comes from the shallowest instrument during the winter and is very close to the East Flower Garden Bank. Proximity to the bank suggests that interaction of the flow with the bank may be a local source of large-scale turbulence. These three factors—proximity to the surface, energetic input during the winter, and proximity to the bank—probably account for the undiminished low-frequency noise in the record.

Periodic and nearly periodic fluctuations appear in the velocity spectra as peaks in the spectral density which rise above the baseline noise at the frequency of the oscillation. These periodic fluctuations are due to perturbations imposed on the mean flow by the tide, inertial currents, and what appear to be quasi-geostrophic shelf waves. The periods and frequencies of the major tidal constituents and the local inertial period for the latitude of the Flower Garden Banks are listed in Table 5.2.

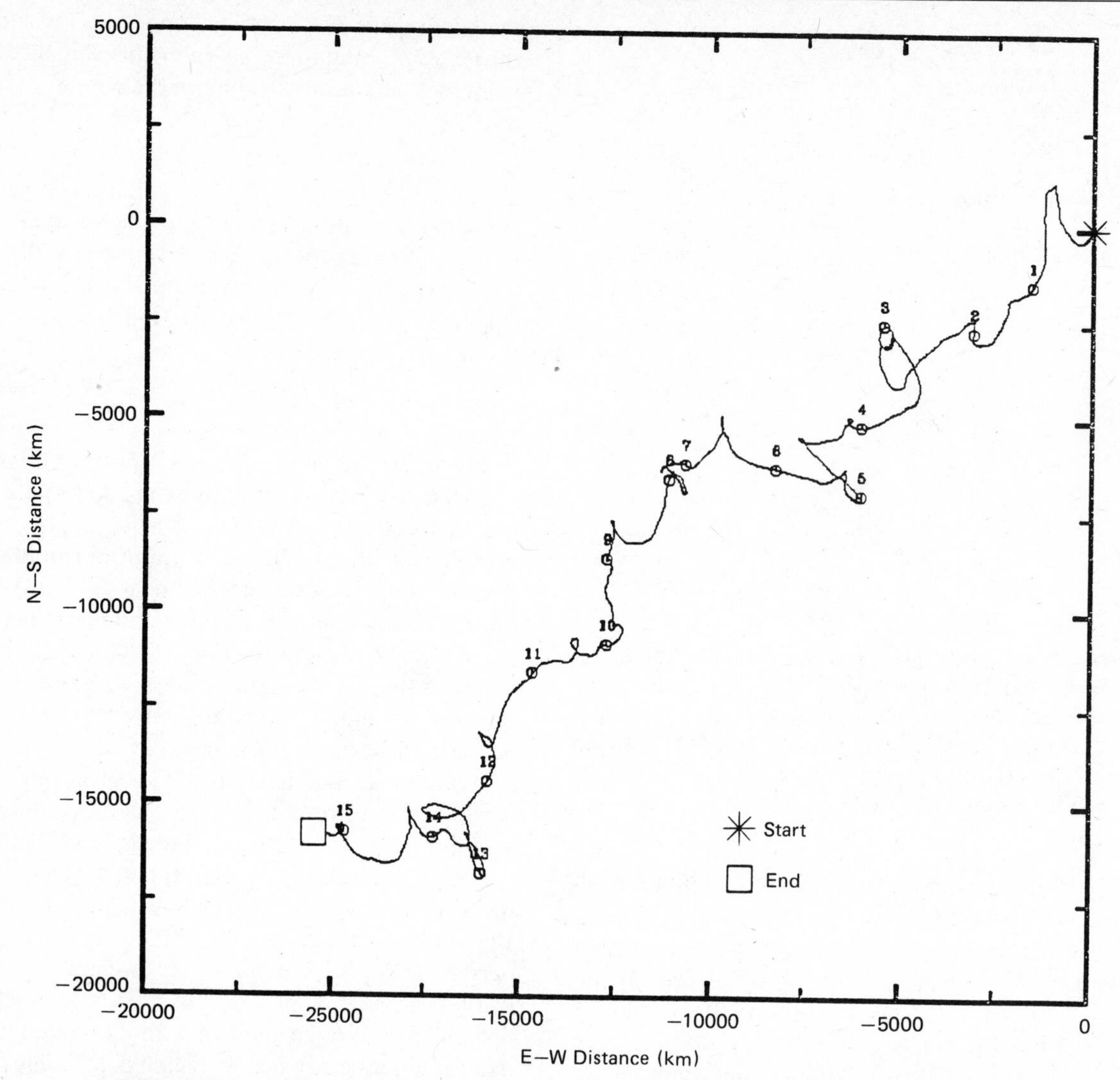

Figure 5.14. Progressive vector diagram, winds at NDBO Buoy 42002, 25 October 1980–February 1981.

Tidal Currents. The tides at Galveston Bay, Texas, are classified as mixed diurnal; that is, there are two highs and two lows in the water height each lunar day (24.84 hours) but with a large inequality between the two highs or the two lows. The amplitudes of the three major constituents (K_1, O_1, M_2) range from 9 to 12 cm (McGrail et al., 1982a).

In all the spectra of the currents near the Flower Garden Banks, shown in Figures 5.16 to 5.24, there are prominent peaks at and close to the inertial frequency and the K_1 and O_1 tidal frequencies (Table 5.2). The peaks near 2 cpd are due to the S_2 and M_2 semidiurnal tidal constituents. Because of the small frequency difference between the inertial oscillations and the O_1 tidal constituent, it is impossible to separate them in the autospectra.

A few basic features of the tidal currents can be derived from the velocity spectra. The diurnal and semidiurnal tidal currents rotate clockwise at all depths. The v (north–south) velocity components of the semidiurnal tidal currents are significantly larger than the u (east–west) components, which indicates that the semidiurnal tide is elliptical, with the major axis oriented north–south. The ellipticity is enhanced near the bottom. The tidal orbits of the K_1 tide are more nearly circular, however, even near the bottom. Because of the difference in bandwidths of the different spectra, which are a function of the lengths

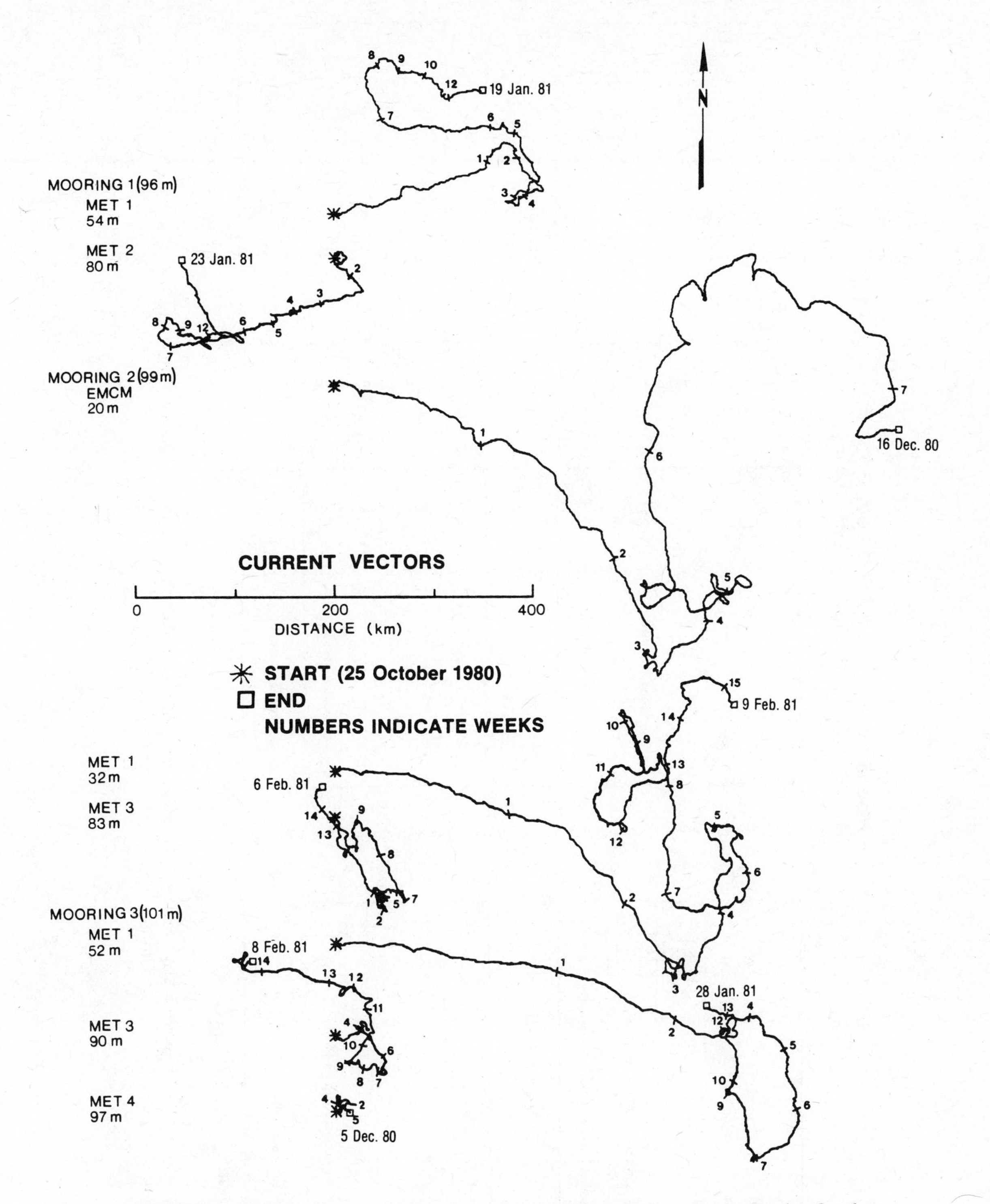

Figure 5.15. Progressive vector diagrams, currents at the Flower Garden Banks, October 1980–February 1981.

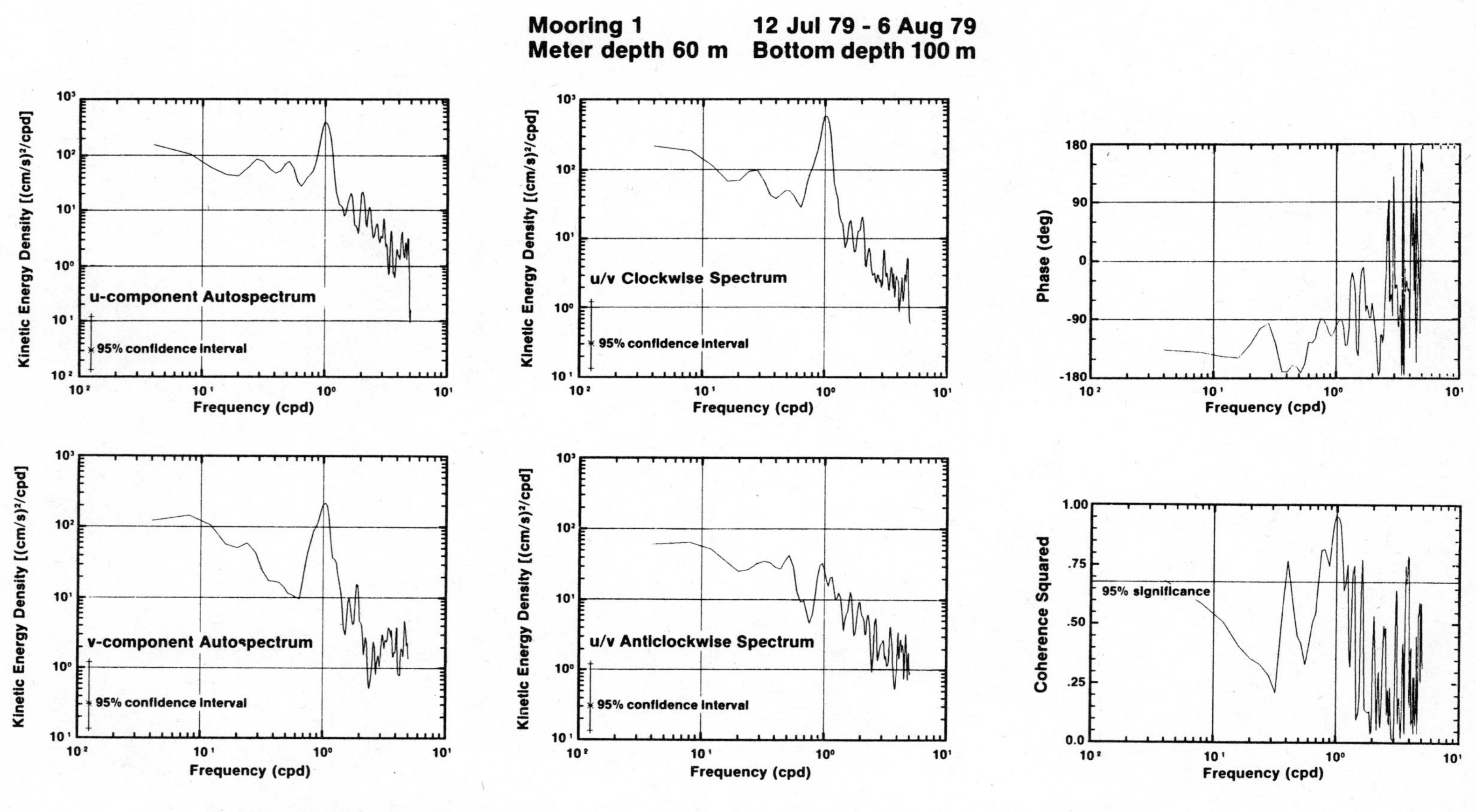

Figure 5.16. Spectral representations of the mid-depth currents at mooring 1 illustrate summer conditions.

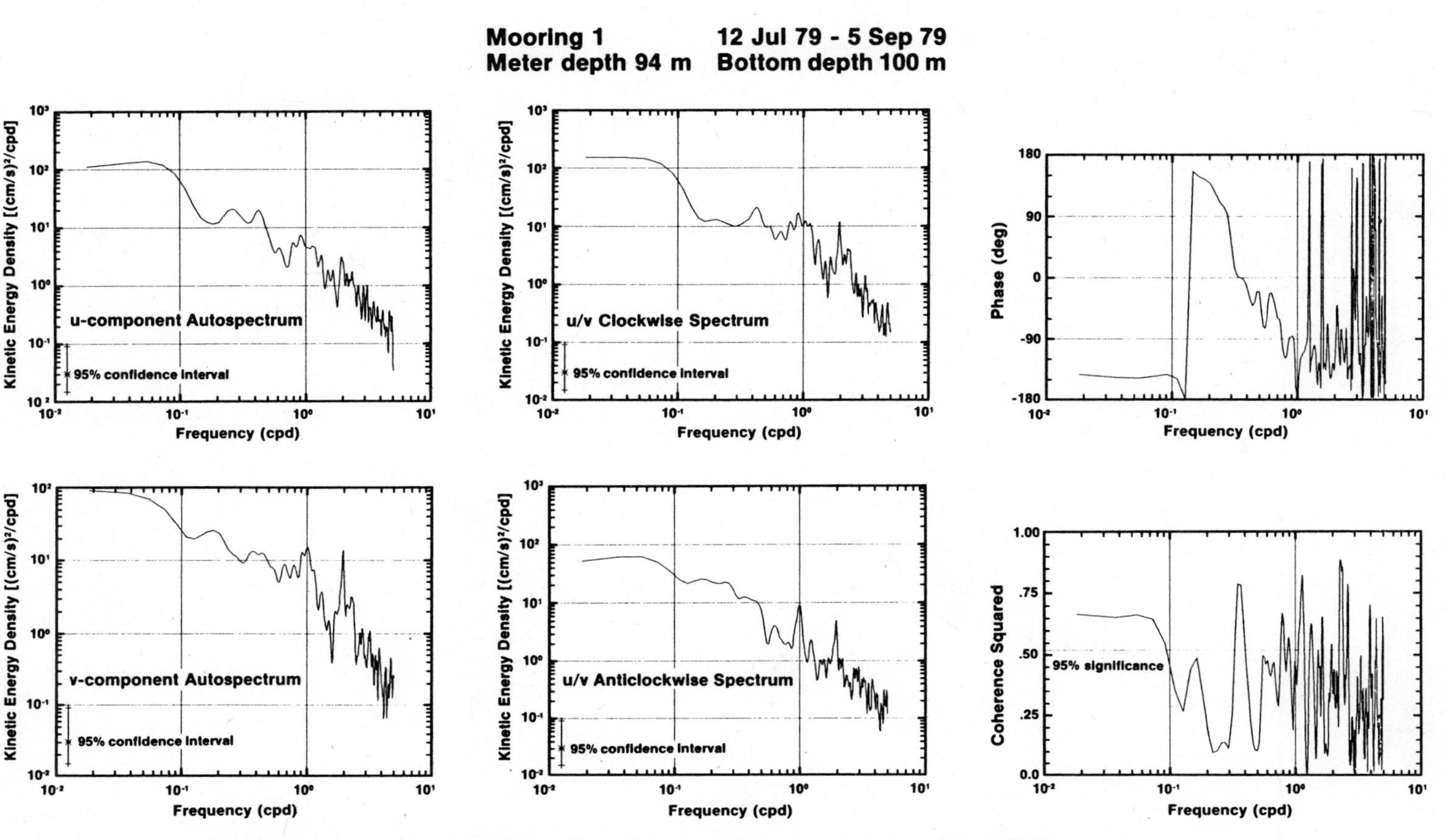

Figure 5.17. Spectral representation of the near-bottom currents at mooring illustrate summer conditions.

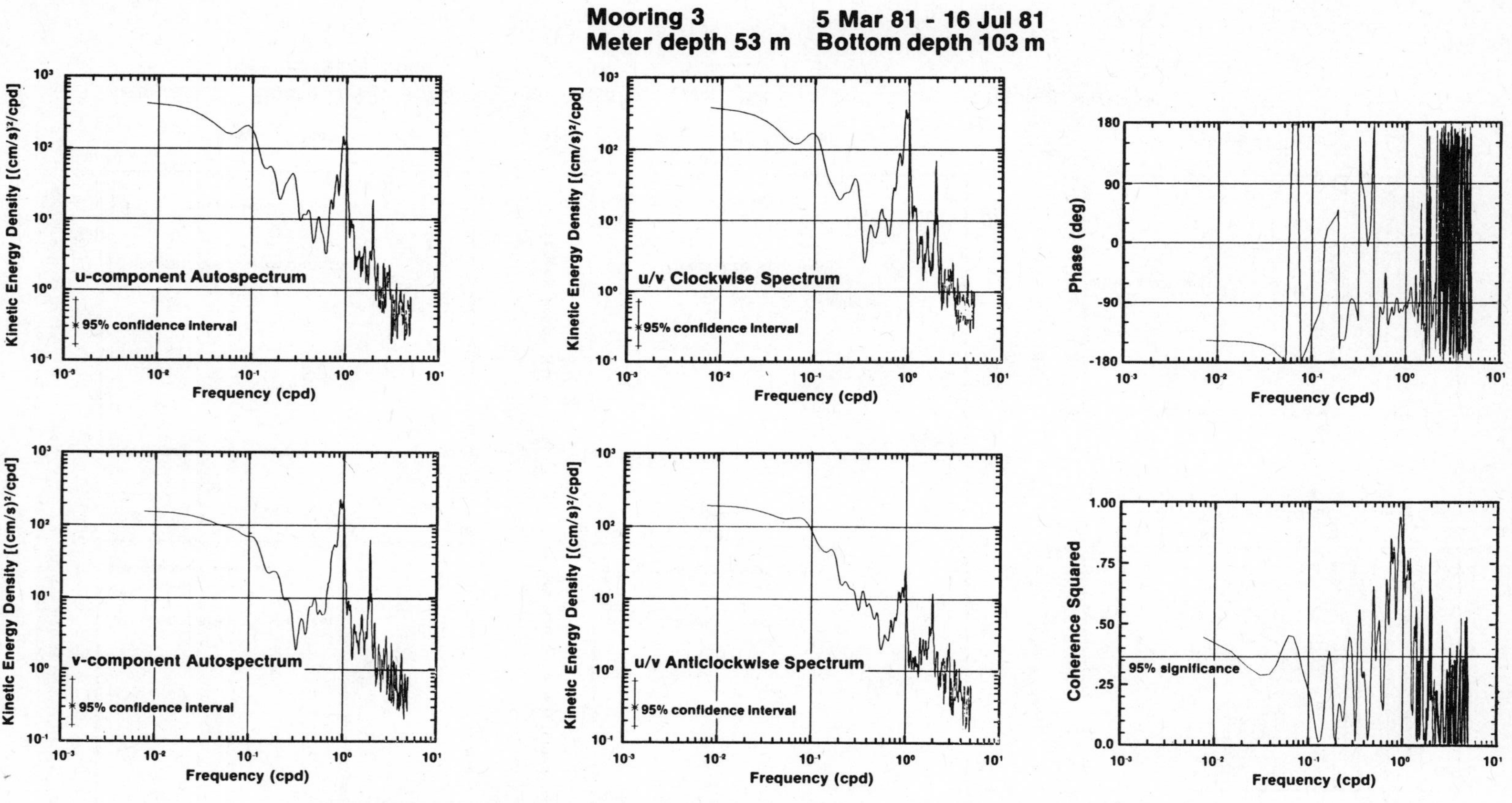

Figure 5.18. Spectral representations of the mid-d epth currents at mooring 3 illustrate spring/summer conditions.

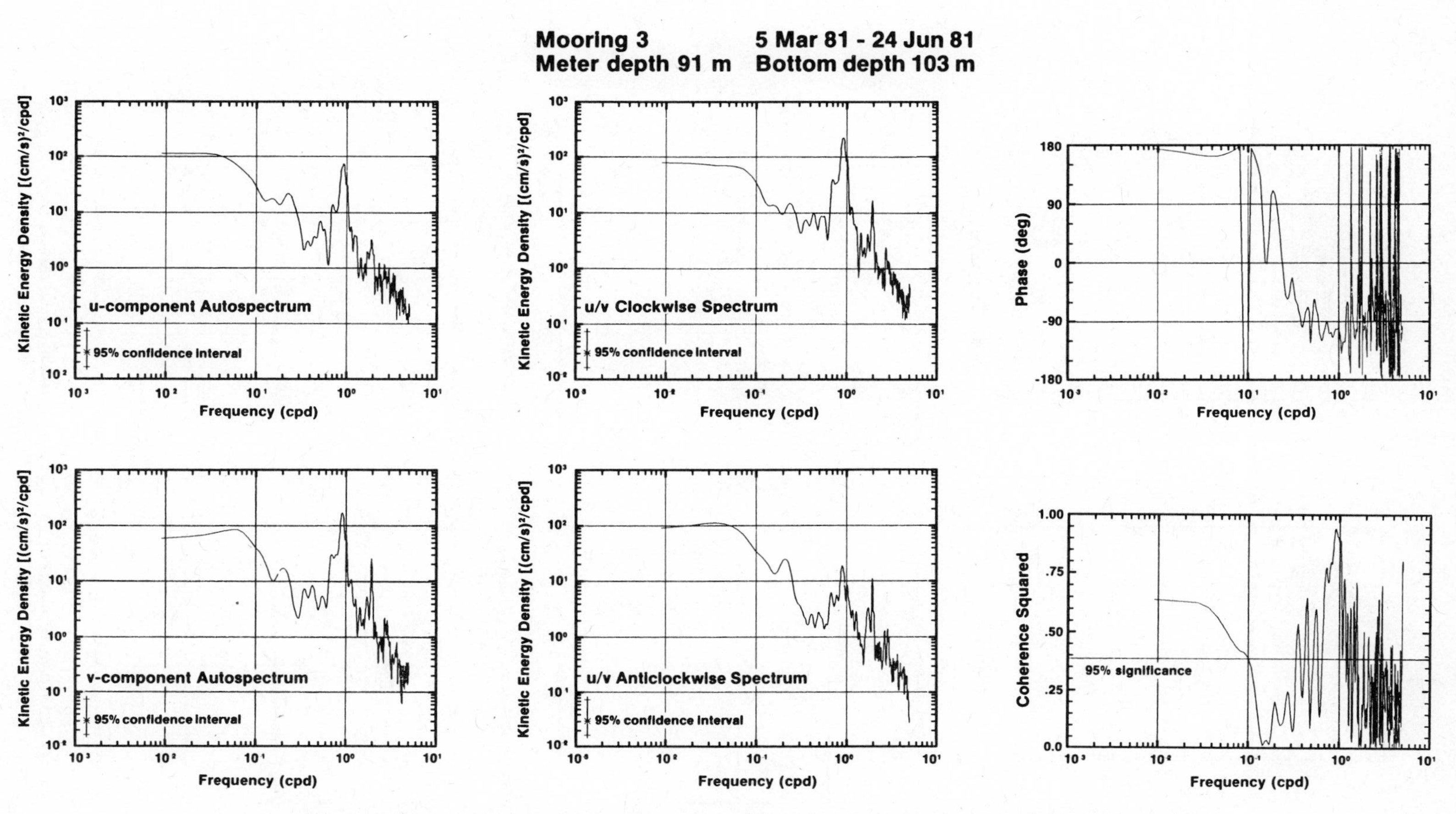

Figure 5.19. Spectral representations of the near-bottom currents at mooring 3 illustrate spring/summer conditions.

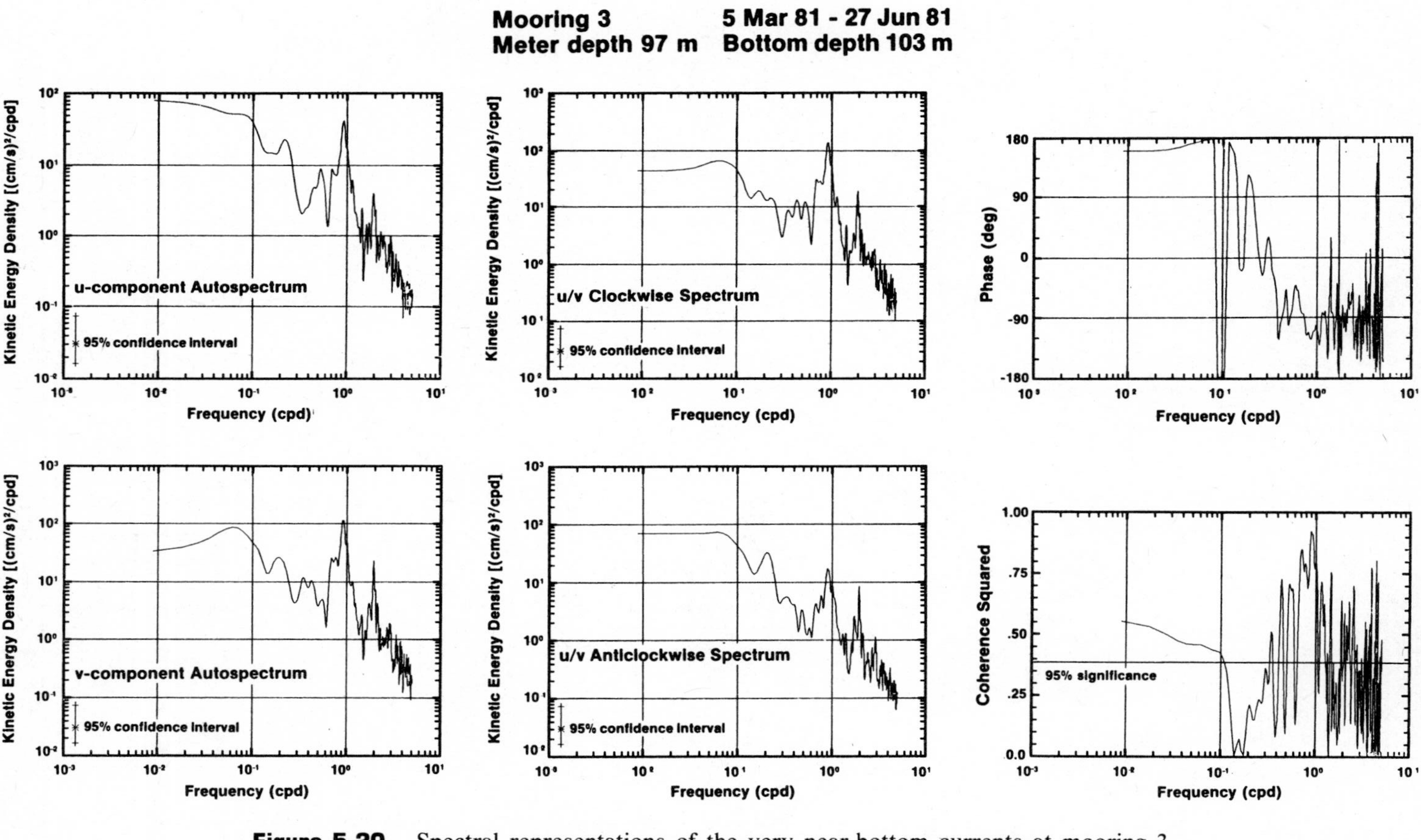

Figure 5.20. Spectral representations of the very near-bottom currents at mooring 3 illustrate spring/summer conditions.

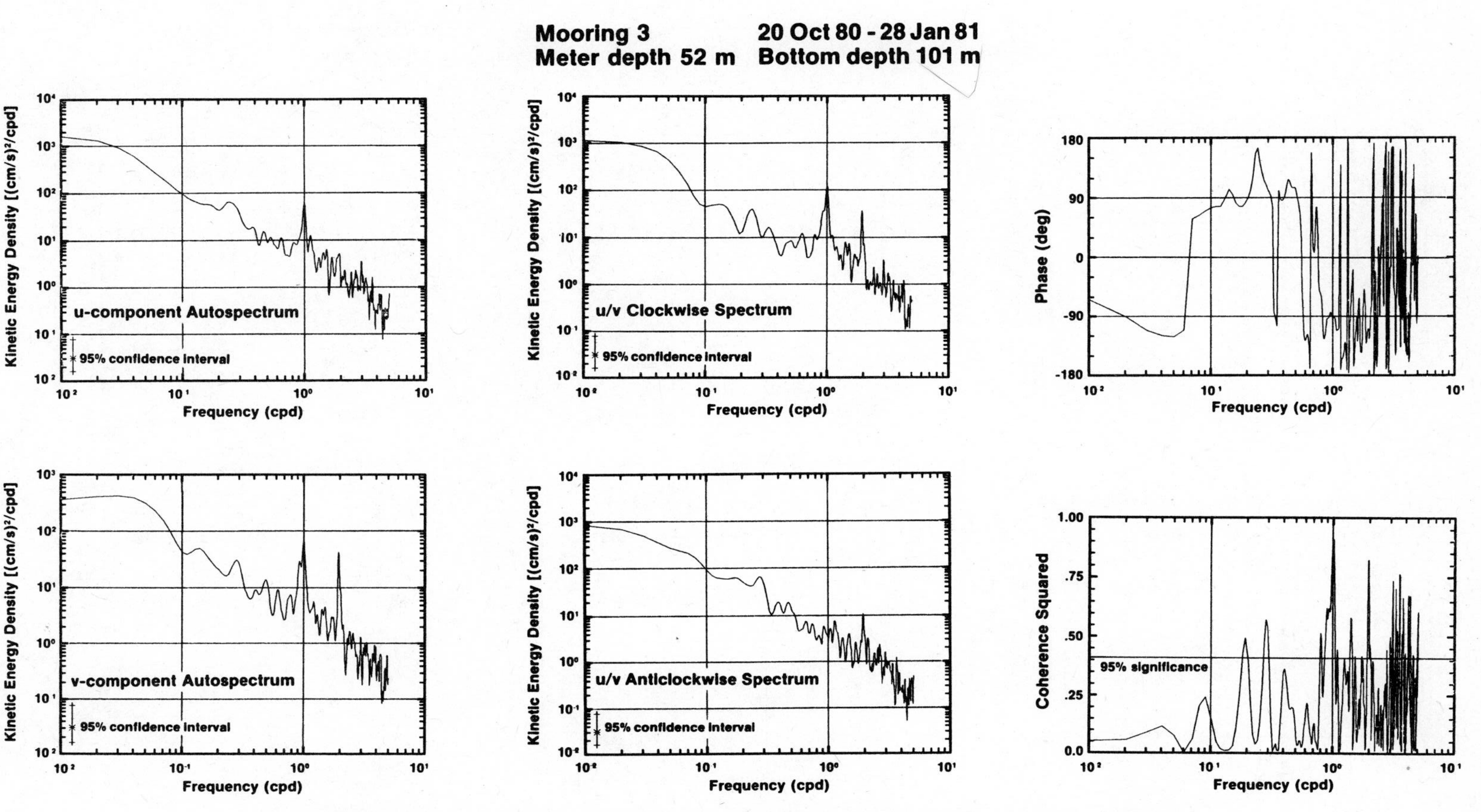

Figure 5.21. Spectral representations of the mid-depth currents at mooring 3 illustrate winter conditions.

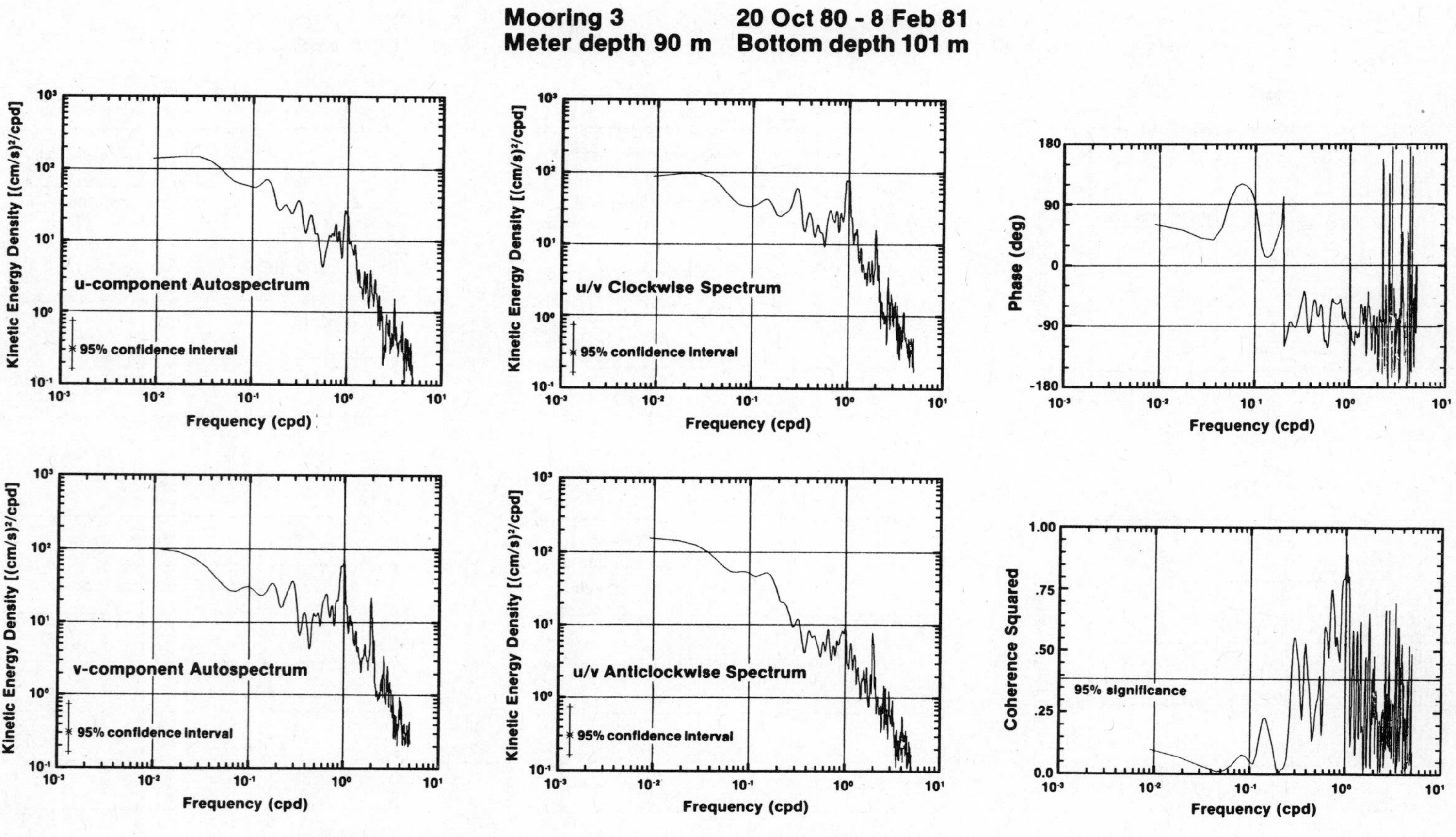

Figure 5.22. Spectral representations of the near-bottom currents at mooring 3 illustrate winter conditions.

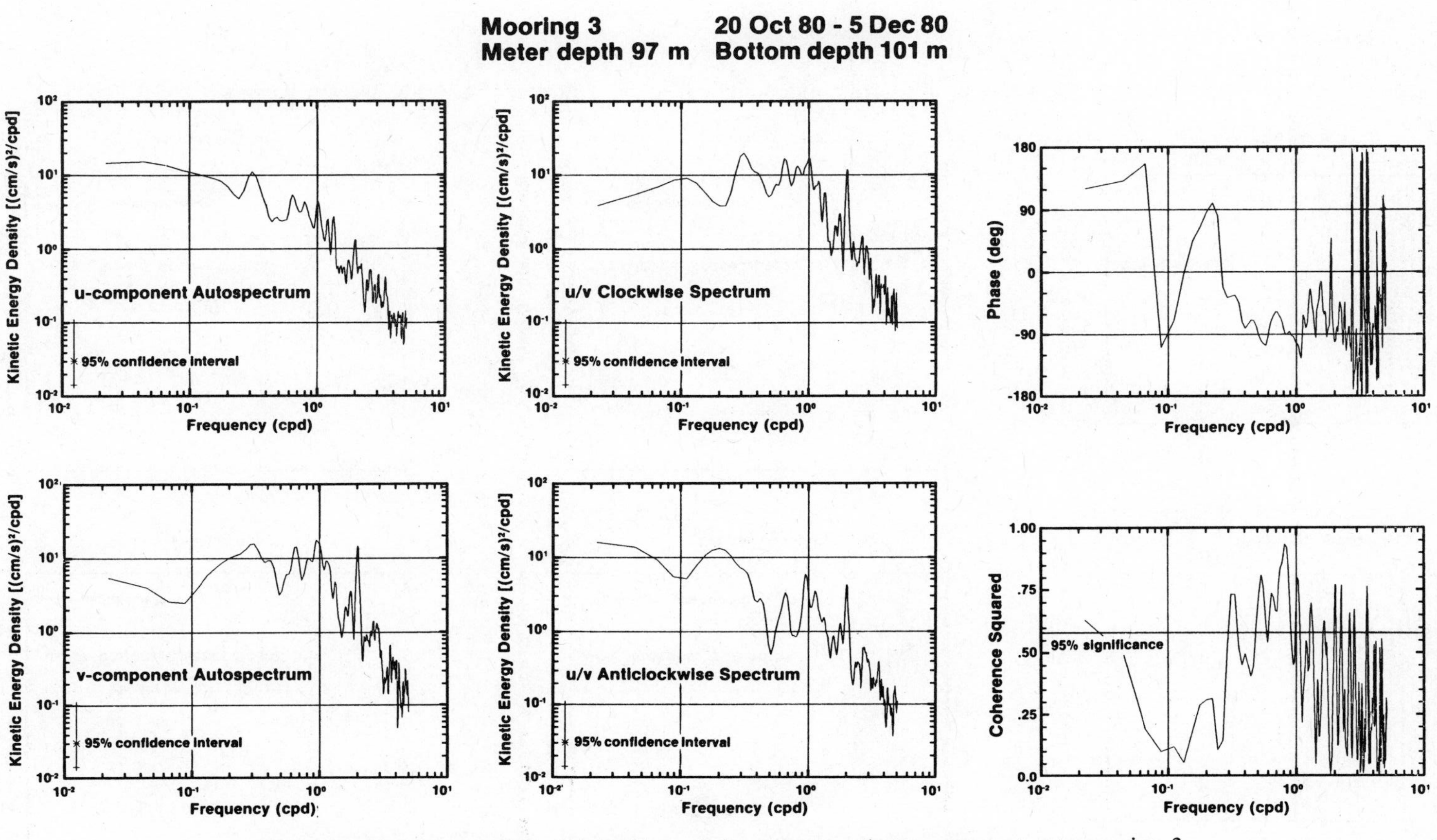

Figure 5.23. Spectral representations of the very near-bottom currents at mooring 3 illustrate winter conditions.

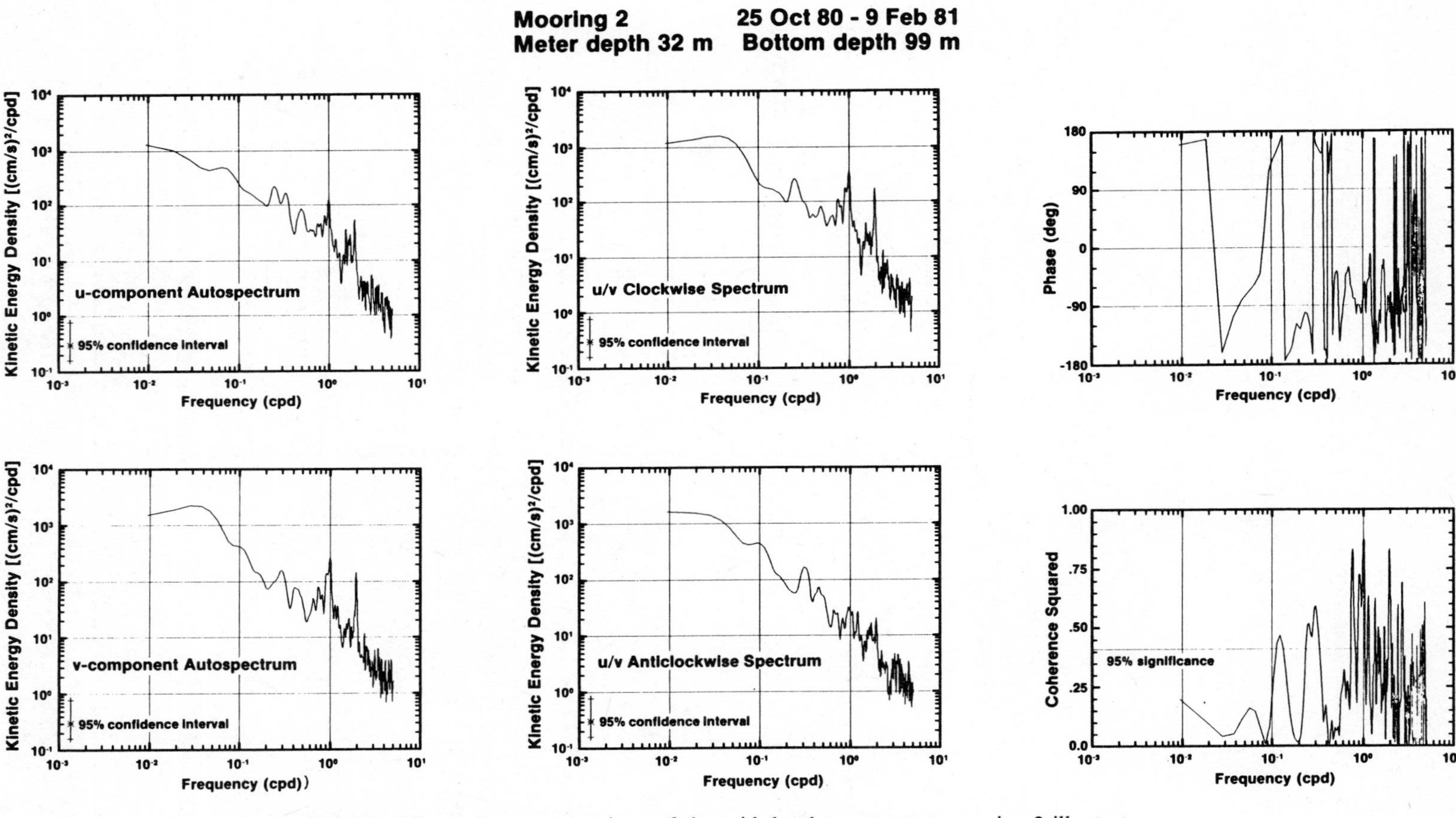

Figure 5.24. Spectral representations of the mid-depth currents at mooring 2 illustrate winter conditions.

TABLE 5.2. Theoretical Frequency and Period of Major Tidal Constituents and Inertial Currents

Darwin Name of Tidal Harmonic	Period (h)	Frequency (cpd)
S_2	12.00	2.0000
M_2	12.42	1.9324
K_1	23.93	1.0029
O_1	25.82	0.9295
Inertial currents at 27°55′N latitude	25.56	0.9390

of the time series and the amount of smoothing applied to the spectra, it is difficult to access from them the changes in amplitude with depth or season. Amplitudes and phases of the tidal currents have therefore been computed by harmonic analysis of 29-day segments of the velocity time-series records.

Table 5.3 provides a typical set of amplitudes of the currents computed by harmonic analysis for the K_1 and M_2 tidal constituents (combined baroclinic and barotropic). The S_2 amplitudes tend to be substantially smaller. Tidal current speeds throughout the year generally range from 1 to 3 cm/s for the M_2 constituent and 2 to 5 cm/s, but sometimes as high as 9 cm/s, for the K_1 constituent.

All three constituents of the tidal currents (K_1, M_2, and S_2) were strongly baroclinic. Harmonic analysis of a large number of 29-day current record segments shows a large temporal variation in the phase and amplitude that would not occur if the tides were barotropic. The K_1 tidal currents are two to three times larger in the summer than in winter. The other two constituents have no apparent seasonal modulation but are characterized by a scatter in their amplitudes and phases in relation to the tide potential.

There are also significant reductions in the amplitudes of the tidal currents near the bottom. These

TABLE 5.3. Tidal Current Amplitudes (in cm/s) at Mooring 3 for 6 March to 4 April 1981

		Tidal Constituent			
		K_1		M_2	
Meter No.	Depth (m)	u	v	u	v
1	53	2.2	3.2	1.4	2.7
2	64	1.8	2.2	0.7	2.4
3	91	1.2	0.8	0.1	2.2
4	97	0.3	0.3	0.2	1.9

reductions are due to attenuation by bottom friction and to the baroclinicity.

Because the tides have a significant baroclinic component, the amplitude of the vertical displacements within the water column was estimated. To do this the same harmonic analysis used on the u and v components of velocity was applied to the temperature records. Estimates of vertical displacements were then made by dividing the temperature fluctuations by the average vertical temperature gradient appropriate for the time of year and depth of the instrument from which the temperature record was obtained. The K_1 displacements were primarily between 1 and 4 m, with a few between 8 and 10 m in winter. The larger amplitudes are suspect because it is likely that horizontal advection in the cross-shelf direction contributes to the winter temperature fluctuations. The M_2 and S_2 also possess larger winter displacements but average between 0.25 and 1.5 m for the M_2 and between 0.1 and 1.2 for the S_2 constituents.

Inertial Currents. Inertial currents are those that oscillate at or near the local inertial frequency:

$$\omega_I = 2\ \pi \sin \phi$$

where π is the angular velocity of the earth at the equator, in radians per second (7.27×10^{-5}/second) and ϕ is the latitude of the observations. Dynamically, the Coriolis acceleration $v\omega_I$ is balanced by the centrifugal acceleration v^2/L, where L is the radius of the inertial circle.

These currents are usually initiated by an impulse of the wind at the sea surface. Basically, what happens is that the water is accelerated up to some mean velocity. The Coriolis acceleration then deflects the current to the right (in the northern hemisphere) in an ever tighter spiral until the centrifugal acceleration of the curving flow balances the Coriolis acceleration. In a current meter record an inertial oscillation would be seen as a clockwise rotation of the velocity vectors with a period of $2\pi/\omega_I$. As shown in Table 5.2, the inertial period at the Flower Garden Banks is 25.56 hours.

The inertial oscillations observed in the current meter records occur episodically in conjunction with hurricanes and major winter storms. They reach maximum velocities of 20 to 30 cm/s near the surface and decay toward the bottom except at the location of mooring 2 (Figure 5.1). The phase relationship between the surface and bottom changes from sea-

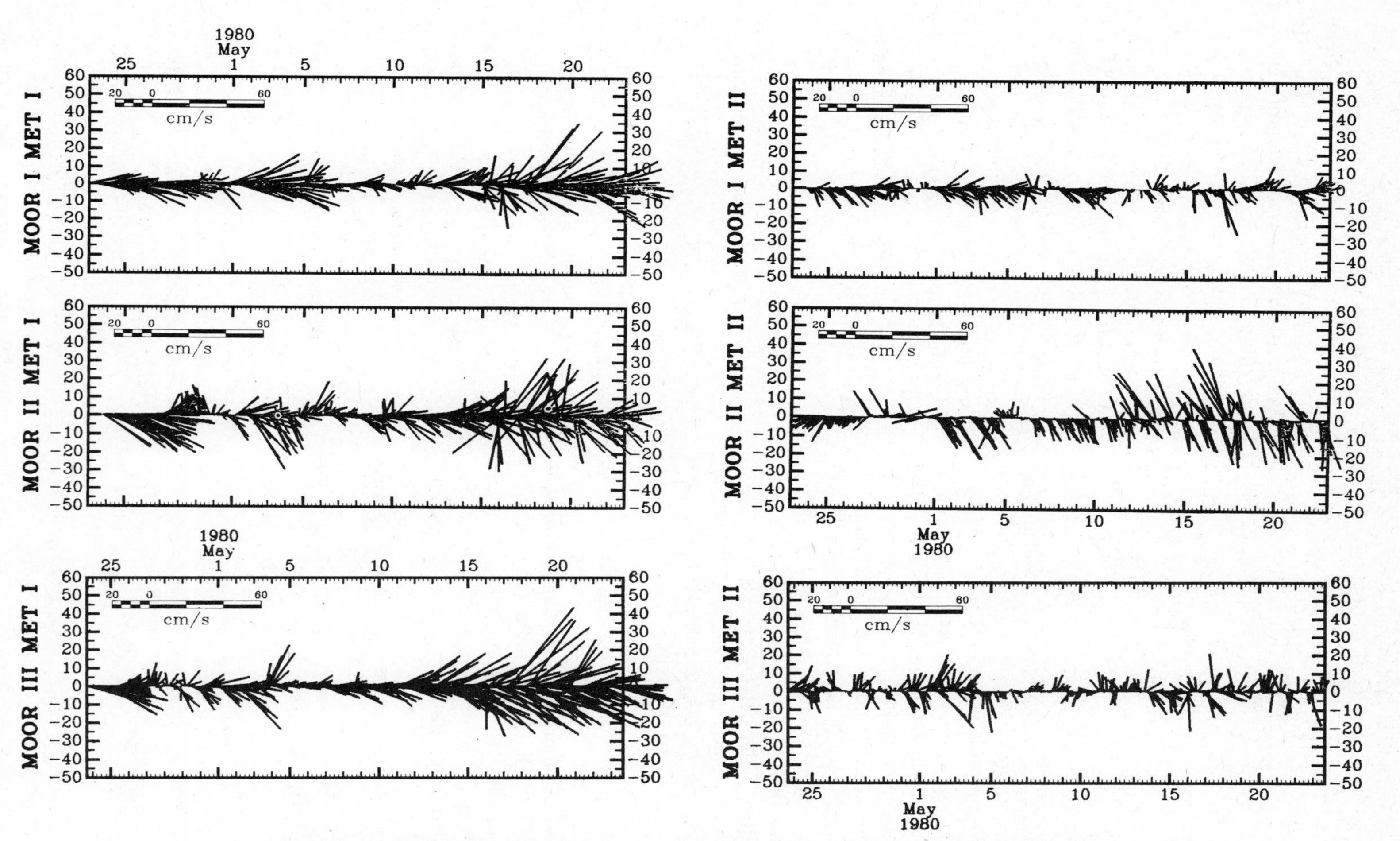

Figure 5.25. Time series of horizontal currents at the upper and lower meters of moorings 1,2, and 3. Strong inertial currents are observed on 15–19 May 1980.

son to season, apparently as a result of variations in stratification and in the wind field generating the inertial currents. The oscillations appear to persist for about 3 to 5 days after initiation but occasionally last as long as 10 days. Figure 5.25 shows a period of strong inertial oscillations at mid-depth and bottom for three moorings during May 1980.

Shelf Waves. When flow on the continental shelf is perturbed at a frequency lower than inertial, waves can be generated that are trapped on the continental shelf by vorticity. These waves travel with their crest normal to the coast and are right-bounded; that is, they are constrained to propagate in only one direction such that an observer standing on the crest of the wave and facing in the direction of propagation would have his right shoulder to the coast (in the northern hemisphere). They are quasi-geostrophic in the sense that, within a few percent, the dynamic balance is between the pressure gradient and the Coriolis acceleration. Shelf waves may be barotropic (surface waves) or baroclinic (internal waves) and they decay exponentially in the offshore direction. Perhaps the most thorough discourse on shelf waves is given by LeBlond and Mysak (1978).

The dispersion relationship for shelf waves (speed of propagation in relation to frequency and wavelength) is governed by the slope of the shelf and the latitude for barotropic waves, complicated by the nature of the density distribution for baroclinic waves. Stated differently, for any given wavelength, latitude, and shelf slope there is only one possible frequency (for each mode) of the barotropic wave.

The most common frequencies for what appear to be shelf waves in the Flower Garden current meter data are 33 hours, 2 days, and 4 days. All have clockwise rotation of the velocity vectors and relatively good coherence between the *u* and *v* velocity components.

The shelf wave at the two-day period is intermittent and seems to be generated by storms. The coherency between the *v* components of velocity

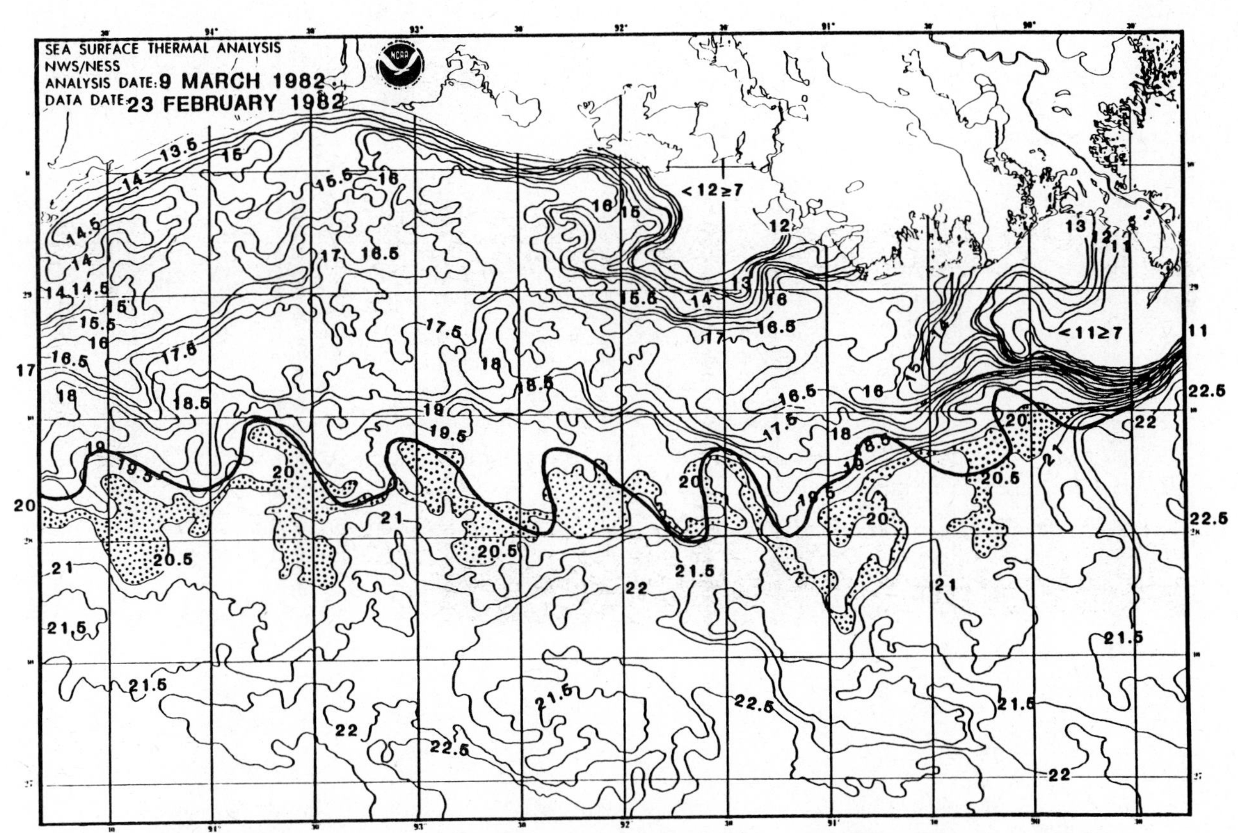

Figure 5.26. Wavelike perturbations in the sea-surface isotherms at the shelf edge outlined with a heavy line. The wavelength of these distributions is approximately 80 km. These are probably shelf waves with a two-day period.

was determined for meter 1 on moorings 1 and 3 (see Figure 5.51). At the 2-day period mooring 1 leads mooring 3 by approximately 100°, which means that the meters are about 28% of a wavelength apart for a 2-day wave. Because the moorings are approximately 22 km apart in the east–west direction, the wave must have a wavelength of about 79 km. In the sea surface thermal data for 23 February 1982 there are wavelike features all the way from 90 to 95°W between about 28° and 28°30′N (Figure 5.26). These waves have an approximate wavelength of 83 km, which is just 5% greater than the estimate from the spectral analysis.

Perhaps one possibility is that these waves are generated by the topographic disturbance of the Mississippi Canyon, then propagate to the west; but it is then not clear why they should be generated only during periods of storms. The amplitude of the current oscillation occasionally rises as high as 15 cm/s but normally is on the order of 4 cm/s.

The 4-day oscillation is very near the 3- to 5-day spectral peak found in the wind spectra which represent cold frontal passages. This could be a directly forced oscillation, but it appears more likely that it is a free shelf wave excited by impulses of the wind stress around this preferred frequency.

Hurricanes and Tropical Storms

During the deployment of current meters at the Flower Garden Banks, several hurricanes and tropical storms passed close enough to generate a recognizable response in the water column. In 1979 three of the five hurricanes that entered the Gulf of Mexico came relatively close to the banks: Bob (9–11 July), Claudette (21–24 July), and Elena (29 August–1 September). In 1980 Hurricane Allen passed close to the banks, causing damage to the reefs and monitoring equipment. Although Allen was the severest storm, the best records of a storm passage were made from Claudette. The response to Allen was recorded by only a single current meter, near the bottom on mooring 2, as discussed under "Orographic Effects."

The normal eastern flow of the shelf appears to have been reversed by Hurricane Bob. The early portion of the current meter records (Figure 5.27) reflects the reversal of the current and the "spin down" or decay of the flow field set up by Bob. Particularly noticeable are the strong inertial oscillations and the 2-day shelf waves at mid-depth. Near-bottom currents were relatively strong (> 20 cm/s) and toward the southeast with weak inertial oscillations. The 2-day shelf waves were present in the near-bottom flow but were very subtle.

Claudette passed directly over the NDBO buoy 42002 and the Flower Garden Banks (see Figure 2.13 for location of the buoy in relation to the banks). By the time Claudette passed over this area she had been reduced to a tropical storm with maximum recorded winds of only 20 m/s (39 knots). After making landfall the storm became nearly stationary. Unusually strong southerly winds persisted at buoy 42002 through 29 July. The recorded wind at NDBO 42002 for 13 July through 1 August 1979 is shown in Figure 5.27 with the current and temperature response at mid-depth and near-bottom at mooring 1.

The approach of Claudette was heralded by a wind shift from southeast to northeast in the early morning hours of 21 July. As the storm approached, the wind vectors swung in a counterclockwise direction and increased in magnitude. When the storm passed over the winds rotated to southwesterly and increased further in magnitude.

Little effect from Claudette was seen at the Flower Garden Banks at 60-m depth until late on 22 July when the current started to rotate counterclockwise toward the southwest and to increase in speed. When the storm passed near midnight (local time) on 23 July the current abruptly turned toward the northwest at 60 m (Figure 5.27) and accelerated to speeds of more than 60 cm/s. This could not have been a direct frictional response because the current vectors would have been oriented in the direction of the wind or to the right of it. Inertial oscillations started to grow on 23 July, reached a maximum on 28 July, then decayed rapidly.

Near the bottom the first response to Claudette was a weak rotation of the current late on 23 July, then a major pulse (> 20 cm/s) to the northeast, followed by a sudden reversal about mid-day on 25 July. By early on 26 July the mean flow at the bottom had become very weak (about 2 cm/s) in the onshore direction. It was modulated by inertial oscillations of about 10 cm/s amplitude through 28 July. The major onshore pulse at the bottom was accompanied by a sudden drop in temperature which was probably caused by cold water upwelling over the shelf break near the bottom.

Because the banks represent major obstructions to the flow, it is quite likely that the currents would have been accelerated across the reef by the Venturi effect. Flow at the level of the reef would also probably have been directly driven by the wind stress.

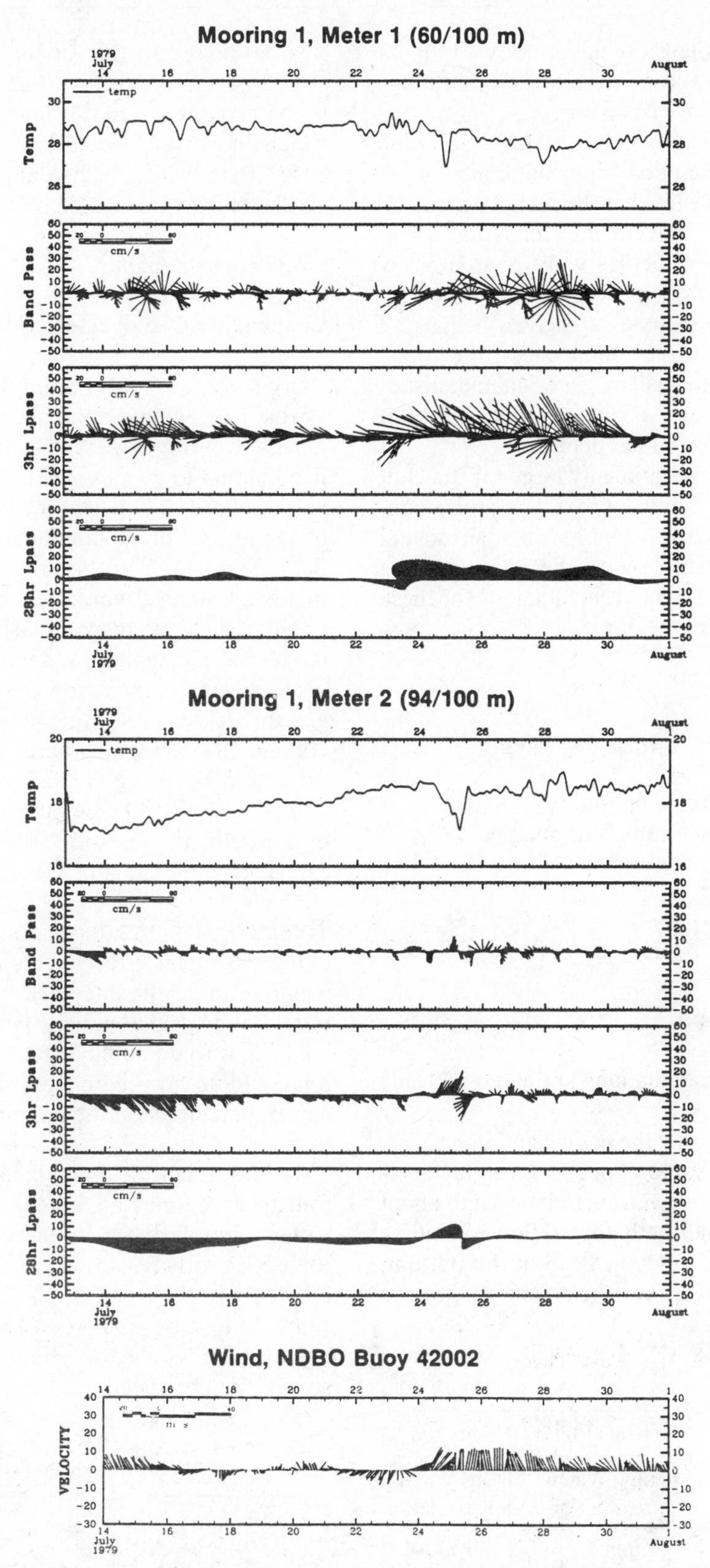

Figure 5.27. Wind and current meter records during the passage of Tropical Storm Claudette.

The currents would therefore have followed the direction of the wind vectors more closely, although leading them in a clockwise sense.

Surface gravity waves generated by the wind stress would also have been important contributors to the effects felt at the level of the reef. To estimate the maximum oscillatory velocities likely to be generated, wave orbital velocities were computed by Airy wave theory for deep- and shallow-water approximations. Both estimates are given because it is not clear that the waves could make the transformation from deep to shallow water characteristics over the steep slopes of the banks. Only deep-water estimates are given for 100-m depths.

Two waves, both exceptionally large for the Gulf of Mexico, were chosen to approximate storm conditions. The first was a wave with a 10-s period and a 5-m height (2.5-m amplitude); the second was a 12-s wave with a 7-m height. The equations for these calculations are the following:

Deep water

$$|u_{max}| = a\sigma e^{-kz}$$

where a = wave amplitude
σ = wave radian frequency
k = wave number

Shallow water

$$|u_{max}| = \frac{a\sigma}{kz}$$

The results of these calculations are given in Table 5.4.

In general, then, during the passage of a hurricane over the banks we should expect a three-layer response. A direct wind-driven current down to about the thermocline, an indirectly forced flow extending from the thermocline to about 10 m of the bottom, and a near-bottom flow, probably compensatory to a large degree to that at the surface. In addition, strong oscillatory velocities would be experienced at the level of the reef from surface gravity waves, which, however, would have little effect near the base of the bank. It also appears that the inertial oscillations are intensified on the bank and penetrate to the bottom with greater magnitude than at sites away from the bank.

TABLE 5.4. Results of Airy Wave Theory Calculations

Water Depth (m)	Wave Period (s)	Wave Height (m)	Maximal Orbital Velocity (cm/s) Shallow Water Approximation	Maximal Orbital Velocity (cm/s) Deep Water Approximation
20	10	5	175	70
20	12	7	245	105
100	10	5	—	1
100	12	7	—	11

Orographic Distortions of the Flow

"Orographic" comes from the Greek base words "oros," meaning mountain, and "graphos," which means to write or describe. Orographic effects, as they pertain to geophysical fluid dynamics, are those perturbations of flow that are caused by the presence of mountains or mountainlike features in the flow field. The nature of the flow about an obstacle depends on several variables, among which are the speed of the flow, the vertical and horizontal scale of the object, the depth of water in relation to the height of the obstacle, the turbulent viscosity of the fluid, and the degree of stratification present in the water column. Reviews of orographic effects have been presented by Hogg (1980) and Baines and Davies (1980). Hogg (1980) reviews the theoretical aspects of the field; Baines and Davies (1980) review laboratory investigations of these phenomena.

Theoretical Considerations. An understanding of the orographic effects at the Flower Garden Banks is important to the interpretation of their biotic and sedimentological zonation. Water at the base of the banks is turbid and cold (on the order of 17°C) and if it could be carried up onto the reefs it would have a profound effect on the population dynamics of the banks.

In both theoretical and laboratory investigations four flow regimes have been studied: irrotational without stratification; irrotational with stratification; rotational without stratification; and rotating stratified.

In the first we could consider a two-dimensional object such as a half-cylinder at the bottom in a steady flow of speed u.

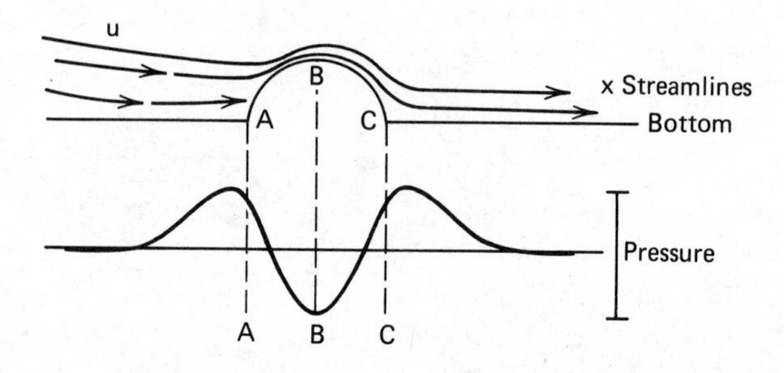

The Bernoulli equation $P + 1/2\ (\rho u^2) = K$ (where P is pressure, ρ is density, u is the x component of velocity, and K is a constant) may be differentiated with respect to x to give

$$\frac{\partial P}{\partial x} + \rho u \frac{\partial u}{\partial x} = 0$$

As the flow approaches the obstacle it is blocked; $u\ \partial u/\partial x$ becomes nonzero and negative (u decreases) and the pressure rises to a maximum at point A. The flow is then accelerated over the object, thus causing $u\ \partial u/\partial x$ to become large and positive, $\partial P/\partial x$ becomes negative, and P reaches a minimum at B. The flow then decelerates in the lee of the object to create a second pressure maximum. The velocity and pressure return to ambient in the lee of the half-cylinder. In essence, the kinetic energy of the flow is converted to potential energy and back again. All this assumes that the fluid is inviscid, extends upward infinitely and cannot move in the y direction (horizontal direction perpendicular to x).

If the flow is viscid, the water particles traveling over the surface of the object lose energy to friction, do not reach the same maximum u at B, and the pressure in the lee of the object drops below ambient. This causes the flow to reverse in the lee of the object and vortices are formed. Also, the lowest level water in front of the object may not have sufficient kinetic energy to rise over the object in the presence of friction. If the water also has a finite depth, then all the water approaching the obstacle from upstream must be accommodated by accelerating it through the reduced depth over the object. In a viscid fluid the flow through the reduced aperture cannot accommodate all of the flow from the original cross section and a surface wave forms on the upstream side of the object. This causes an adverse pressure gradient and further reduces the rise of the near-bottom water particles in front of the object. To say it differently, the near-bottom flow can become blocked.

If the flow is three-dimensional and the fluid is free to move in the y direction, and if the flow is essentially unrestricted on either side of the object, most of the excess fluid blocked by the object will flow around rather than over it. Because both Flower Garden Banks occupy about 80% of the water depth and because their rough surfaces generate considerable turbulence (therefore making the eddy viscosity large), we would not expect the near-bottom waters to rise very far up their flanks.

Stratification can also reduce the vertical excursion of water impinging on an obstacle. Now we return to the two-dimensional model for a moment. The kinetic energy of the oncoming water is converted into potential energy as it moves vertically, raising the isopycnals and creating an internal wave. The relationship is

$$\rho \frac{u^2}{2} = gh^2 \frac{\partial \rho}{\partial z}$$

where g is the acceleration of gravity and h is the height above equilibrium that a water particle is raised. For any given velocity and stratification that height is

$$h = \left[\rho \frac{u^2}{2} (g \frac{\partial \rho}{\partial z})^{-1}\right]^{1/2}$$

McGrail and Horne (1981) showed that for a 100-cm/s flow and stratification comparable to that for April at the Flower Garden Banks ($\partial \rho/\partial z = 2 \times 10^{-7}$ g/cm^4), h is only 50.5 m. This is the worst possible case approach because the velocity is large and the stratification low. As in the unstratified model, the introduction of viscosity significantly reduces the actual permissible rise of a water particle. The aperture at the top of the bank would have to accommodate approximately 20 m of water that normally flows at that depth plus the 50 m of water displaced vertically from below. Because the fluid is incompressible, the velocity through the aperture (U_a) would be

$$U_a = U\left(1 + \frac{h}{a}\right)$$

where a is the distance between the top of the bank and the water surface. By using the 100 cm/s value for U, U_a would be 350 cm/s. If friction reduced this by only 29%, U_a would be 250 cm/s and h would be reduced to 30 m. Now, if the flow is permitted to be three-dimensional, the rise would be significantly reduced as the kinetic energy is drained off by horizontal flow around the bank. In a rotating homogeneous flow field the flow is said to be geostrophic. The x and y components of flow are then

$$+ vf = -\frac{1}{\rho}\frac{\partial P}{\partial x}$$

$$- uf = -\frac{1}{\rho}\frac{\partial P}{\partial y}$$

where v = the y north–south component of flow)
u = the x (east–west component of flow)
f = the Coriolis parameter $2\omega \sin \phi$

Under these conditions the flow is constrained to be two-dimensional in the x–y plane; that is, the vertical velocity w is identically zero everywhere (Hogg, 1980) and the flow is depth-independent. When this flow encounters a bump on the bottom the flow at all levels diverges and flows around the bump as if it were a cylinder reaching to the surface. This type of flow is called a Taylor column. For the column to form, the flow around the bump must remain reasonably geostrophic and the bump must be relatively large. For the flow to remain geostrophic, the horizontal scale of the bump must be large enough so that the Rossby number ε remains $<<1$:

$$\varepsilon = \frac{U}{Lf}$$

where U = a characteristic velocity
L = a length scale
f = the Coriolis parameter

If the obstacle has a small radius L or the flow has a high velocity u, then ε becomes large, thus signifying that the inertial accelerations are nonnegligible and the flow is ageostrophic. The vertical scale of the obstacle is the ratio of the height of the object h to the depth of the water. Hogg (1980) combined the vertical scale and the inverse of ε into a single parameter β_T:

$$\beta_T = \frac{h}{H_o}\varepsilon^{-1}$$

According to Hogg (1980), a Taylor column can form only when β_T is larger than some number of the order 1. Using numbers appropriate to the Flower Garden Banks (u = 20 cm/s), $h/H_o = 0.8$, $L = 4 \times 10$ cm, and $f = 7 \times 10$/s), β_T is equal to 1.12. The number is large enough to indicate that rotational effects are important but too small to suggest unequivocally that all vertical motion would be suppressed.

When both rotation and stratification are strong, a modification of the Taylor column takes place and cones of deadwater form over the obstacle. The criterion Hogg (1980) gives implies that the cones at the Flower Garden Banks in summer would be more steeply inclined than the sides of the banks. Therefore, the cones could not form. The strength of the stratification, however, is so great that little vertical motion is possible as the flow encounters the bank.

In summary, from both theory and order of magnitude estimates, we would expect the flow to diverge around the banks with a modest vertical excursion (on the order of 10 m) on the point of the bank at which the flow diverges. Weak vortices and lee waves (internal) probably form on the downstream side of the banks where the flow separates.

Observations. Several types of measurement carried out at the Flower Garden Banks in the period from January 1978 through July 1981 were specifically designed to study orographic effects. The results of these measurements, plus a completely serendipitous set, are consistent with the expectations developed in the last section. The flow is accelerated around rather than over the bank. Vertical excursions on the bank are probably limited to approximately 10 m by rotation, friction, and stratification.

The bottom current directions from velocity profiles taken in July 1979 and a 2-day vector mean direction from the lowest meter on mooring 2 are shown in Figure 5.28. By October 1980 the PHISH system was being used. Near-surface (5 m) velocity vectors are shown in Figure 5.29 and the near-bottom (1 m up) are shown in Figure 5.30. It must be remembered that these are instantaneous vectors that contain random fluctuations as well as the longer

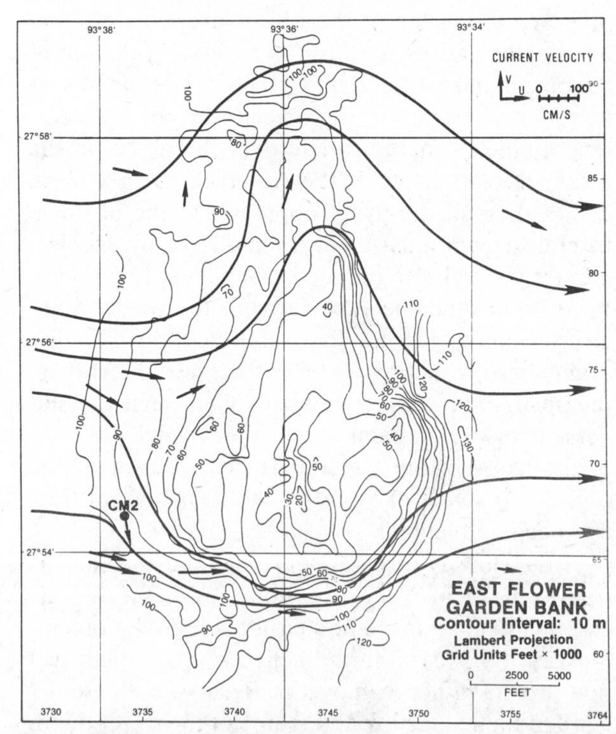

Figure 5.28. Bottom velocity vectors from all July 1979 stations plotted with directional sense only (no magnitude implied). The lines connecting stations of like depth are inferred streamlines. Mean directions for the lowest current meter for 16 to 17 July 1979 are also plotted and demonstrate the consistency of the two types of measurement.

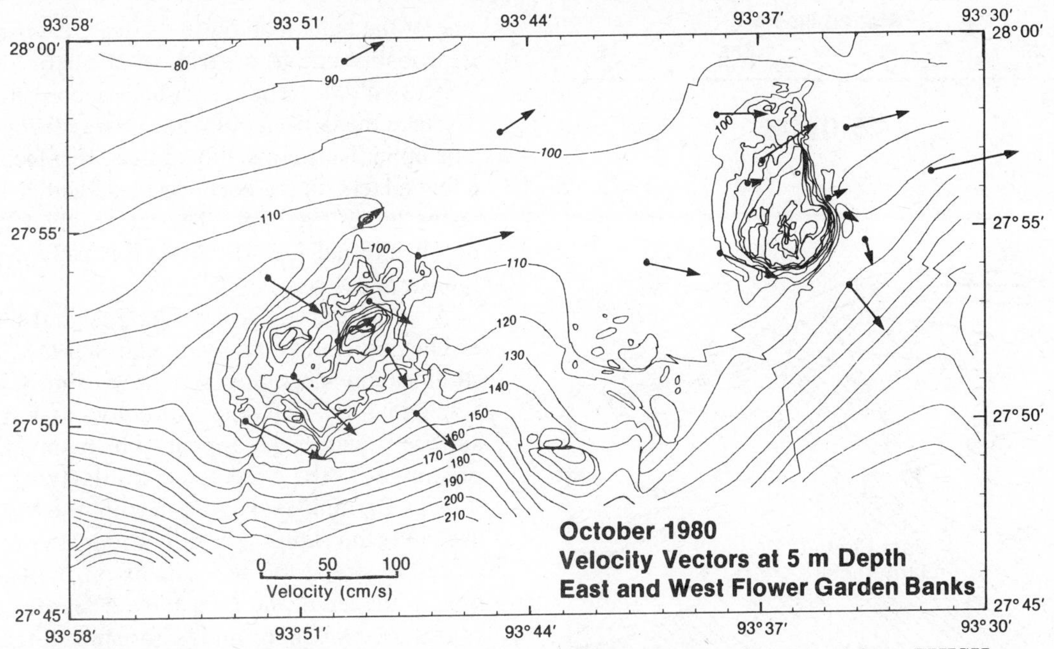

Figure 5.29. Horizontal velocity vectors at a depth of 70 m, determined from PHISH stations taken on the October 1980 BALTIC SEAL cruises.

term mean signal. Note, however, the tendency of the vectors to split around the banks on their western (upstream) side and converge on the downstream side. Near bottom, in the lee of the East Flower Garden Bank, there is some indication of a weak return circulation toward the base of the bank.

Temperature data from some of the same stations shown in Figures 5.29 and 5.30 were used to construct a cross section of the thermal structure on a line from the northwest to southeast of the East Flower Garden Bank (Figure 5.31). It shows that near-bottom isotherms are deeper on the upstream

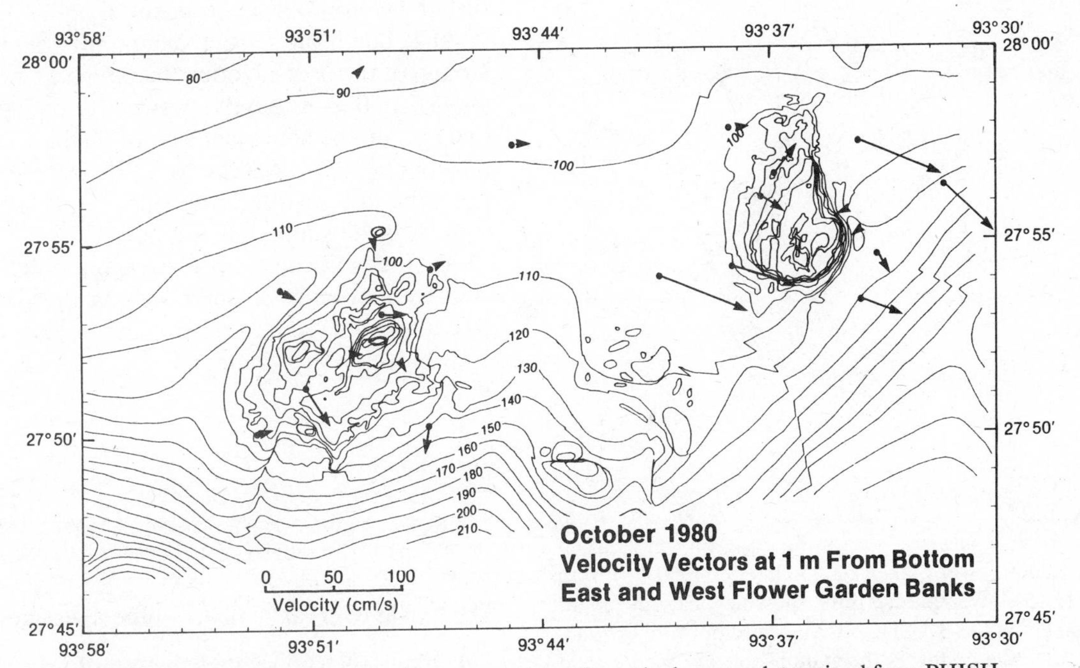

Figure 5.30. Horizontal velocity vectors at 1 m from the bottom, determined from PHISH stations taken on the October 1980 BALTIC SEAL.

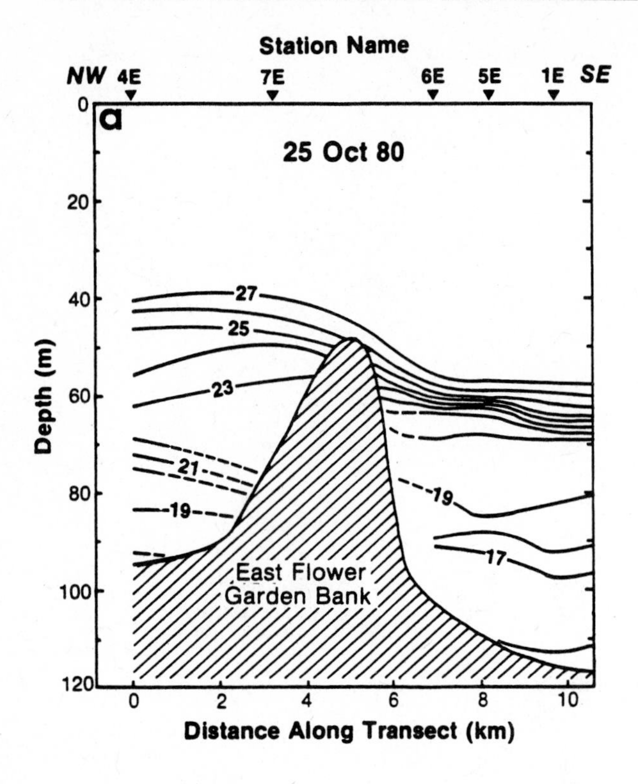

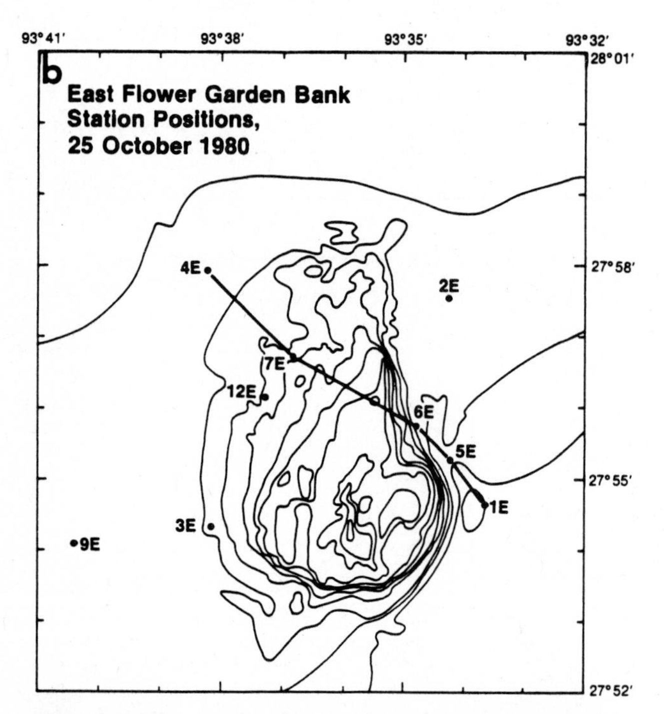

Figure 5.31. *(a)* Vertical section of temperature from northwest to southeast across the East Flower Garden Bank. Stations were taken on 25 October 1980. *(b)* Map of the East Flower Garden Bank showing locations of stations taken on the BALTIC SEAL cruise of October 1980. The heavy line connects the stations used in the vertical section of Figure 31*a*.

side of the bank than on the downstream side, which is probably due to centrifugal accelerations on the northwest side. The centrifugal acceleration would displace the dense cold waters away from the bank. The upper isotherms appear to be displaced upward a few meters on the northwest side with a sharp drop in the lee of the bank. The latter effect appears to be an internal lee wave that severely compresses the thermocline.

A similar survey around the East and West Flower Garden Banks taken in March 1981, when the stratification was rather minimal, is shown in Figures 5.32 and 5.33. The velocity vectors at 5 m from the surface (Figure 5.32) appear to be responding to the presence of the banks, particularly at the East Flower Garden Bank. At 1 m from the bottom (Figure 5.33) the flow is sharply attenuated and a little more tangential to the isobaths on the banks. The flow in the lee of the East Flower Garden Bank appears to be weaker and less organized than that to the west.

In July 1981 another transect across the East Flower Garden Bank from the northwest to southeast was used to construct a thermal cross section (Figure 5.43). This one also has transmissivity values on it. As usual, the flow was from west to east. In this case the lower isotherms rise against the bank on the upstream side, but the upper isotherms, below 40 m, are depressed. Again there appears to be a lee wave downstream of the crest of the bank. Very turbid water ($<$ 40%/m transmissivity) is limited to $\geq$ 80 m depth on the upstream and $\geq$ 90 m downstream of the bank. Notice the plume of dirty water ($<$ 40%/m transmissivity) between about 90 and 110 m depth on the southeast side of the bank. This appears to be turbid water carried from the bottom on the west side of the bank out across the deeper waters to the south and southeast.

Another indication of orographic effects generated by the banks is embedded in the time-series current meter records. From each record the variance tensor **v** was computed:

$$\mathbf{v} = \begin{vmatrix} \overline{u^2} & \overline{uv} \\ \overline{vu} & \overline{v^2} \end{vmatrix}$$

where u = the east–west velocity vector magnitude
v = the north–south velocity vector magnitude

The single overbar indicates time averaging.

It has been shown that the coordinate axes can be rotated so that $\bar{u}\bar{v}$ and $\bar{v}\bar{u}$ become zero. At that

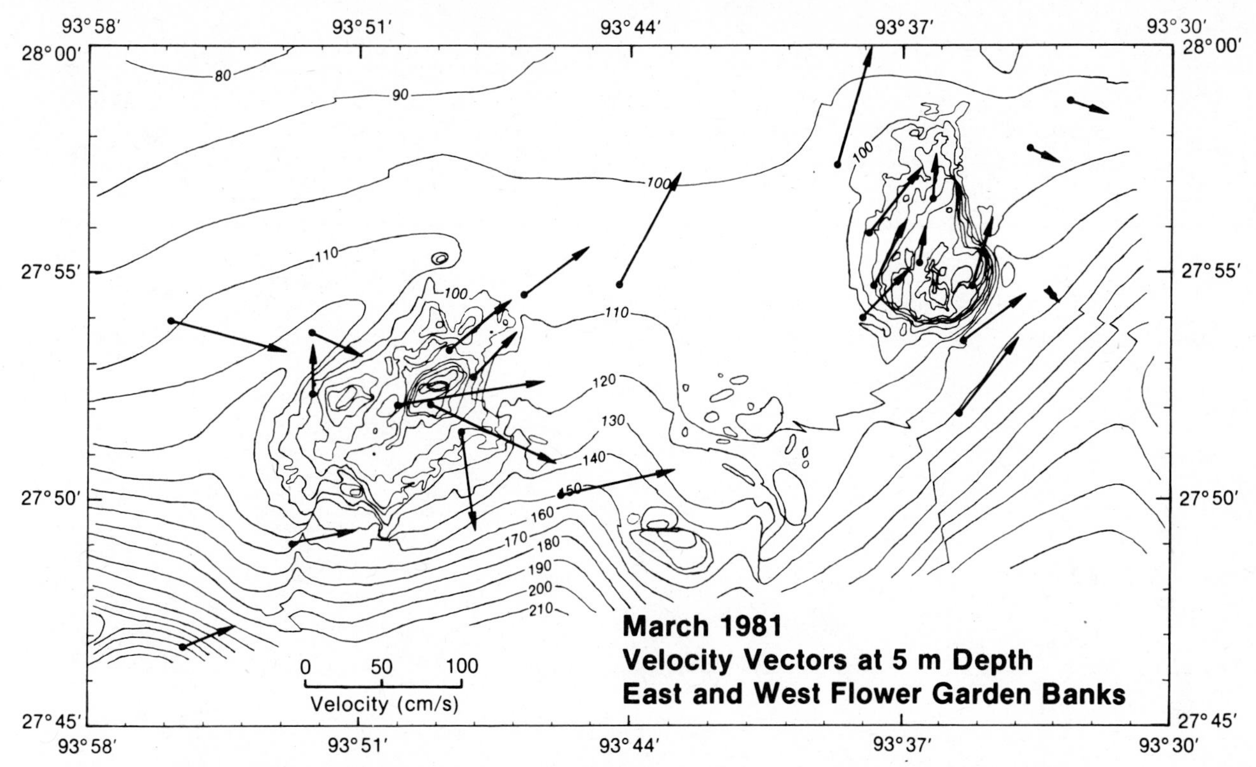

Figure 5.32. Horizontal velocity vectors at a depth of 5 m, determined from PHISH stations taken on the March 1981 GYRE cruise.

orientation $\overline{u_o^2}$ is a maximum and $\overline{v^2}$ a minimum. The axes at this orientation θ are known as the principal axes and $\overline{u_o^2}$ and $\overline{v_o^2}$ are the characteristics which can be computed as follows:

The characteristics are the eigenvalues λ of the variance tensor **v** and are found from the determinant of

$$\begin{vmatrix} \overline{u^2} - \lambda & \overline{uv} \\ \overline{vu} & \overline{v^2} - \lambda \end{vmatrix} = 0$$

The orientation of the principal axis θ is

$$\theta = \frac{1}{2}\tan^{-1}\left(\frac{2\overline{uv}}{\overline{u^2} - \overline{v^2}}\right)$$

Monthly averages of the autocovariance $\overline{u^2}$ and $\overline{v^2}$ and the cross covariances uv were computed for all current meter records that contained good quality data. From these values the following was found:

1. θ, the orientation of the major axis of variance $\overline{u_o^2}$; $\theta = 0$ is north.
2. $\varepsilon = \overline{u_o^2}/\overline{v_o^2}$ is the ratio of the maximum to minimum variance, which is a measure of the ellipticity of variance. The greater the value of ε, the more strongly the variance of the flow is aligned with the principal aixs. The closer ε is to 1, the more incoherent the variance is with respect to direction.

Plots of θ versus ε were prepared for all moorings and all deployments. The plots for all the meters on the first five deployments of mooring 2 (the mooring closest to the bank; see Figure 5.1) are shown in Figure 5.34; the plots for all meters on the last deployment (Figure 5.35) were separated because the last deployment of mooring 2 was moved for the purpose of determining the effectiveness of topographic steering. On each plot the orientation of the local isobaths is shown with the angles subtended by the East Flower Garden Bank. In general, the records from shallow meters possessed much smaller ε values, which indicated that the orographic effects, though still present, diminished upward. Near-bottom instruments recorded flow that was sharply polarized along the isobaths, as evidenced by values over 12.

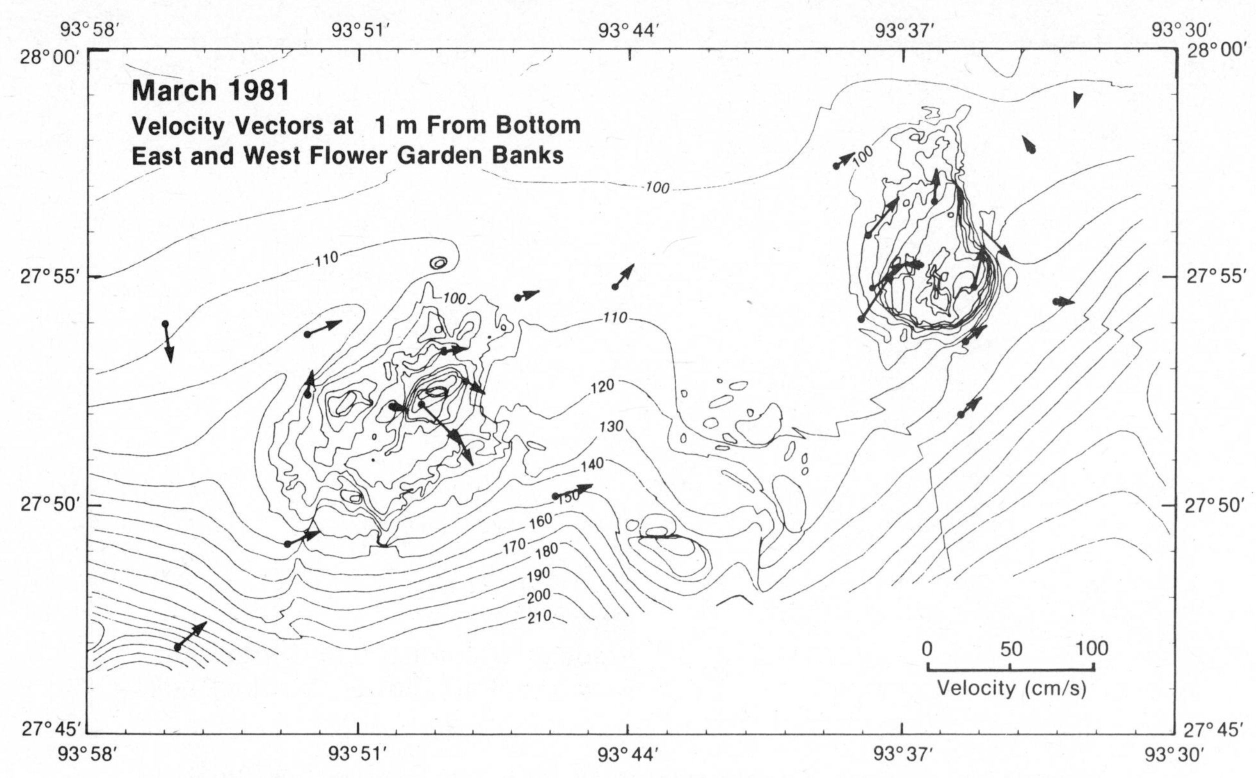

Figure 5.33. Horizontal velocity vectors at 1 m from the bottom, determined from PHISH stations taken on the March 1981 GYRE cruise.

Table 5.5 lists the average direction θ for each deployment at mooring 2, the standard deviation of the direction θ, the orientation of the local isobath (θ isobath), and the difference between the Δθ. Two records from shallow meters with small ε values were excluded from the averages because θ in such cases has little meaning. The Δθ varies from 5 to 24° but is always positive, meaning that the variance is slightly to the right of the isobath or away from the bank. Considering the location of the mooring, it is possible that this difference is caused by centrifugal acceleration.

For comparison θ versus ε for the other moorings is shown in Figure 5.35. Notice that the other moorings the ε values were much smaller and less clustered about the orientation of the local isobaths, which is expected because of their greater distances from the banks in relation to mooring 2. What was not expected was the disinclination of the flow at any of these sites to go toward the banks. This is illustrated in Figure 5.36, which shows the range of orientations of the principal axis at each mooring in relation to the bottom topography.

In summary, the station data and time series current meter data show that the flow goes around the bank horizontally with only low amplitude vertical motions.

An unexpected confirmation of this hypothesis appeared when side-scan sonar records from the East and West Flower Garden Banks were examined. On each of these banks sediment in the Algal Nodule Zone and *Amphistegina* Sand Zone is heaped into giant sand or gravel waves which have amplitudes of 1 to 2 m and wavelengths of 10 to 30 m. The orientation and location of these features are shown in (Figure 5.37) for the West Flower Garden Bank and in (Figure 5.38) East Flower Garden Bank.

Formation of bedforms of this size and in sediment of this coarseness would require high velocity flow. The gravel waves are most prevalent on the west side of both banks. This distribution, with the orientation of the crests and asymmetry of their form, indicates that the flow forming them comes from the west. With a few exceptions the crests of the bedforms are nearly perpendicular to the local isobaths. These exceptions occur on the upstream side of the East Flower Garden Bank, where the flow bifurcates (Figure 5.38). South of this region the bedforms indicate southerly isobath-parallel flow, and north of it they indicate northerly isobath-

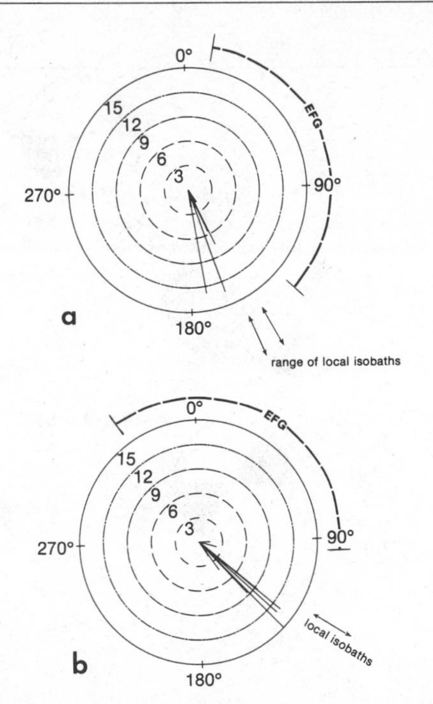

Figure 5.34. Ratio (ε) of the variance along the major axis to the variance along the minor axis, plotted in the direction θ of the major axis for *(a)* all meters on mooring 2 deployed before March 1981 and *(b)* all meters on mooring 2 for the March to July 1981 deployment.

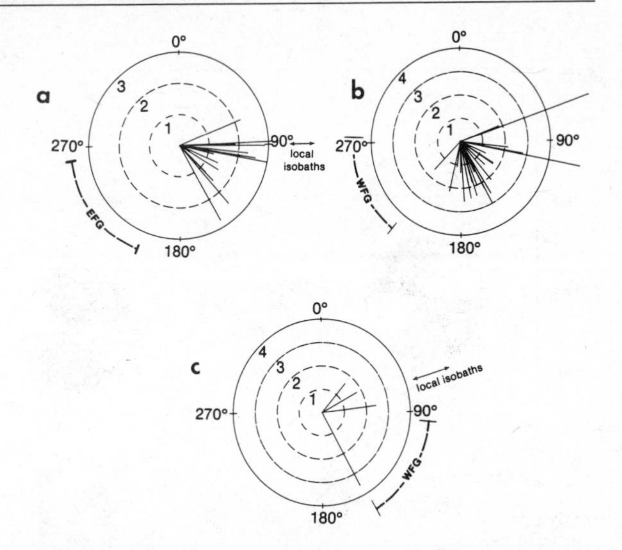

Figure 5.35. Ratio (ε) of the variance along the major axis to the variance along the minor axis, plotted in the direction θ of the major axis for *(a)* all meters on mooring, *(b)* all meters on mooring 3, and *(c)* all meters on mooring 4.

parallel flow. On the West Flower Garden Bank (Figure 5.37) crests of the bedforms are parallel to the isobaths on the downstream (eastern) side of the bank, suggesting a separation point.

The significance of these features is that their formation requires exceptionally high velocities and their orientation is such that even during this extraordinary flow the currents must remain essentially parallel to the isobaths.

In addition to deforming the mean flow field, it has been demonstrated (Baines and Davies, 1980) that the banks may trap long waves, evidence of which is given at the Flower Garden Banks by the strong inertial signal in the bottom current meter records at mooring 2. It is only at this site, adjacent to the bank, that inertial oscillations are as strong near the bottom as they are in the upper water column. A segment of the current meter record from 5 m above the bottom at mooring 2 during Hurricane Allen is shown in Figure 5.39.

The strong polarization of the flow along the isobaths, even as it oscillates in response to a hurricane, is clearly shown. Curiously, the temperature oscillations during this time have a period exactly one-half that of the currents. The vertical lines in Figure 5.39 run through the zero crossings of the velocity

TABLE 5.5. Average Direction of Major Axis of Variance Compared with Local Isobath Direction

Deployment	Average Direction ($\underline{\theta}$)	Local Isobath Direction ($\theta_{isobath}$)	Standard Deviation of Direction	Direction ($\underline{\theta} - \theta_{isobath}$)	Number of Estimates
1	170.1	156.4	—	13.7°	1
2	166.4	149.4	7.14	17°	2
3	173.5	149.4	—	24°	1
4	157.0	152.0	3.7	5°	3
5	158.2	152.0	7.82	6°	6
6	136.9	122.8	6.9	14°	7

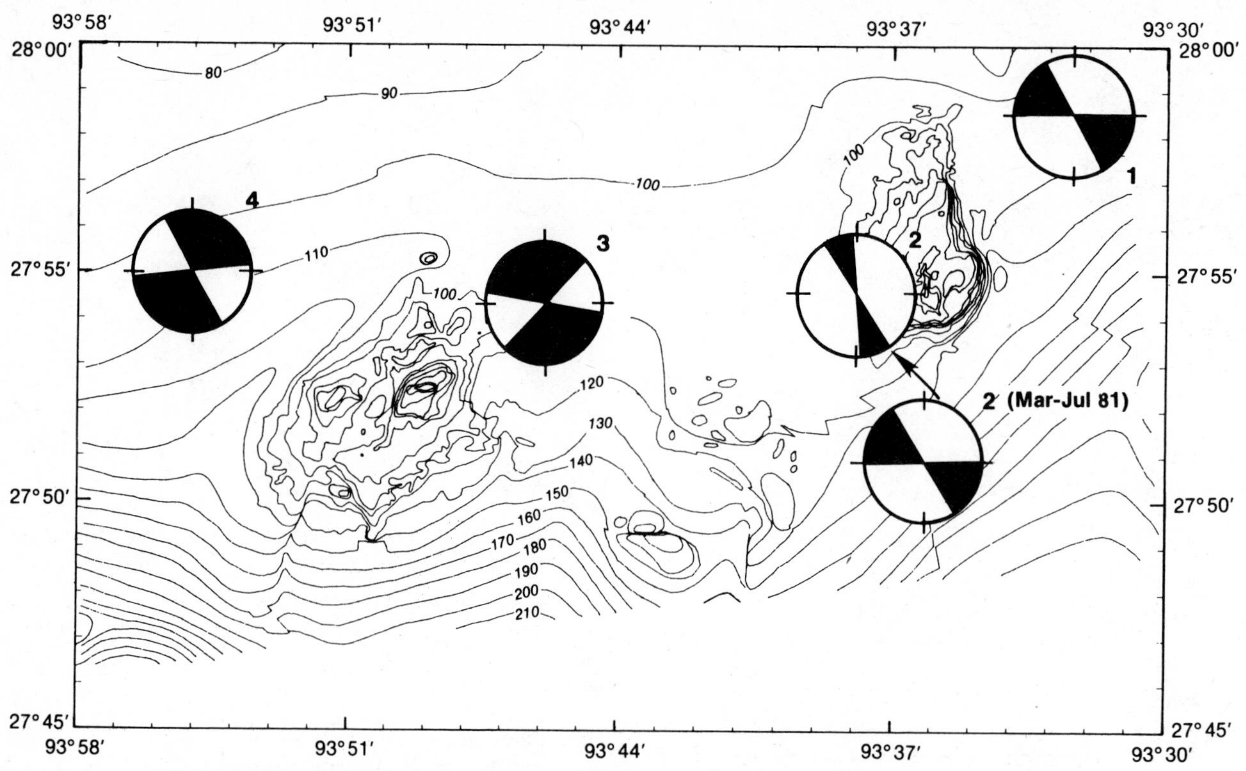

Figure 5.36. Shaded region in each circle shows the range of directions of the major axis of velocity variances for most 1-month velocity segments at each mooring (see text for exceptions). Each circle is centered at the mooring position, except for the March to July 1981 deployment of mooring 2. The arrow tip indicates the position of mooring 2 during March through July 1981.

vectors and extend through the temperature record to show that the zero crossings (two per cycle) correspond to temperature minima. These complex relationships are difficult to interpret with respect to the structure and propagation of the waves on the bank. It does appear, however, that the oscillations are concentrated on the slopes of the bank.

The accelerations that attend flow diverging around an obstacle, combined with the trapping of inertial oscillations on the slopes of the banks, should make bottom velocities on the bank surfaces significantly higher than velocities on the adjacent seafloor. Weak flow in the lee of the banks, particularly the East Flower Garden Bank, may contribute to the deposition of fine sediment on the seafloor.

BOUNDARY LAYERS AND SUSPENDED SEDIMENT

When a viscous fluid flows over any bounding surface, the molecules of the fluid adhere to those of the surface so that on that surface the relative velocity of the fluid becomes zero. This is known as the "no slip" condition. Beyond the immediate bounding surface, the fluid is retarded because of internal friction and the presence of the boundary. At some distance δ the flow is unaffected by the presence of the boundary and may, for practical purposes, be considered inviscid. Within the distance δ from the bounding surface the dynamics of the flow are dominated by viscous and turbulent effects. This zone of δ thickness is known as the boundary layer.

To facilitate a discussion of boundary layers it may be useful to introduce the jargon of boundary layer research.

Viscosity (μ). The internal friction of a fluid, caused by molecular attractions, which produces a resistance to flow or deformation of the fluid. The unit of measure is the poise ($g \cdot cm^2/s$). Viscosity is a characteristic of the fluid.

Turbulent or Eddy Viscosity (η). The internal friction of a fluid, caused by the presence of turbulence, which produces a resistance to flow or defor-

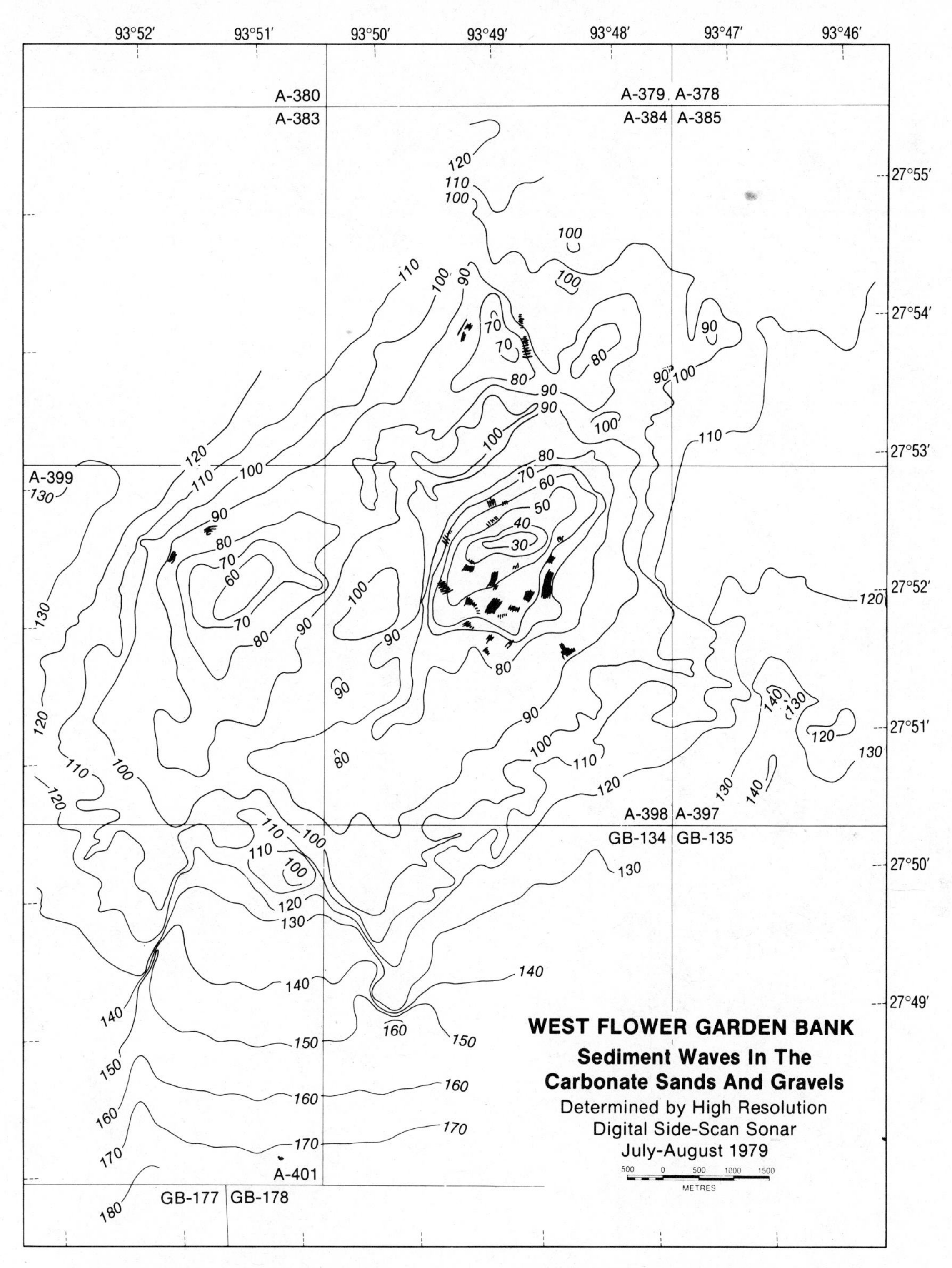

Figure 5.37. Large-scale bedforms on the West Flower Garden Bank, determined from a digital side-scan mosaic. Wavelengths and crest lengths are plotted to scale.

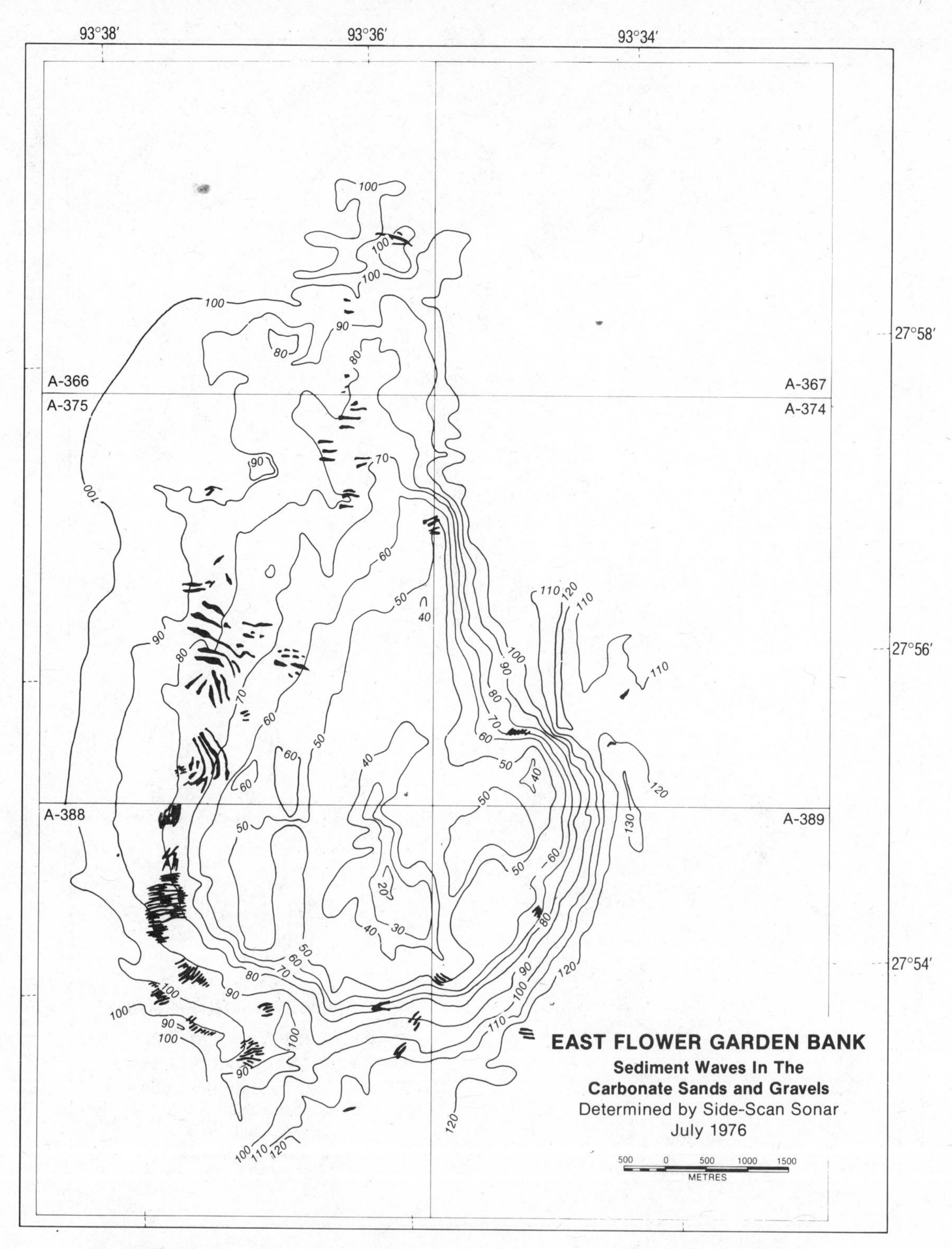

Figure 5.38. Large-scale bedforms on the East Flower Garden Bank, determined from side-scan sonar records. Original interpretation by Stephen Viada. Wavelengths and crest lengths are plotted to scale.

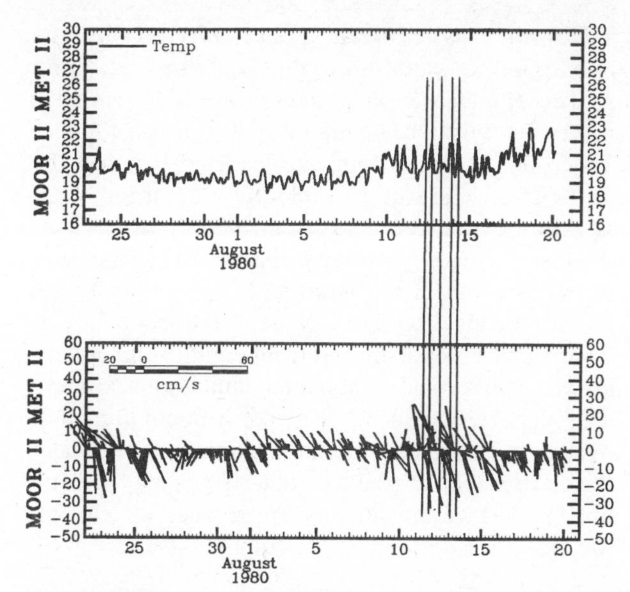

Figure 5.39. Time series of temperature and horizontal velocity from meter 2 of mooring 2, located near the southwest edge of the East Flower Garden Bank at 5 m from the bottom. Hurrican Allen traveled through the Gulf of Mexico during the period of this record (7–10 August 1980).

mation of the fluid. Its magnitude depends on the intensity of turbulence present but its value is always orders of magnitude greater than molecular viscosity (μ).

Shear Stress (τ). The force that causes two contacting surfaces (or layers) to slide on each other, moving apart in opposite directions parallel to the plane of their contact; for example, if you had a deck of playing cards neatly stacked on a table and you placed your finger in the middle of the top card and pushed horizontally, parallel to the table top, you would be exerting a "shear stress" on the cards.

To demonstrate the relationship that exists among these variables consider a small element of fluid at the bottom of a container in which all the fluid is still.

z x time t_0 — a

τ $v_1^2 = \frac{\Delta x}{\Delta t}$ Δz time t_1 $v_0 = 0$ Δx — b

Let t_0 be the initial time. Now assume that something far from the bottom sets the fluid in motion parallel to the bottom. The internal friction (viscosity μ) causes the molecules of fluid to stick together. Those at the top of the element of fluid start to move but those at the bottom stick to the container and do not move. Now, over a period of time, Δt the top of the element of fluid will move a distance Δx but those at the bottom will not have moved at all. However, $\Delta x/\Delta t \neq 0$), a velocity; at the top of the fluid element the velocity is $v \neq 0$ and at the bottom, $v_0 = 0/\Delta t = 0$. Therefore we have a vertical "velocity gradient" in which $v - v_0 = \Delta v$ and the gradient is $\Delta v/\Delta z$. The force required to make the top of the fluid element move in relation to the bottom is the "shear stress" (τ).

Now we can express the relation of these variables mathematically:

$$\tau = \mu \frac{\partial v}{\partial z} \tag{1}$$

From this we understand how viscosity represents a resistance to flow or deformation, because, if τ is held constant and μ is increased in value, $\delta v/\delta z$ must decrease; that is, deformation $\Delta x/\Delta z$ must decrease.

It should also be clear that for a given viscosity the application of a large shear stress will cause a large velocity gradient and the deformation of the fluid element will become so extreme that it will be torn apart. That is exactly how turbulence begins. Notice that the shearing of the fluid element introduces a rotation about the center of the element. Therefore, when the velocity gradient is high, small elements of fluid with a large v are rotated down and small elements of fluid with a small v are rotated upward. The result is that in short order fluid at height Δz is slowed down by the addition of fluid elements with low velocity and fluid close to the bottom is accelerated by the addition of high velocity water from above. Therefore the velocity difference Δv per unit height Δz is decreased by turbulence. For the same shear stress τ the effect of turbulence is to reduce the velocity gradient and increase the apparent viscosity.

This apparent viscosity is called the "eddy" or "turbulent" viscosity (μ) and the turbulent equivalent of (1) is

$$\tau_t = \eta \frac{\partial v}{\partial z} \tag{2}$$

τ_t is sometimes referred to as a Reynolds stress.

Diffusion. The transport or mixing of a substance *through* a medium (e.g., like water). The substance diffused may be a contaminant, such as sedimentary particles, or some solute. It may, on the other hand, be something like heat, kinetic energy, or momentum that is being diffused. Diffusion always goes "down" the concentration gradient of the diffusing substance; that is, diffusion moves from high concentration toward low concentration. As with viscosity, there is molecular diffusion and turbulent diffusion. Molecular diffusion is due to the random motions of molecules. Because the length scales of these motions are small and their speed finite, molecular diffusion is very slow. Turbulent diffusion, however, proceeds much more rapidly because the motions have much larger length scales. To develop an intuitive feel for the difference between the two remember that we diffuse cream and sugar in coffee by stirring it and thereby inducing turbulence. If we waited for molecular diffusion to do the mixing, the coffee would grow cold long before it was completed.

Advection. The transport of a substance from one location to another *in* the medium. The direction of transport is independent of gradients but dependent on the direction of flow.

The difference between diffusion and advection may be seen in the example of dye poured on the surface of the sea. Turbulent diffusion causes the patch of dye to grow in size but diminish in concentration, whereas the mean flow advects the patch away from its point of introduction. So it is that sediment is *diffused* up from the seafloor by turbulence when the shear stress is sufficiently large to cause erosion. The sediment, however, is *advected* from the site of erosion to the site of deposition by the horizontal flow.

In a nonrotating system the velocity profile is given by

$$v(z) = \frac{u^*}{k} \ln\left(\frac{z}{z_o}\right)$$

where $v(z)$ = the horizontal velocity at a height z above the boundary
u^* = $(\tau/\rho)^{1/2}$ is the friction velocity
k = the von Karman constant, assumed to be 0.4
z_o = a characteristic length scale of a rough boundary; it is the distance above the mean bottom at which $v(z)$ goes to zero

This is the classic logarithmic velocity profile. The two important variables in this equation are u^* and z_0. The u^* increases with increasing τ, which, in turn, increases with increasing intensity of turbulence. The rougher the bottom, the larger the z_0 and the farther from the wall the velocity is retarded.

Stratification can also affect the boundary layer thickness by extinguishing turbulence. The measure of the strength of the stratification is typically the Brunt–Vaisala frequency N. It is the frequency at which a particle, displaced from equilibrium and released, would oscillate about its equilibrium position. This particle would act just like a pendulum, first overshooting the equilibrium position as it fell under the acceleration of gravity, then rising past the equilibrium surface on the way up because of buoyant forces. The Brunt–Vaisala frequency N is given by

$$N^2 = -\frac{g}{\rho_o}\frac{\partial\rho}{\partial z}$$

where ρ = density
g = gravitational acceleration

Remember that energy is required to lift the particle above equilibrium or push it down. The larger the density gradient, the more energy required to move the particle vertically. Also remember that in turbulent flow vertical motions are induced by the rotation of water particles under the torque of velocity shear. If the turbulence is very energetic, it will overcome the buoyant forces and mix the stratified water, but if the stratification is very strong, the turbulence will not be energetic enough to displace the water particles and the turbulence will be extinguished. The measure of these competing forces is the Richardson number Ri:

$$\mathrm{Ri} = \frac{N^2}{(\partial v/\partial z)^2}$$

It has been found that at a Richardson number of about 0.25 turbulence is extinguished. If Ri is smaller, it means that $\delta v/\delta z$ is large in relation to buoyancy and turbulence will mix the stratified water. If Ri is greater than 0.25, it means the N is large in relation to the shear forces.

Consider the relation between N and the velocity gradient in the boundary layer of an energetic but stratified flow. At the initial time the density gradient $\delta\rho/\delta z$ would be constant throughout the boundary layer. Near the bottom $\delta v/\delta z$ tends to be large but

TABLE 5.6. Boundary Measurements at West Flower Garden Bank

Measurements	Dive D80-17	Dive D80-18	Dive D80-20
Depth (m)	49	50	75
Bottom type	Algal nodules	Algal nodules	Algal nodules
Mean flow direction	150°	140°	115°
Mean speed (cm/s)/height of emitter (cm)	14.5/130	15.9/130	16.6/150
	12.1/80	15.7/80	16.5/90
	10.8/30	15.1/30	15.6/30
Probability that velocity is independent of height above the bottom	.0001	.06	.03

decreases upward to zero at the level of the free stream. Therefore near the bottom Ri tends to be small and the turbulence mixes the water, rendering its density intermediate between that at the bottom and that at the top of the mixing layer. That intermediate value, however, is greater than the density of the water originally just below the top of the mixing layer. The density gradient at the top of the mixing layer therefore increases. This continues upward from the bottom until the declining velocity and increasing density gradients reach the point at which Ri becomes greater than 0.25. The turbulence is extinguished and no more mixing takes place. That also extinguishes turbulent diffusion in the bottom boundary layer.

The effect of stratification, then, is to compress the boundary layer by extinguishing turbulent diffusion of momentum (and anything else) at the top. This effectively caps the boundary layer.

If the flow is geostrophic, the boundary layer must

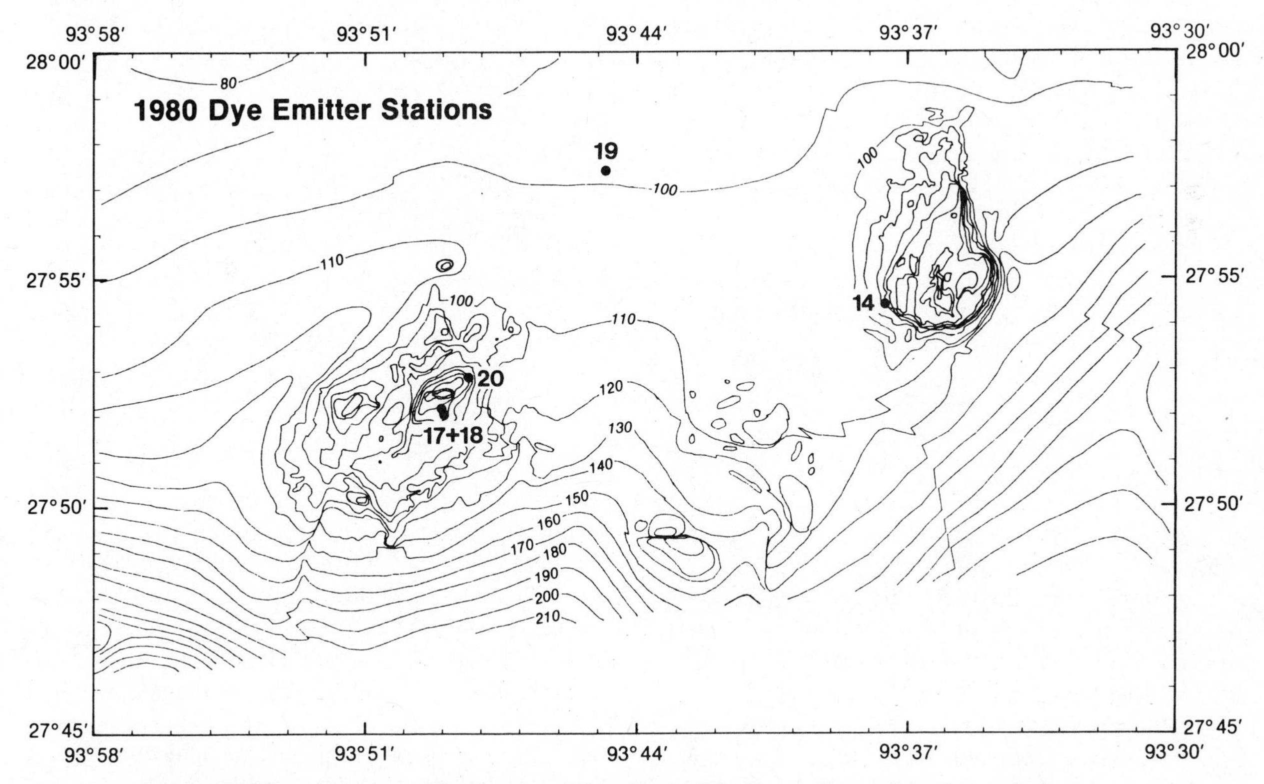

Figure 5.40. Location of the October 1980 submersible dive stations where dye emission experiements were made.

TABLE 5.7. Measured Angular Shearing in the Bottom Boundary Layer

Moorings	Meters	Dates	Depths (m)	Angle
2	2 and 3	July–September, 1979	94/100, 96/100	−14.5°
3[a]	3 and 4	October–December, 1980	90/101, 97/101	+15.1°
3[a]	3 and 4	March–July, 1981	91/103, 97/103	+ 5.7°

[a]Records were low-pass filtered first to remove inertial and higher frequencies.

adjust the relative speed and the angular velocity of the fluid to that of the bounding surface. This layer, called the Ekman boundary layer, consists of two parts; an upper part in which the velocity vectors rotate counterclockwise with depth (to the left of the geostrophic velocity V_g when looking downstream) and a lower part in which the speed is attenuated logarithmically with no additional angular shearing. According to Weatherly (1972), the thickness of the whole layer is $h = ku^*/f$ and the log layer thickness is

$$\delta_{\ln} = \frac{2u^{*2}}{f\,|\,Vg\,|}$$

The angular shearing α is given by $\sin^2\alpha = A^2(u^*/|V_g|)^2$, where $A^2 = 20$. Weatherly (1972), however, observed that most of the veering in his experiment did take place in the log layer. He also said that the effect of stratification is to compress the Ekman layer.

For high velocity flows in particular the boundary layer on the surface of the bank should not be an Ekman layer for the reason that the flow on the bank becomes ageostrophic because of inertial (centrifugal) accelerations. Also, on the upstream side of the bank the water encounters the bank suddenly so that the boundary layer is initiated on the leading edge. The bottom boundary layer therefore is not fully developed over the entire surface of the bank.

From three dye experiments carried out on the bank between 49- and 75-m depth it appears that the boundary layer flow did not fit the classic logarithmic profile. The average velocities at three distances above the bottom are shown in Table 5.6.

Locations of the dive sites are shown in Figure 5.40. The failure of these boundary-layer velocity profiles to assume the classic logarithmic form is probably due to several factors. The theory assumes that the flow is steady, yet it is clear that the velocity at 30 m above the bottom increased by more than 4 cm/s in the four hours between dives D80-17 and D80-18. The bottom at these sites is hydrodynamically rough and inhomogeneous. The algal nodules that cover the surface vary from the size of a green pea to a grapefruit. Also, the gravel waves introduce still another scale of roughness. It may be, therefore, that turbulence due to each scale of these roughness elements is superimposed on the others so that no averaging period is appropriate for all scales.

At any rate, the reduction of the flow velocity from 15 cm/s to zero over just 30 cm should exert large stresses on the bottom and would account for the sphericity of the algal nodules and the transport of sand-sized material (*Amphistegina* tests) out of this zone. Accumulation of silt and clay in this environment would be out of the question.

Assuming that the boundary layer on the open shelf at the base of the banks is less complicated and using Weatherly's (1972) formulas to estimate the thickness of the logarithmic layer, we would obtain the following. From a dye emission experiment in 98 m of water it was estimated that u^* was on the order of 0.3 cm/s in a flow that averaged 27 cm/s at a depth of 61 m. From these values $\delta_{\ln}$ should be on the order of 95 cm and angular shearing α should be on the order of 3^0. If the actual velocity were really 15 cm/s just outside the boundary due to baroclinicity $\delta_{\ln}$ would have been 1.7 m and α would have been about 5^0. The Ekman layer thickness h for either case would have been approximately 17 m.

The Kundu (1976) method was used to compute the mean angular difference between the two meters closest to the bottom on moorings 2 and 3. The results are shown in Table 5.7.

At mooring 2 the angular difference between the two meters is rather large and in the wrong sense for Ekman veering. The lower instrument recorded flow with a clockwise downward spiral, undoubtedly due to centrifugal accelerations in the bottom boundary layer (BBL) as the flow accelerates around the East Flower Garden Bank (see Figure 5.1 for locations).

The records from mooring 3 during both deployments rotate in the correct sense for Ekman veering, but they too could be biased by centrifugal accel-

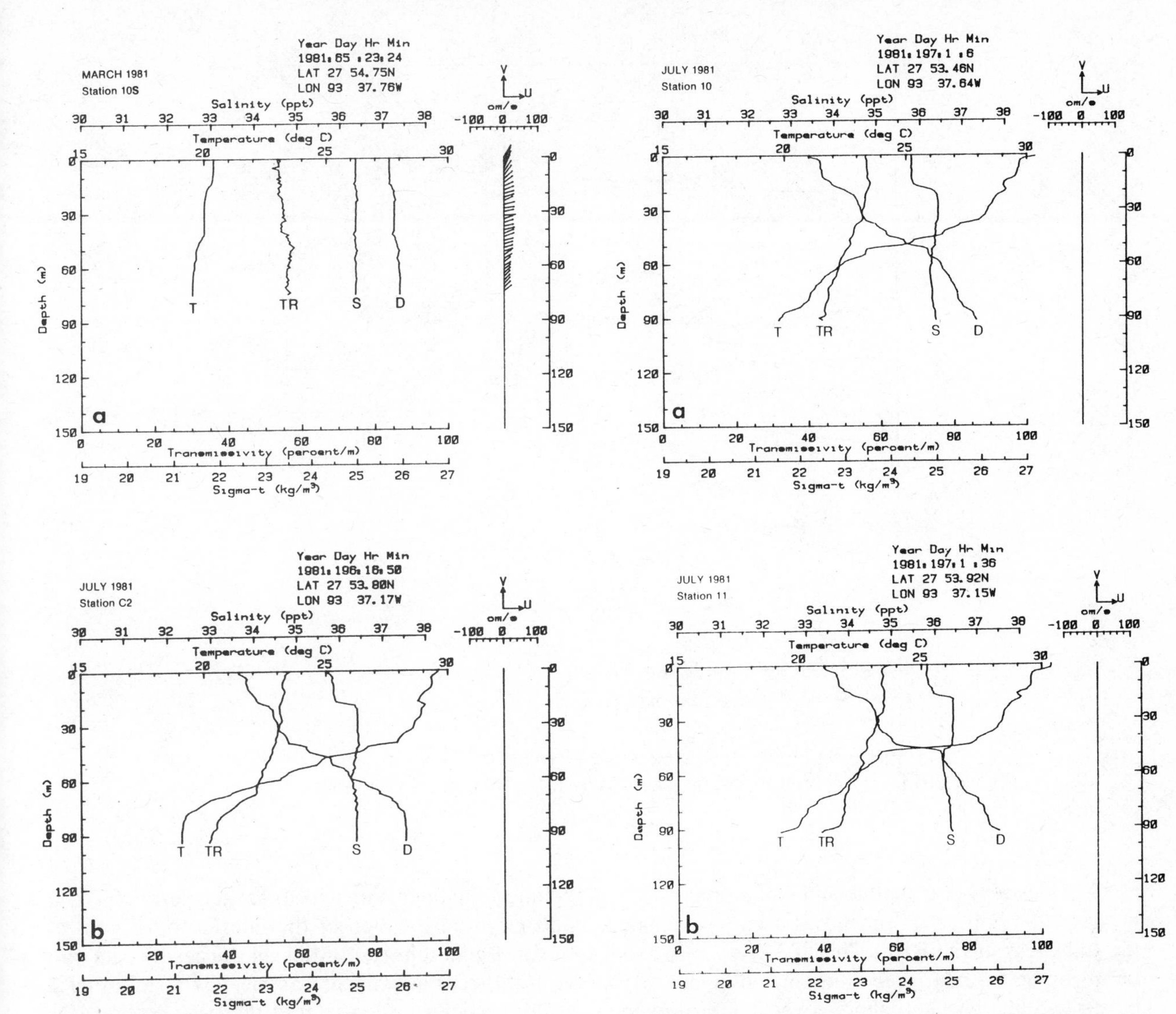

Figure 5.41. Profiles of transmissivity, temperature, salinity, density, and horizontal velocity at stations (*a*) 10S and (*b*) C2. The origin of the velocity vectors is at the depth of the measurement; vectors point in the direction of flow referenced to standard cartographic coordinates (north is toward the top of the page).

Figure 5.42. Profiles of transmissivity, temperature, salinity, density, and horizontal velocity at stations (*a*) 10 and (*b*) 11. The origin of the velocity vectors is at the depth of the measurement; vectors point in the direction of flow referenced to standard cartographic coordinates (north is toward the top of the page).

erations around the northeastern side of the West Flower Garden Bank. The possibility that this veering at mooring 3 is at least partly an Ekman-type turning cannot be dismissed, however.

The estimates of $\delta_{\ln}$ and h suggest that for all deployments none of the instruments was in the log layer and only those within 10 m of the bottom were likely to have been in the Ekman layer. This is somewhat reassuring because the mean monthly vectors (Figure 5.13) at mid-depth are oriented just south of east. An Ekman BBL flow under these vectors should turn counterclockwise with depth, but instruments 11 to 18 m above the bottom recorded flow that was rotated clockwise, or more southerly. In view of the thinness of the Ekman BBL this rotation must be attributed to the baroclinicity of the flow rather than to some peculiarity of the BBL flow.

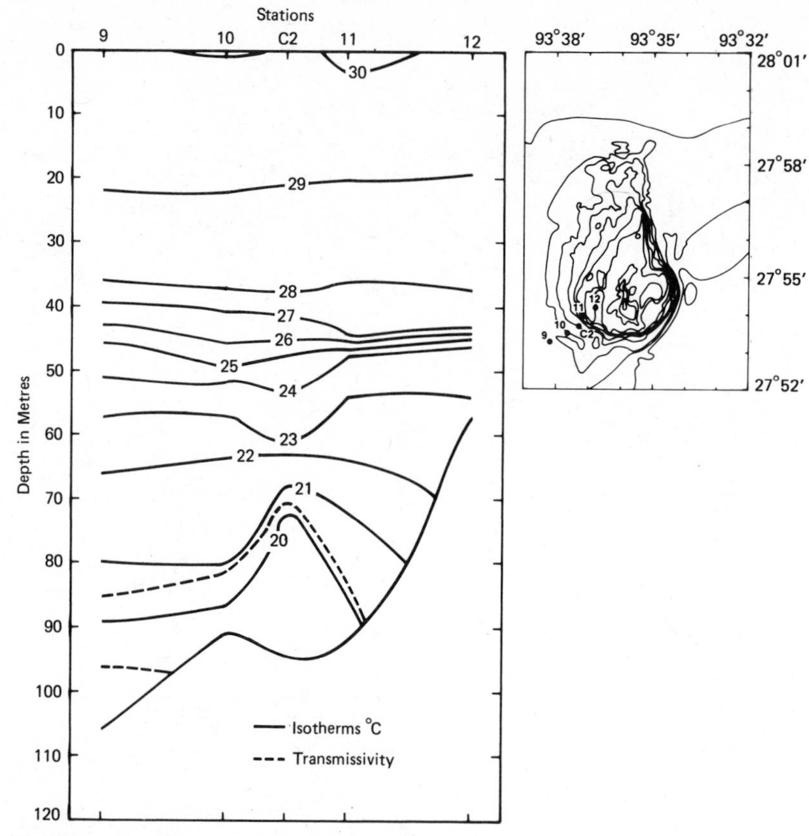

Figure 5.43. Temperature and transmissivity, July 1981, on a southwest transect at the East Flower Garden Bank.

The thickness of isohaline and isothermal mixed layers at the bottom is sometimes used to estimate the thickness of the BBL. That is a risky business on a sloping boundary on which mixed layers may be stacked by oblique offshore advection. However, all of the stations taken in the vicinity of the Flower Garden Banks on the October 1980, March 1981, and July 1981 cruises were examined for bottom mixed layers. In October and July fewer than 40% of the stations had bottom mixed layers. The average thickness in July and October was approximately 6 m, with a standard deviation of 4.1 in October and 3.4 m in July. In March 46% of the stations had a bottom mixed layer. The average thickness was 10 $\pm$ 5.1 m. A total of 73 stations were taken during these three surveys.

All but three of the March stations with thick bottom mixed layers ($\geqslant$ 10 m) were taken on or within 1 km of the banks. These thickened layers appear to owe their existence to orographic effects rather than to increased turbulence and mixing at the bottom. These effects include convergence, baroclinic adjustments to centrifugal accelerations, and bottom intensification of the internal tides and internal oscillations (IO). Michael Carnes (in McGrail et al., 1982a) estimated that the IOs at mooring 2 after Hurricane Allen caused the near-bottom isotherms to oscillate over a vertical distance of 10 to 15 m.

Station 10S (Figure 5.41) from the March survey exemplifies this thickening of the bottom mixed layer. The station was taken on the western slope of the East Flower Garden Bank (Figure 5.7) during a period of strong flow from the southwest. Flow at that location would have been undergoing considerable acceleration (V^2/r) as it encountered the bank. Using 30 cm/s for V and 4.5 km for the radius of curvature yields 2×10^{-3}cm/s. The Coriolis acceleration (Vf) for the same velocity flow at this location would have been 2.1×10^{-3}cm/s.

A comparable station from the July survey is C2 (Figure 5.41), which was taken on the southwest slope of the East Flower Garden Bank (Figure 5.8). Stations 10 and 11 (Figure 5.42) were taken on either

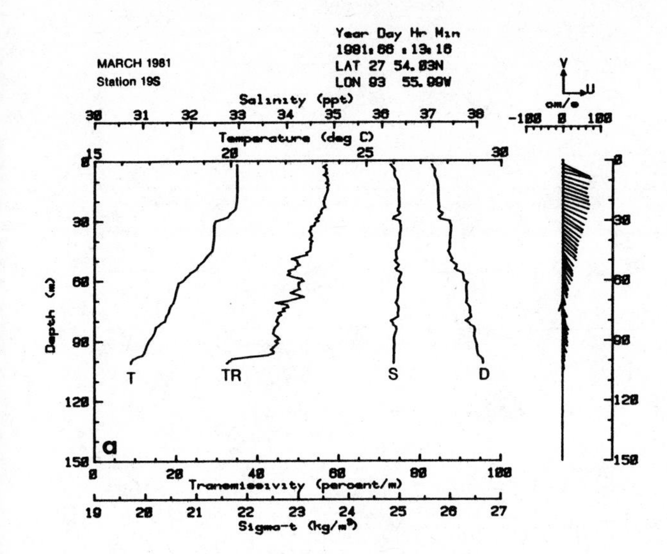

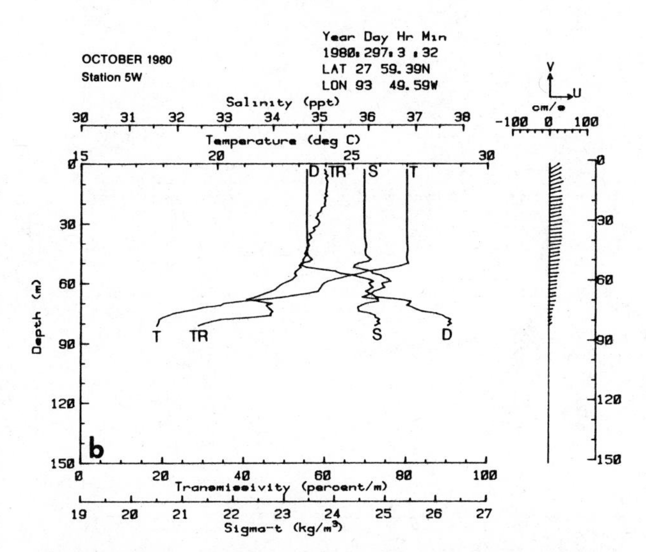

Figure 5.44. Profiles of transmissivity, temperature, salinity, density, and horizontal velocity at stations (*a*) 19S and (*b*) 5W. The origin of the velocity vectors is at the depth of the measurement; vectors point in direction of flow referenced to standard cartographic coordinates (north is toward the top of the page).

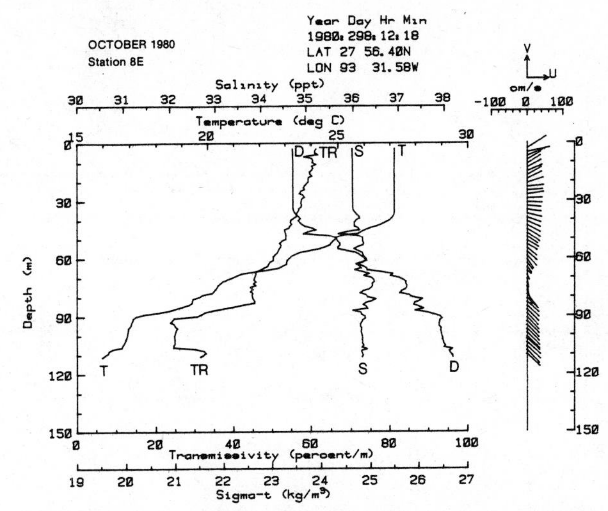

Figure 5.45. Profiles of transmissivity, temperature, salinity, density, and horizontal velocity at station 8E. The origin of the velocity vectors is at the depth of the measurement; vectors point in the direction of flow referenced to standard cartographic coordinates (north is toward the top of the page).

side of C2 but approximately 7.5 to 8 hours later. Notice that neither of these stations has thick bottom mixed layers. Figure 5.43 shows a cross section of stations 9 through 12, including C2, which runs from southwest toward the northeast (Figure 5.9). It is not clear whether the eight-hour difference in time from C2 to 11 resulted in sampling at different phases of the internal tide or whether orographic effects account for the thickening of the bottom mixed layer. However, a local increase in turbulent vertical mixing can be ruled out altogether because that would not have caused the downward flexure of the isotherms in the lower thermocline and it would not have been restricted to this one location.

The bottom mixed layer in station 19S (Figure 5.44) of the March survey is more in line with the scale of δ_{ln} expected from the estimates of 1 to 2 m made above. This station is well away from the influence of the banks and it is likely that the bottom mixed layer sampled here is due to turbulent mixing in the boundary layer. Notice the sharp transmissivity gradient in the mixed layer. It suggests active resuspension and upward diffusion of sediment from the bottom that would be commensurate with such mixing.

Station 5W (Figure 5.44) from the October 1980 survey was taken about 8 km north of the West Flower Garden Bank (Figure 5.6) and should have been beyond the influence of the bank. The water within 2m of the bottom is not completely isothermal but the influence of mixing is clearly present. Again the transmissivity spike in this layer implies active resuspension and upward diffusion due to the turbulence in the bottom boundary layer.

Many of the stations taken during these cruises have steplike structures in their near-bottom thermal structures which are associated with increased turbidity. An excellent example is station 8E (Figure 5.45) of the October 1980 cruise. This association

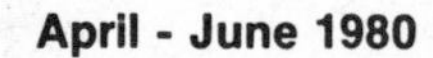

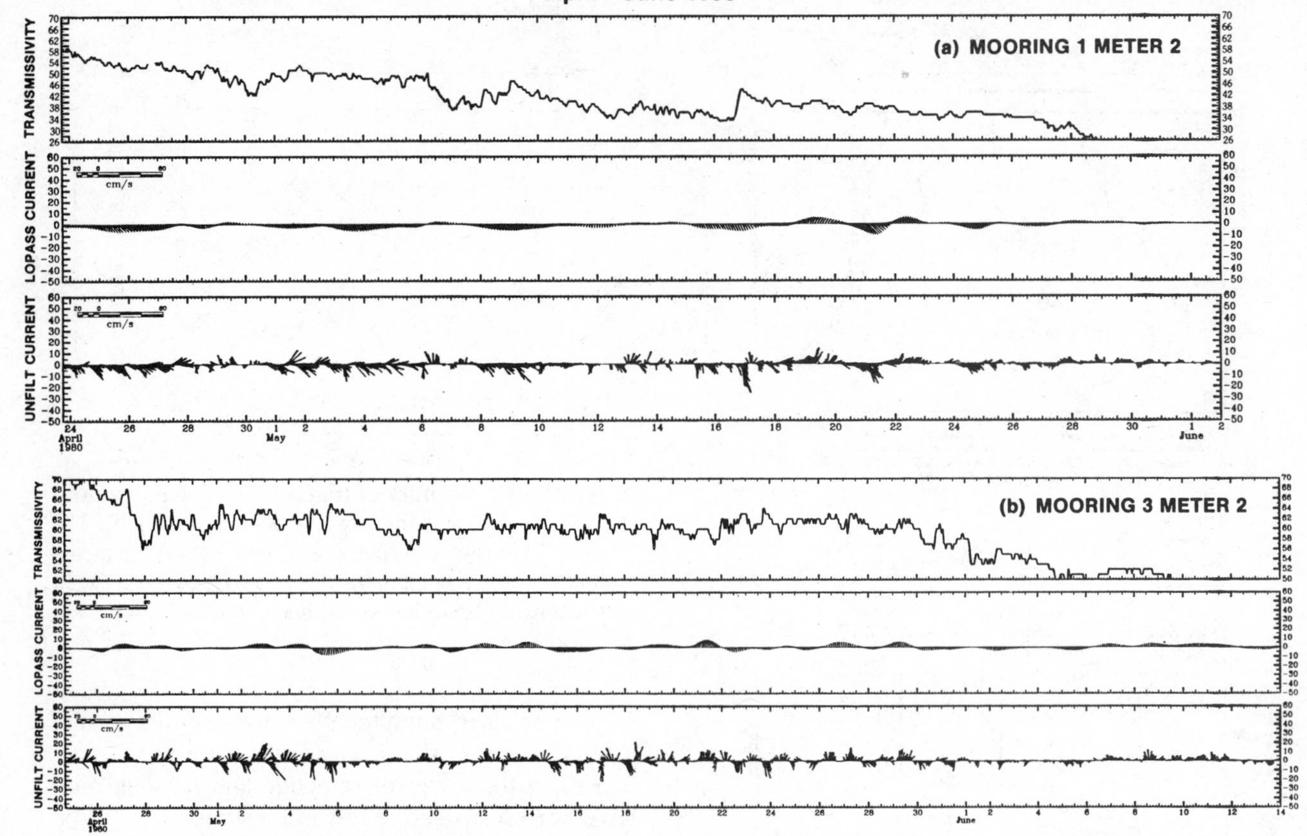

Figure 5.46. Raw velocity and transmissivity (XMS) time series records for April to June 1980 from (*a*) mooring 1 at 95-m depth (bottom depth 99 m), and (*b*) mooring 3 at 100-m depth (bottom depth 109 m). See Figure 5.1 for mooring locations.

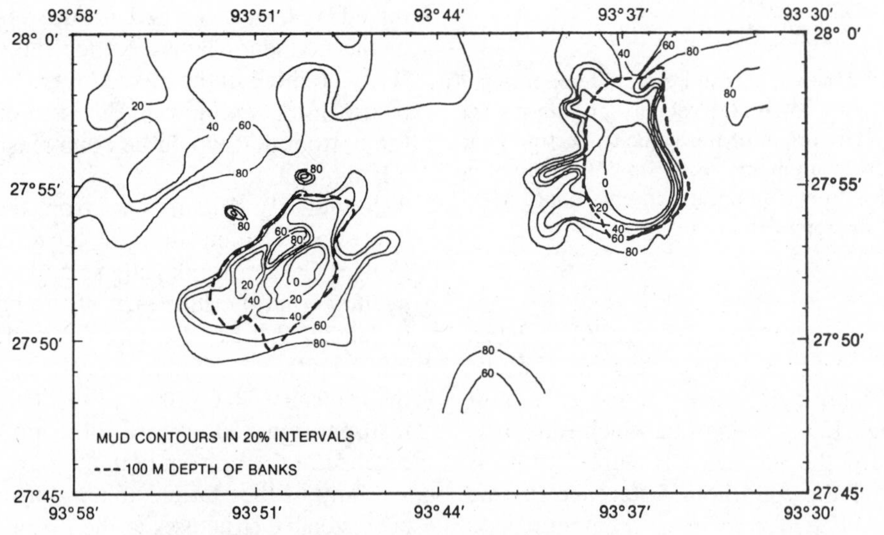

Figure 5.47. Mud contours in 20% intervals.

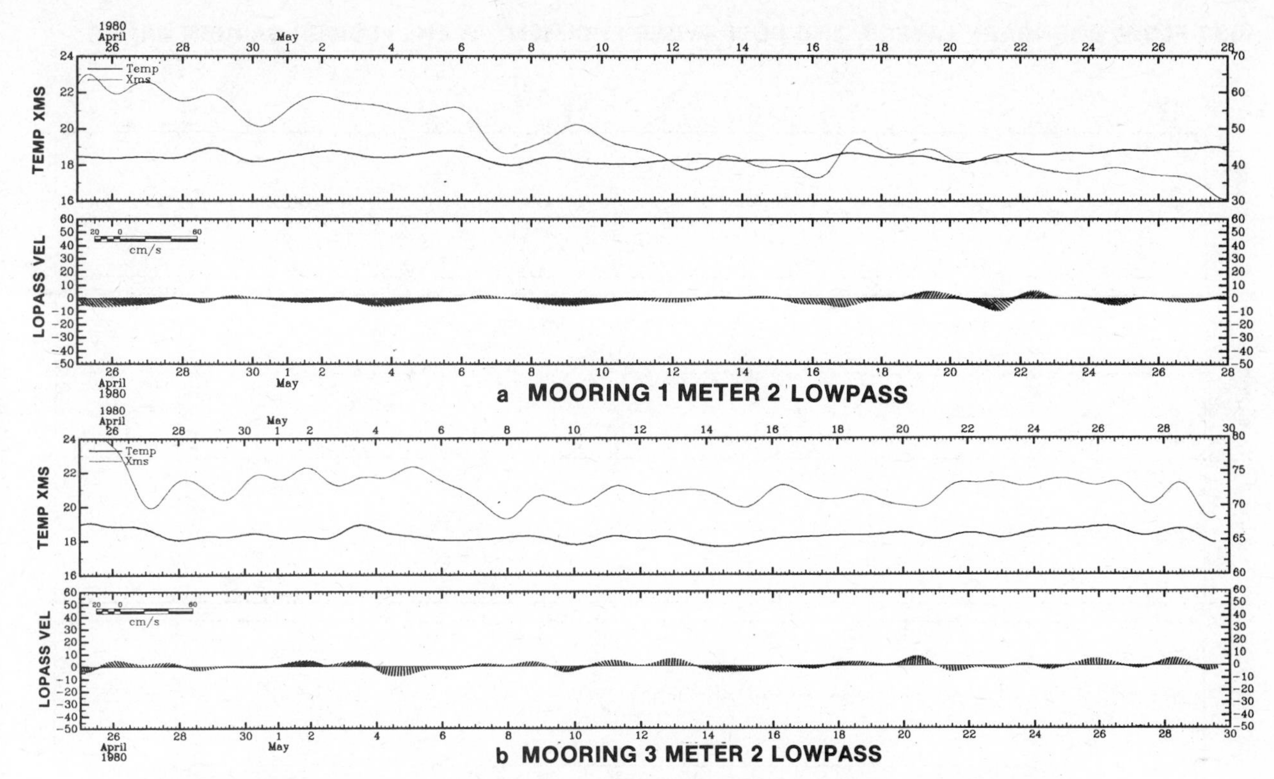

Figure 5.48. Time series of temperature (°C), transmissivity (%/m), and horizontal velocity (stick-plot representation in normal map orientations) low-passed with a 28-hour filter near the bottom on (*a*) mooring 1 at 95-m depth (bottom depth 99 m) and (*b*) mooring 3 at 100-m depth, 109 m) near the Flower Garden Banks.

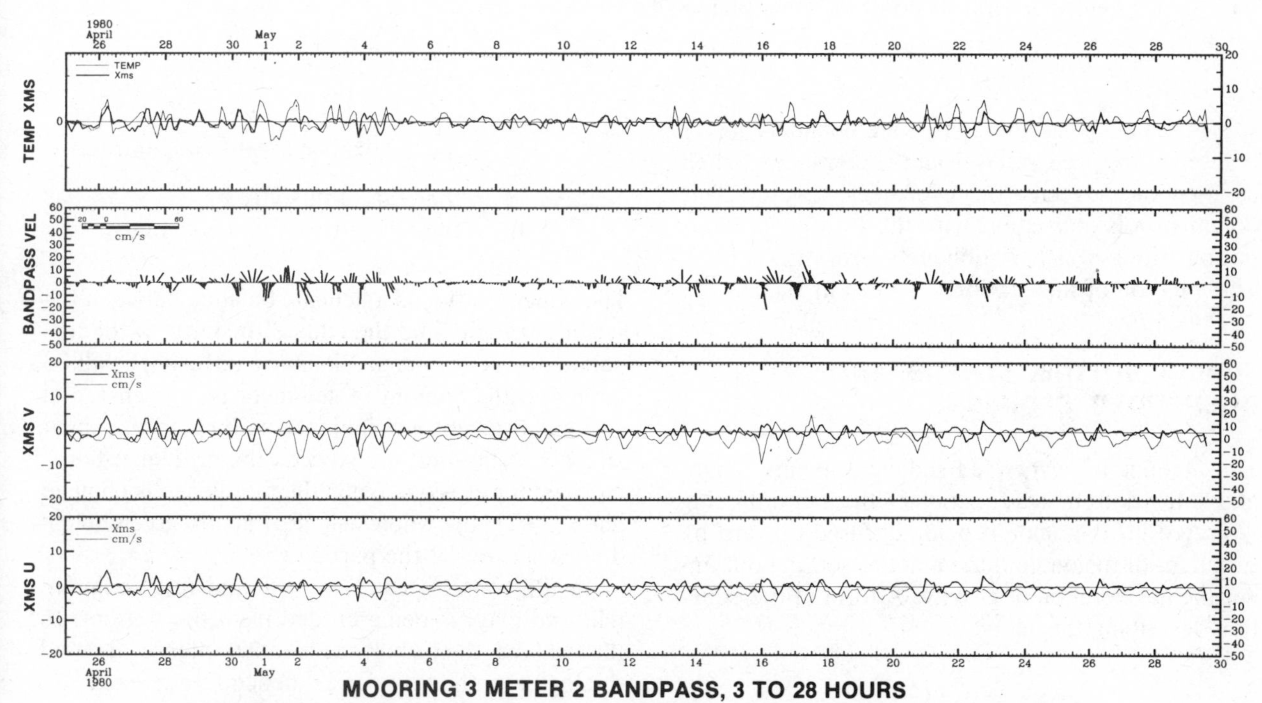

Figure 5.49. Bandpassed time series of temperature, transmissivity, and horizontal velocity, both stick plots, and individual orthogonal components near the bottom on mooring 3, northeast of the West Flower Garden Bank. The bandpass eliminates components of the current with periods outside the range of 3 to 28 hours.

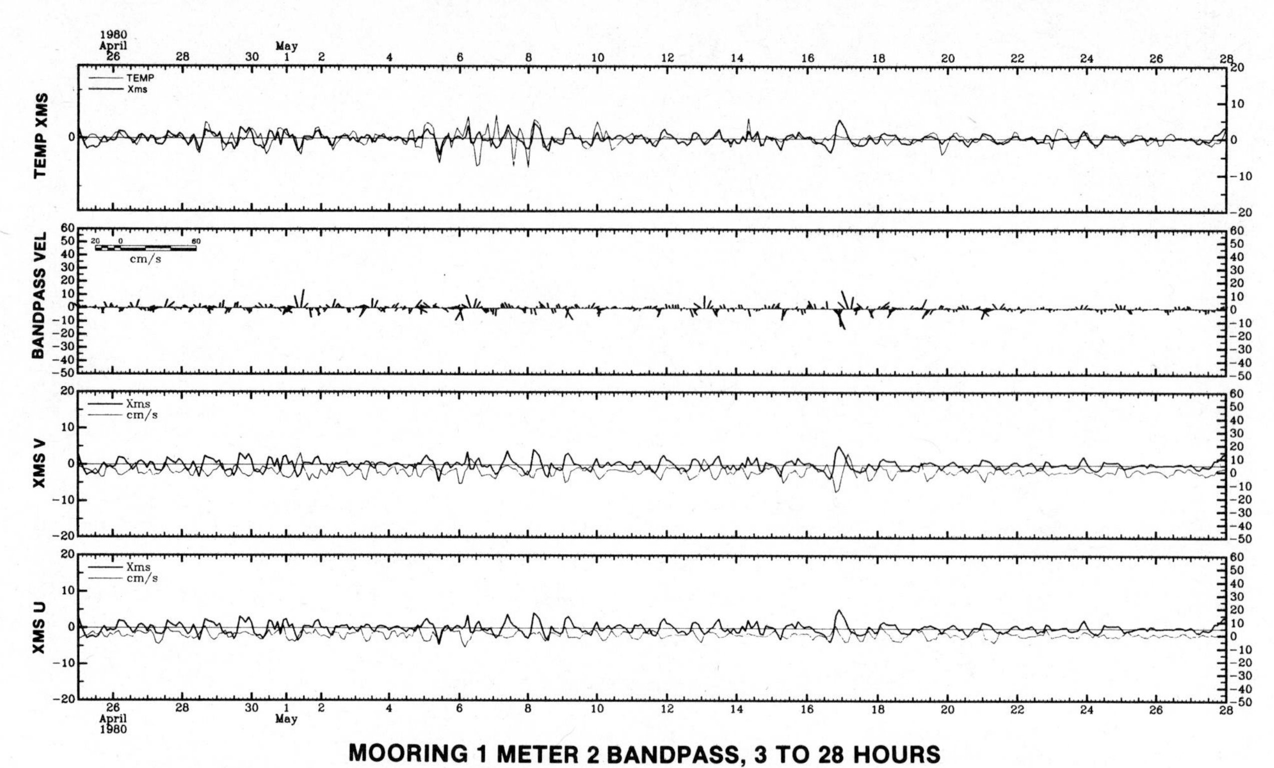

Figure 5.50. Bandpassed time series of temperature, transmissivity, and horizontal velocity, both stick plots and individual orthogonal components near the bottom on mooring 1, northeast of the East Flower Garden Bank. The bandpass eliminates components of the current with periods outside the range of 3 to 28 hours.

strongly implies that these steps are boundary layers that have been separated from the continental shelf upslope and advected out over deeper water. This conclusion is consistent with the southwest near-bottom flow, which is obliquely cross-isobathyal. The location of this station is shown in Figure 5.6.

TURBID BOTTOM LAYERS AND SUSPENDED SEDIMENT

The amount of suspended sediment at any height above the bottom is a function of the amount being advected in, the amount being diffused upward by turbulent diffusion, and the amount settling out under the acceleration of gravity. A form of the equation describing this is

$$C(z) = \frac{E}{W\rho}\left(\frac{\partial c}{\partial z}\right)$$

where C(z) = the sediment concentration
ρ = the fluid density
$\frac{\partial c}{\partial z}$ = the vertical sediment concentration
E = the eddy diffusivity
W = the fall velocity of the sediment

The concentration is a delicate balance between the settling velocity and the eddy diffusivity. If the turbulence is very intense, the eddy diffusivity will be large and the amount of sediment in suspension at any given height will increase as long as the shear stress τ at the bottom exceeds the critical value τ_c required for erosion. When the turbulence is damped and $E/W<1$, the water will clear at any given height due to settling of the particles.

In the simplest case, $\tau > \tau_c$, and fine sediment (silt and clay) is being eroded from the bottom. In the log layer the turbulence is rather intense. Beyond the log layer the turbulence, hence E, decrease rapidly, but W remains a constant and near the top of the log layer there is as much sediment lost to settling as gained by upward diffusion. The sediment cannot penetrate above the point at which $W > E$.

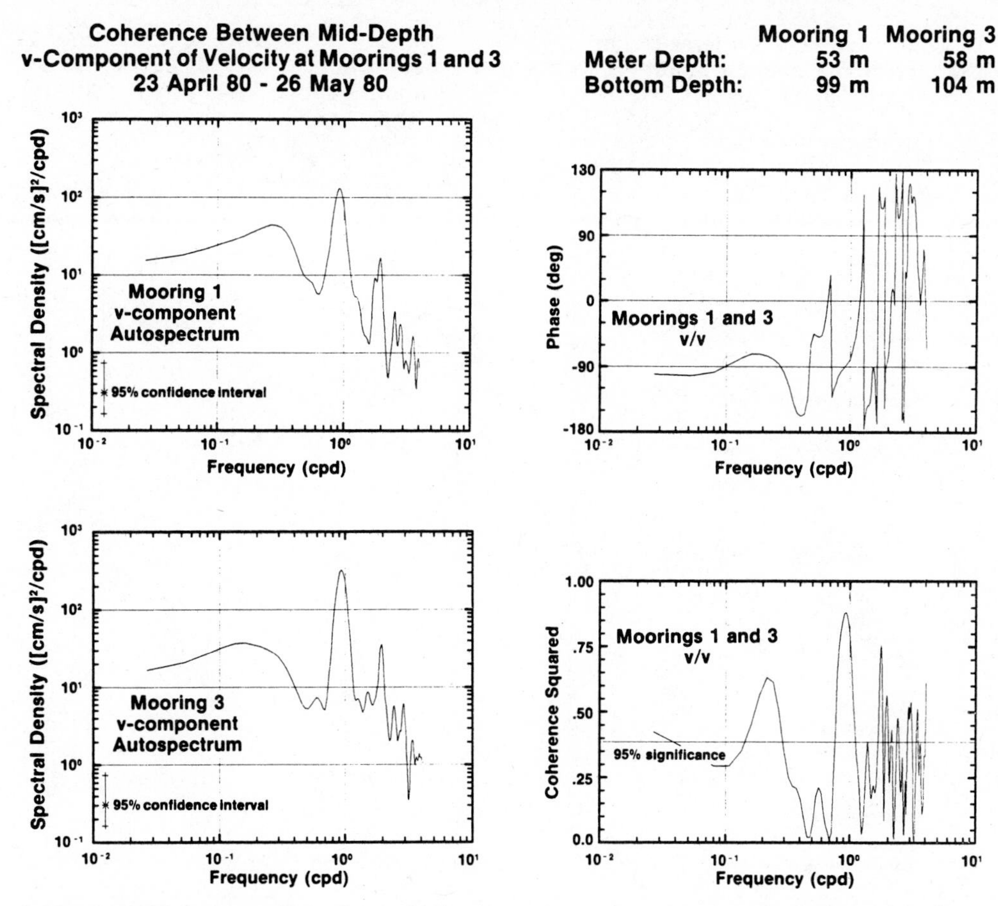

Figure 5.51. Comparison of the mid-depth, north–south component of velocity at mooring 1 with that at mooring 3.

This equation can be integrated from a distance z_1 above the bottom to some other distance z to yield

$$\ln \frac{C(z)}{C(z_1)} = -\rho \frac{W}{E}(z - z_1)$$

or

$$C(z) = e^{(-\rho W/E)(z - z_1)}\, C(z_1)$$

where $C(z_1)$ is the concentration at Z_1 (after Shepard, 1963).

Now if z is a substantial distance and E becomes vanishingly small, as in the case above, the concentration at z also becomes vanishingly small. All of this is predicated on a steady-state condition.

The transmissivity profile in station 19S (Figure 5.44) is illustrative of what happens during resuspension of fine sediment in a stratified fluid. Near the bottom (z_1), $C(z_1)$ is relatively large. Clay particles are therefore brought together frequently and tend to flocculate so that W is relatively large. On the other hand, $E >> W$ in the mixed layer, where turbulence is undamped. The decrease in C with respect to z is therefore relatively small. Above the mixed layer E becomes very small as the turbulence is extinguished by stratification. Therefore W/E becomes large and C decreases rapidly with respect to increasing height above the bottom, as shown in Figure 5.44.

If a turbid, mixed bottom boundary layer, like that in Figure 5.44, becomes separated from the bottom, there is no longer sediment available from the bottom to replace that lost by diffusion to the water above, in which case z_1 is measured from the base of the separated layer and $C(z_1)$ becomes a function of time. If E remains large, as is often the case, $C(z_1)$ decreases, $\delta C/\delta z$ becomes small, and the fine sediment becomes rather uniformly distributed through the layer. The transmissivity profile in station 8E (Figure 5.45) is illustrative of this point. In

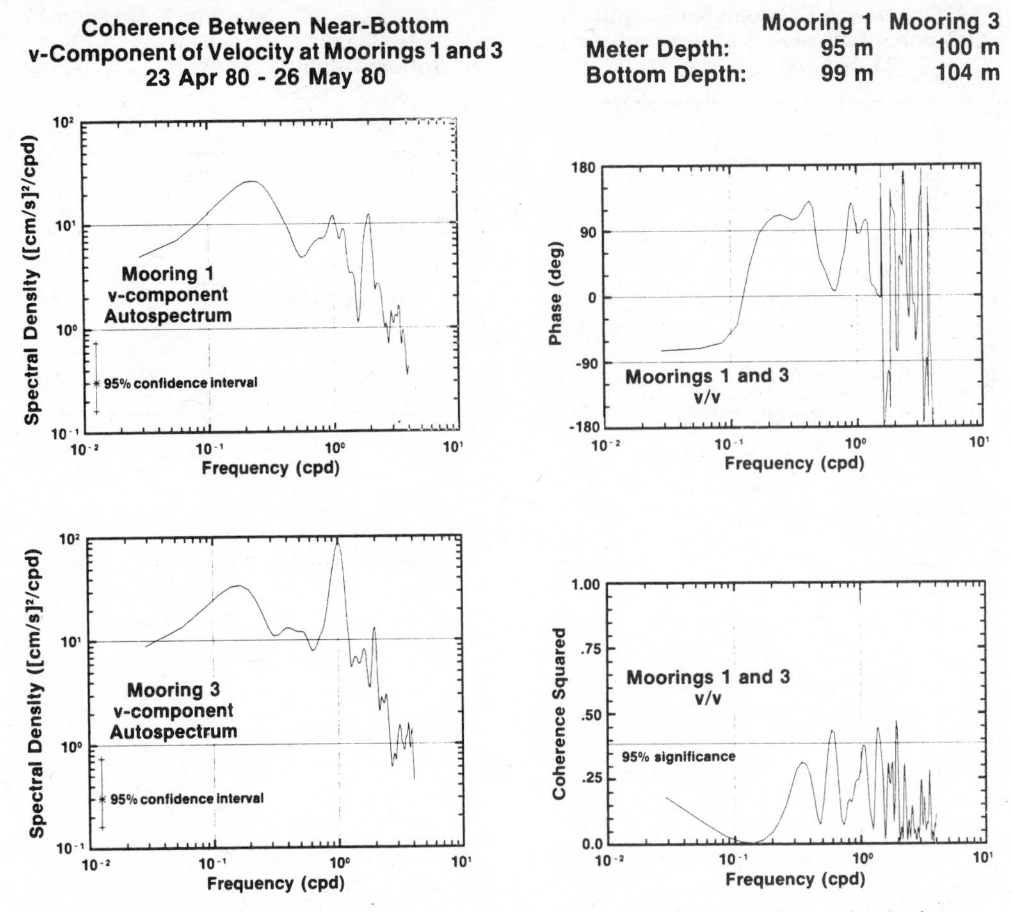

Figure 5.52. Comparison of the near-bottom, north–south component of velocity at mooring 1 with that at mooring 3.

this case there appear to be at least two separated layers between 105 and 90 m depth.

On the other hand, if E becomes small in the separated layer and if the layer has flowed out across water of significantly greater density, then $C(z_1)$ can increase with time as the sediment settles onto the density interface. At the interface W is diminished because of the increased fluid density. Sediment in this separated layer may be advected over distances of several tens of kilometers before becoming so diffused that it is undetectable in the transmissivity profiles.

From the profiles, dye emission studies, and theoretical estimates it appears that δ_{1n} on the open shelf around the base of the bank seldom, if ever, exceeds 3m. It also appears that the local penetration height of the resuspended sediment is limited to that layer. Variations in the concentration of suspended sediment above that height would therefore be a function of advective processes that carry separated boundary layers and associated sediment out over the shelf break.

This hypothesis is consistent with the time series transmissivity and velocity records from the moorings in the vicinity of the banks. Examples of records from moorings 1 and 3 are shown in Figure 5.46. Both meters were 5m above the bottom and the water was consistently clearer at mooring 3 than at mooring 1. This discrepancy may be related to the fact that the bottom to the northwest (upstream) of mooring 3 has less silt and clay in it than that to the northwest of mooring 1 (Figure 5.47).

The velocity, temperature, and transmissivity records for these two instruments were filtered to remove signals with periods equal to or shorter than inertial (Figure 5.48). These show that at low frequencies the transmissivity (sediment concentration) is unrelated to speed; that is, some of the highest speed flow is associated with clear water. The records were then high-pass filtered so that the high

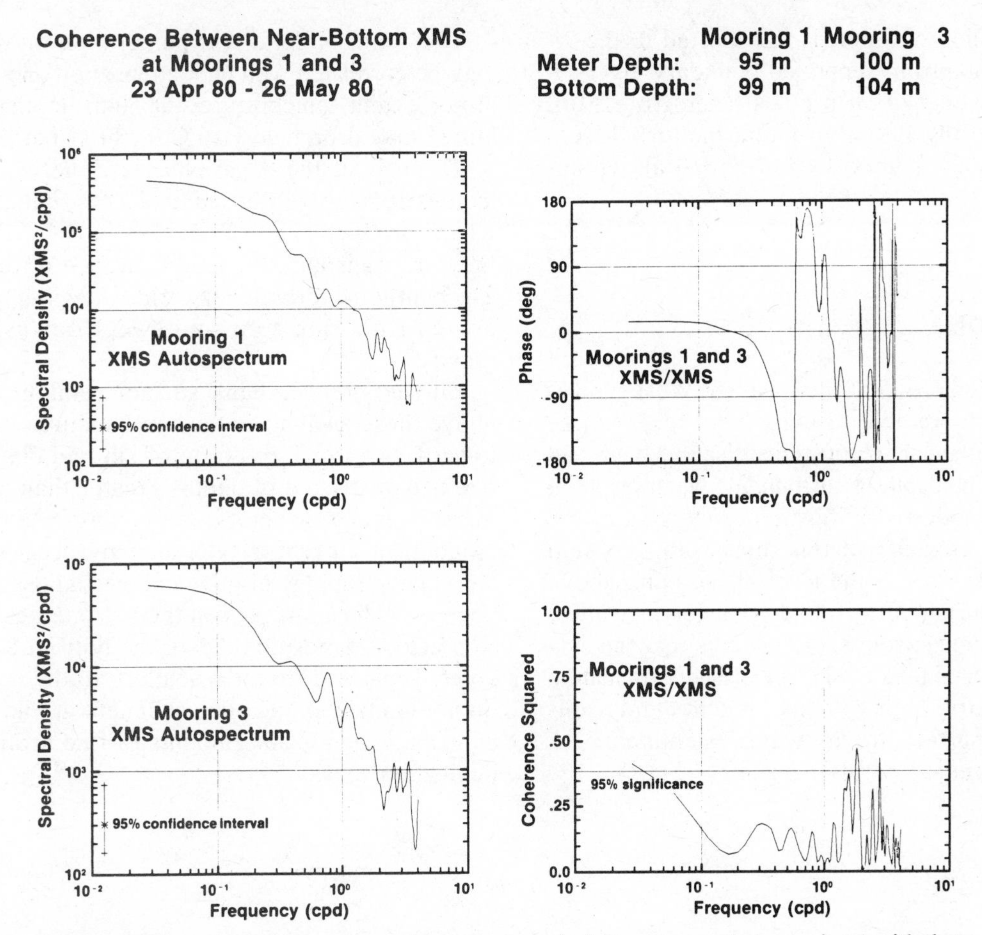

Figure 5.53. Comparison of near-bottom transmissivity measured at mooring 1 with that at mooring 3.

amplitude, high frequency current, and transmissivity signals could be examined (Figures 5.49 and 5.50). These records show that the transmissivity is well correlated with temperature and nearly in phase. Lower temperatures are, in general, also accompanied by lower transmissivities. The transmissivity and velocity components are not quite so well correlated and there is a phase lag between them.

All of this suggests that at 5 m above the bottom changes in transmissivity at high frequencies (inertial and greater) are related to vertical motions associated with tides and inertial oscillations. This may also be true to some extent in the low-frequency signals.

The spectra for the *v* components of velocity and transmissivity from the moorings discussed are shown in Figures 5.51 to 5.53. The *v* components of velocity for the upper meters (Figure 5.51) are coherent in a variety of frequencies, including tidal, inertial, and a broad band encompassing periods of about three to seven days. Notice the strong peaks in those spectra near 1 cpd. The complementary spectra for the lower meters on moorings 1 and 3 (Figure 5.52) are much less coherent, perhaps because the bottom meter at mooring 3 is closer to a bank than the meter at mooring 1. See, for example, how large the signal at 1 cpd is at mooring 3 compared with that at mooring 1. The transmissivity records from these instruments are even more poorly correlated than the *v* components of velocity (Figure 5.53). In fact, they are significantly correlated only at the semidiurnal frequency. Neither record shows the large peak just below 1 cpd which dominates the velocity spectra of both upper meters and the lower meter of mooring 3.

A review of all the moored transmissivity and current meter records indicates that during cold-front passages, southerly flow causes the transmissivity to decrease. This implies that sediment, resuspended by combined action of wind-driven current and waves in shallower water, is transported to the vicinity of the banks in separated boundary layers.

A similar study of all profiles taken from April 1979 through July 1981 shows that the turbid, separated boundary layers do not extend above approximately 75 m.

CONCLUSIONS

The reefs at the crests of the East and West Flower Garden Banks are able to exist there because the banks are bathed in warm, clear, saline water all year round. The depth of surrounding waters is great enough that modest stratification exists year round and the heat capacity of this thick water column keeps temperatures at the level of the bank above 19°C, even during cold frontal passages in winter. The great distance from shore ensures that the surface waters remain relatively free of light-attenuating sediment. Fairly high velocity currents and oscillations from surface gravity waves maintain active circulation over the corals.

The depth of coral reef penetration on the bank may be correlated with light attenuation and to some minor extent temperature. At 50-m depth temperatures may drop below 19°C for brief periods.

Throughout the Algal Nodule Zone current velocities remain high because of orographic effects, which produces large bedforms in the nodules as they are transported over the surface of the bank. The benthonic foraminifers, *Amphistegina*, are winnowed from this zone and swept downslope by gravity.

Currents on the bank surface remain elevated above those away from the bank but are attenuated toward its base. Deposition of silt and clay on the banks is restricted to depths greater than about 80 m because of two factors. Above this level the combination of current acceleration by orographic effects, plus trapping of tidal and inertial oscillations, keeps the shear stresses on the bottom high enough to preclude deposition. Also, the bottom boundary layers separated from the seafloor to the northwest do not contain enough fine sediment at this level to contribute significant amounts of fine sediment by deposition.

6

ZONATION, ABUNDANCE, AND GROWTH OF REEF BUILDERS AT THE FLOWER GARDEN BANKS

HABITAT DESCRIPTION AND ZONATION

The East and West Flower Garden Banks (EFG and WFG), approximately 100 NM southeast of Galveston, Texas, are two of the numerous elevations of the seafloor (banks) in the northwestern Gulf of Mexico produced by intrusion of salt plugs from the Jurassic, Louann evaporite deposits 15 km below the seafloor (Rezak, 1981). Both banks are capped by what are currently considered to be the northernmost thriving tropical coral reefs on the eastern coast of North America. The northern limit of Bahamian reefs is some 20 miles south of the latitude of the Flower Garden Banks. Reefs of the Bermuda Islands ae nearly 300 miles north of the Flower Gardens' latitude but they are situated 570 miles off Cape Hatteras, North Carolina. The ecology of the Bermudian and Bahamian island systems is influenced greatly by the warm Gulf Stream waters that surround them, and like the Flower Gardens these islands harbor elements of the typical Caribbean reef biota.

Within the Gulf of Mexico, the Flower Garden Banks appear to be elements of a discontinuous arc of reefal structures that occur on the continental shelf. Aside from some low diversity reefs on neighboring banks, the coral reefs closest to the Flower Gardens are off Cabo Rojo, about 60 miles south of Tampico, Mexico (Figure 3.1) (Villalobos, 1971). Moore (1958) listed 43 species of Caribbean reef invertebrates from Cabo Rojo, many of which are common at the Flower Gardens. However, certain abundant corals typical of emergent reefs, such as *Acropora palmata* (Elkhorn coral) and *A. cervicornis* (Staghorn coral), do not occur at the submerged Flower Garden reefs. Shallow-water octocorals (sea fans and sea whips), which surprisingly are absent from the Flower Gardens, are present at Cabo Rojo and other reefs several miles south near Isla de Lobos (Chamberlain, 1966; Rigby and McIntyre, 1966). Reportedly, octocorals are abundant on Alacran reef (Kornicker et al., 1959) and other reefs on the Yucatan continental shelf. Coral reefs off the city of Veracruz were reported by Heilprin (1890), and Logan (1969) described the physiography of all reefs and hard banks on the Yucatan Shelf in some detail (Figure 3.1).

Coral reefs occur in the Tortugas and Florida

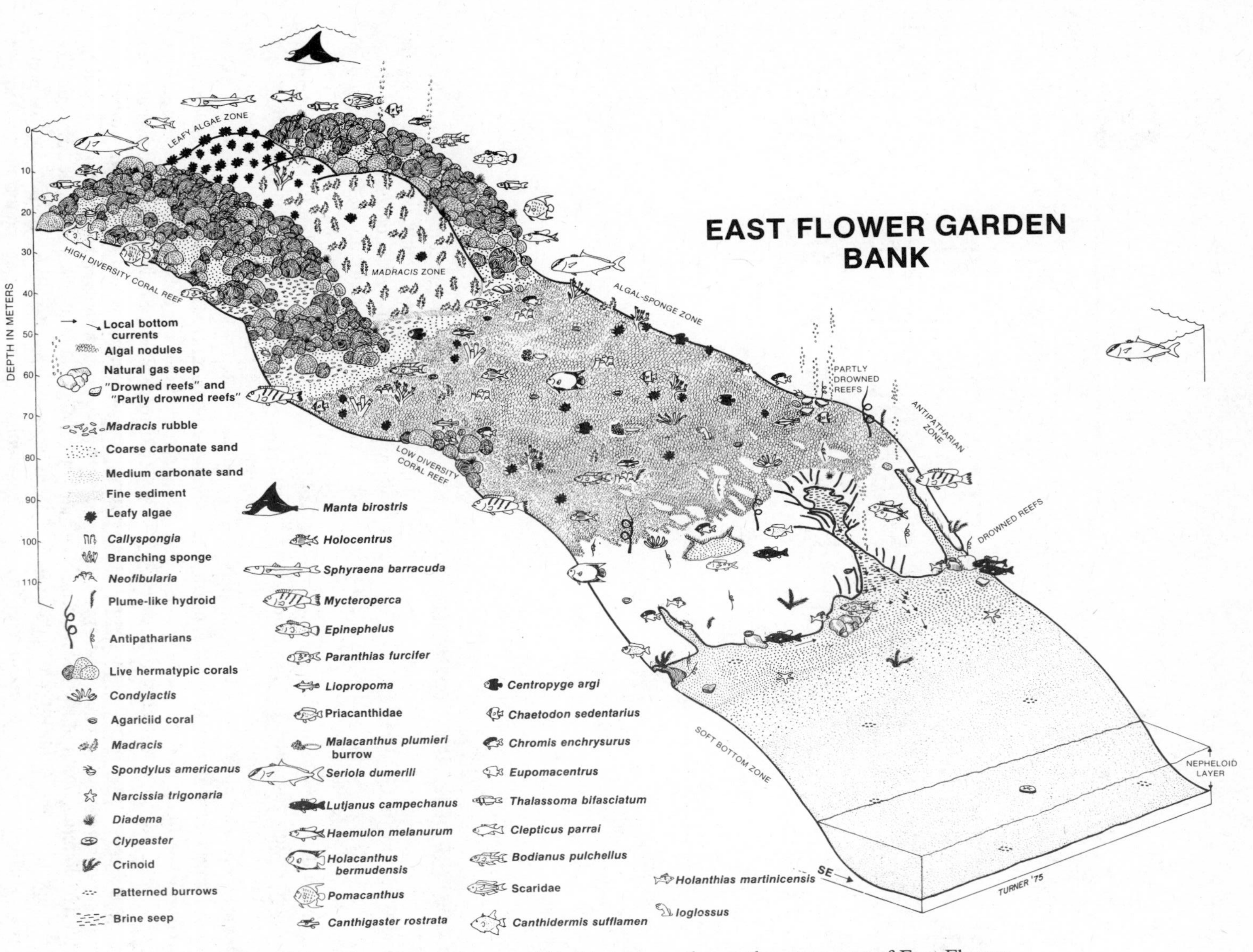

Figure 6.1. Conspicuous biota and biotic zones on the southeastern part of East Flower Garden Bank. High-diversity coral reefs = *Diploria-Montastrea-Porites* Zone. Low-diversity coral reefs = *Stephanocoenia-Millepora* Zone. This is also representative of the West Flower Garden Bank. From Bright, Kraemer, Minnery, and Viada (1984) by permission of *Bulletin of Marine Science*.

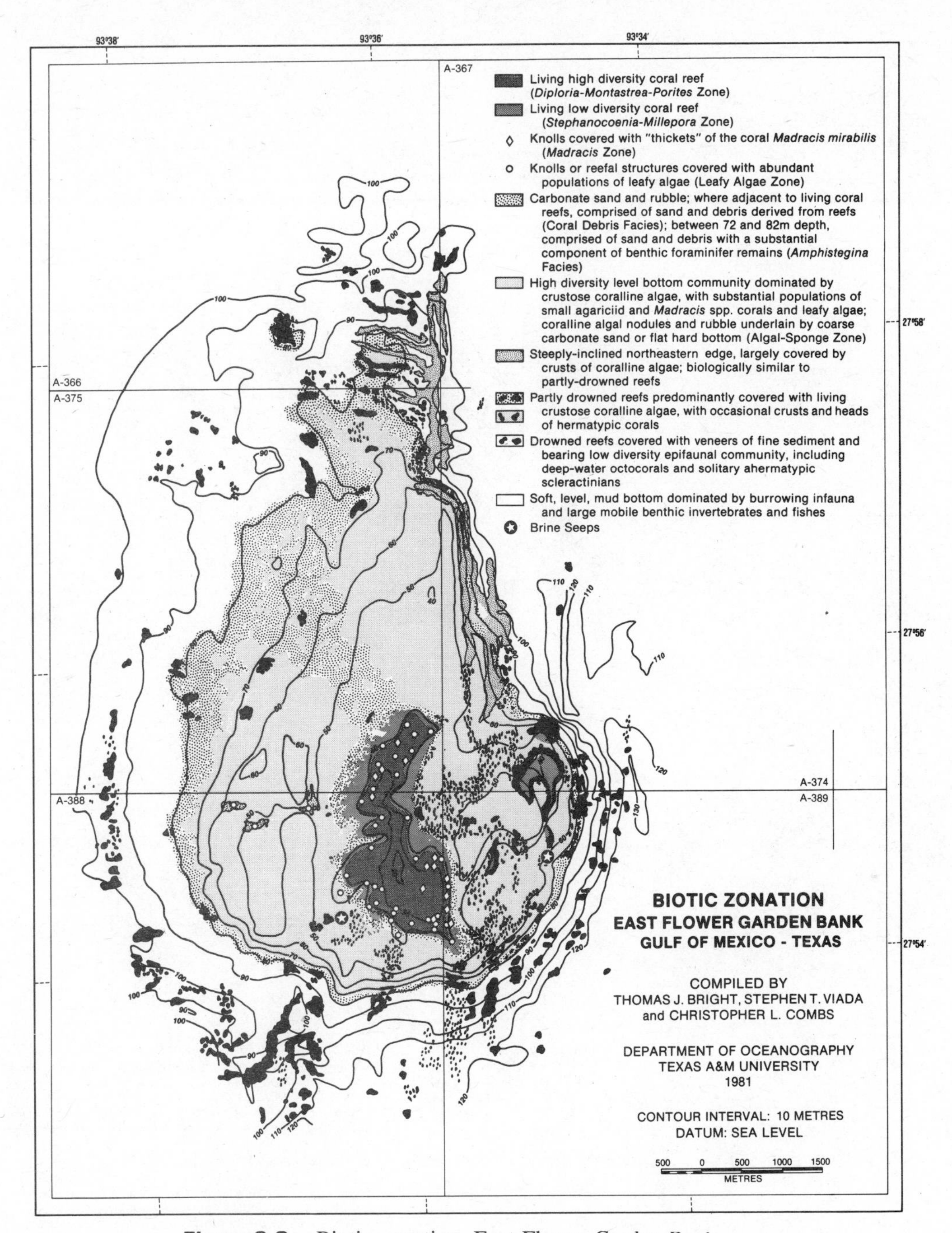

Figure 6.2. Biotic zonation, East Flower Garden Bank.

Keys in the Eastern Gulf of Mexico. Elements of the Caribbean reef biota occupy hard bottoms and "patch reefs" from Tampa Bay to Sanibel Island on the West Florida Shelf (Joyce and Williams, 1969; Smith, 1976). Jordan (1952) described aspects of biota from the Florida Middle Ground, approximately 300 square miles of reef formations off Appalachicola Bay, Florida. Grimm and Hopkins (1977) indicate octocoral predominance at the Florida Middle Ground above 28 m, with dominance shifting to the hermatypic corals between 28 and 36 m. Zonation at the Florida Middle Ground differs considerably from that of the aforementioned coral reefs, even though the dominant organisms are components of the Caribbean biota.

The Gulf of Mexico is therefore ringed by a com-

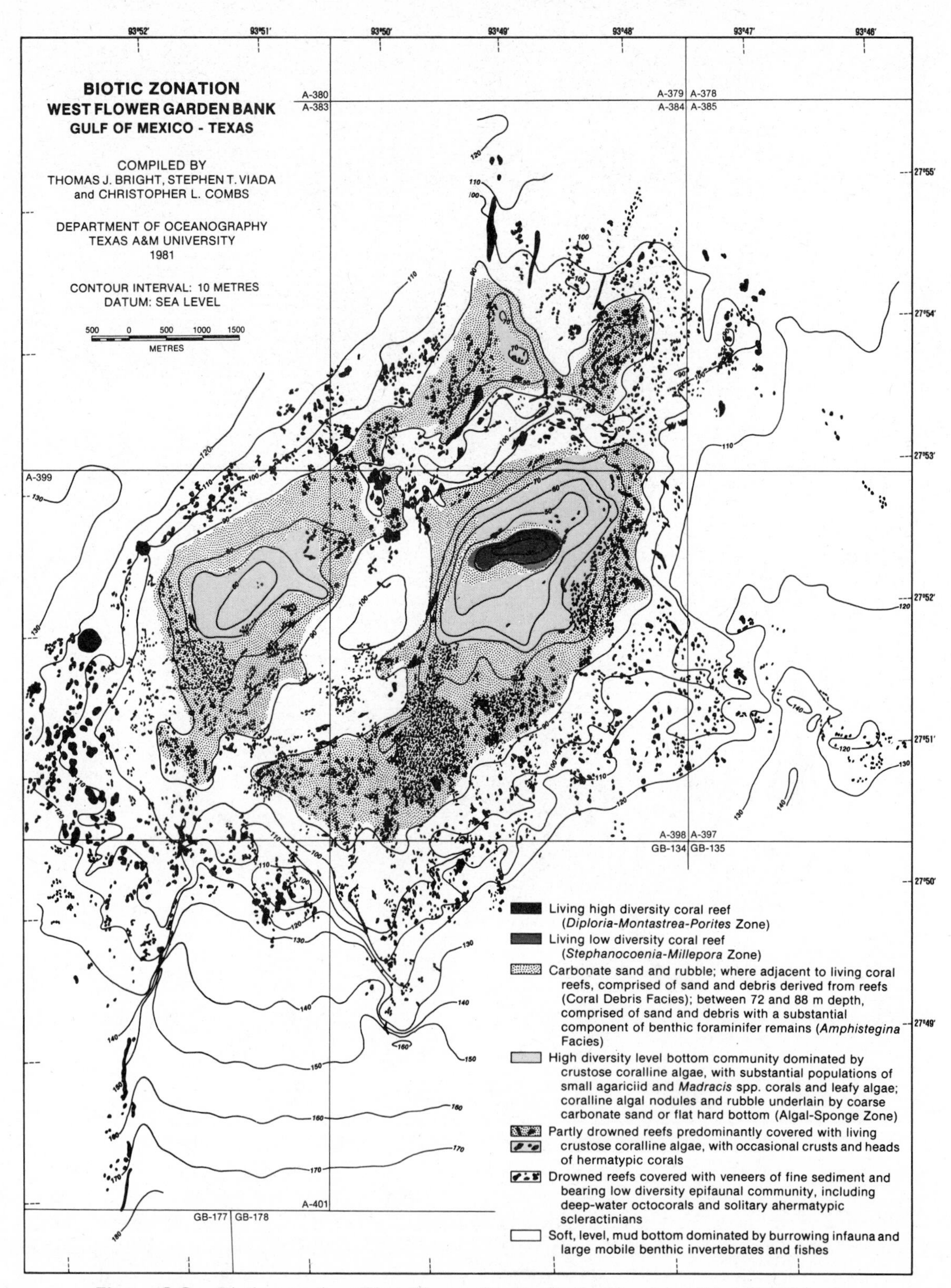

Figure 6.3. Biotic zonation, West Flower Garden Bank. The three "dots" within the 60-m contour on the westernmost knoll represent low-diversity coral reefs.

bination of thriving coral reefs and scattered hard banks and patches that bear elements of the Caribbean reef biota. To what extent the Caribbean biota are represented at the Flower Garden Banks with respect to all taxonomic groups is not yet known. Bright and Pequegnat (1974) reported 253 invertebrate species and 103 fishes from the West Flower Garden. Bright and Rezak (1978a) listed more than 30 species of benthic algae from the East Flower Garden. The list of species has been substantially expanded by additional research sponsored by the U.S. Bureau of Land Management (Bright and Rezak, 1976; Bright and Rezak, 1978b) (See Appendix I).

It is not surprising that thriving reefs occur at the locality of the Flower Garden Banks, where environmental conditions are consistent with the existence of coral reefs (Stoddart, 1969). Positioned at the edge of the continental shelf, with high vertical relief and surrounding depths of 100 to 140 m, the upper portions of the banks are exposed almost continually to tropical-subtropical, clear, oceanic water. In comparison to waters closer to shore, clarity is exceptional. Between the surface and 70-m depth, transmissivity is generally above 75%/m for white light (McGrail et al., 1982a) and sunlight penetration is substantial. Salinities typically remain above 35 ppt but sometimes can be as low as 32 ppt at the surface and 34 ppt at the top of the bank (20-m depth). Above 25-m depth water temperature varies annually between 18 and 32°C (Abbott, 1975; Etter and Cochrane, 1975). Under these conditions the banks have acquired biologically generated carbonate "caps" above approximately 100-m depth, which harbor characteristically Caribbean benthic communities dominated by coralline algae and hermatypic corals (Bright et al., 1974; Bright, 1977; Rezak, 1977).

Biotic zonation is distinct, stringently depth-related, and nearly identical on both banks (Figures 6.1–6.3). The characteristics and limits of biotic zones on the banks have been defined primarily on the basis of direct visual observations and selective sampling performed by using the Texas A&M research submersible DIAPHUS and scientific SCUBA diving. Such techniques result in basically qualitative judgments. In pursuit of objective insight that may reveal subjectively overlooked subtleties in zonation, we abstracted and analyzed a quasi-quantitative compilation of algae and invertebrate occurrence and abundance observsations taken from videotaped records of all submersible transects. Raw abundance and depth-of-occurrence values were derived from recorded sightings of the various taxa (except fishes). Each sighting was weighted (ranked 1, 5, 10, or 20), according to the observer's (Bright's) assessment of taxon abundance made at the time. Raw abundances per taxon were then combined within 5-m depth intervals for each bank and divided by total abundance of all taxa combined within the depth interval. In this way data were converted to relative abundance of each taxon within each depth interval at each bank.

Bray–Curtis cluster techniques were then used to explore the relations between the 5-m depth intervals. The attributes used were taxon type and square root of relative abundance within intervals. Cluster results (Figures 6.4 and 6.5) indicate strong segregation of taxon type and abundance by depth, which implies logical, depth-related biotic zonation on the banks.

The cluster analyses agree with and confirm the authors' qualitative definition of benthic communities and zones at the Flower Gardens. The direct qualitative observations, however, after all provide a more complete and reliable basis on which to consider the subtleties of variation in community structure and distribution on the banks.

Coral Reefs

Submerged coral reefs constitute the shallowest and largest reefal structures on the East and West Flower Garden Banks and occupy the crests of the banks down to 52-m depth in places (Figures 6.1–6.3). The main reef tops generally vary from 18 to 28 m, but 15-m depths are common and an 11-m depth has been encountered at the East Flower Garden (top of a large coral head). The reefs are made up of closely spaced or crowded coral heads up to 3 m in diameter and height (Figures 6.6 and 6.7). "Patches" of sand or carbonate gravel occur among the frequently cavernous coral heads, which show evidence of substantial internal and surficial bioerosion.

Two biotic zones are recognizable on the coral reefs: a high-diversity assemblage (16 hermatypic coral species) limited to depths of less than 36 m (*Diploria-Montastrea-Porites* Zone), and a comparatively low-diversity assemblage (approximately 12 hermatypic coral species) between 36 and 52 m (*Stephanocoenia-Millepora* Zone) (Table 6.1).* The

*Tables 6.1 to 6.7 from Bright, Kraemer, Minnery, and Viada (1984) by permission of *Bulletin of Marine Science*.

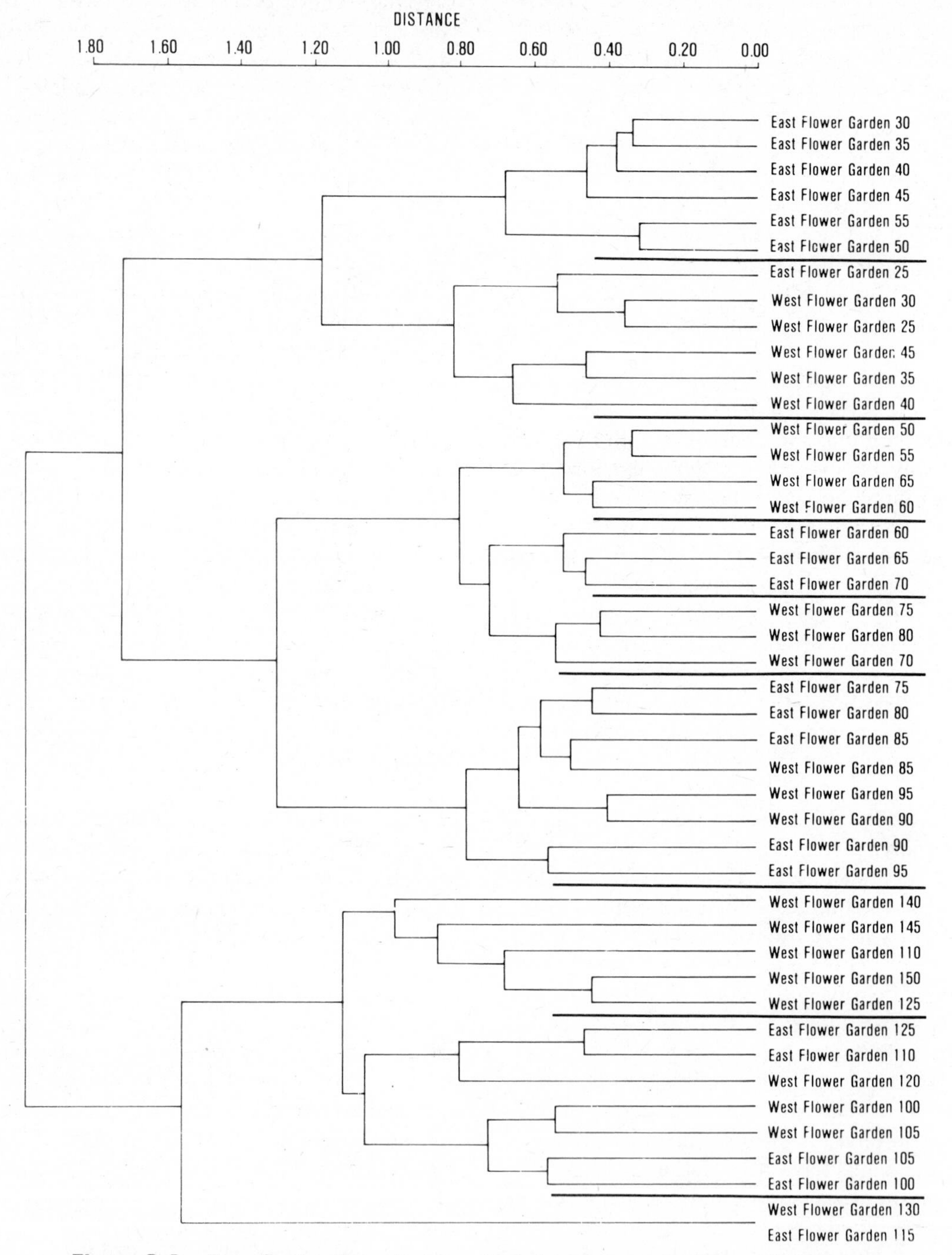

Figure 6.4. Bray-Curtis cluster for 5-m depth intervals at the East and West Flower Garden Banks together. Attributes are relative square roots of abundance for invertebrates and algae. The deepest depth in each 5-m interval is indicated for each entity in the clusters (i.e., "West Flower Garden 30" = the depth interval 26 to 30 m at the West Flower Garden Bank).

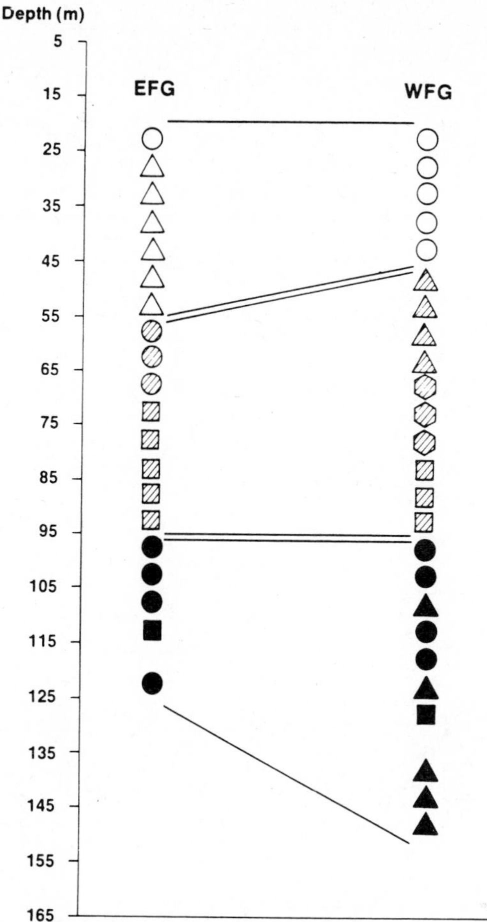

Figure 6.5. Depth-zone relationships derived from interpretation of Bray–Curtis clusters (Figure 6.4) for East and West Flower Garden invertebrates and algae. Horizontal and diagonal lines separate major biotic zones as defined by the custer analysis. Identically shaped geometric figures (circles, squares, etc.) indicate locations and depth ranges that clustered together (were most similar in terms of species composition and abundance) within the major zones.

terms "high diversity" and "low diversity" as used here relate to the northwestern Gulf of Mexico only and pertain to coral diversity. Coral diversities on reefs in other parts of the Caribbean and southern Gulf of Mexico are substantially greater.

Diploria-Montastrea-Porites Zone. More is known of the *Diploria-Montastrea-Porites* community than any other at the Flower Garden Banks because it is accessible to research divers using SCUBA. Edwards (1971) considered that all coral reefs at the West Flower Garden belonged to this zone, thereby implying an hierarchy of coral dominance (percent cover) similar to that described by

Figure 6.6. Typical Flower Garden coral-reef top (21-m depth).

Figure 6.7. Brain corals and reef fishes (*Chromis* sp.; *Bodianus rufus*, Spanish hogfish; *Thalassoma bifasciatum*, Bluehead) at 20-m depth on the West Flower Garden).

Logan (1969) for submerged reefs on the Yucatan Shelf, southwestern Gulf of Mexico. However, subsequent studies by Bright et al. (1974, 1984), Tresslar (1974), Viada (1980), and Kraemer (1982) show conclusively that *Montastrea annularis* is the dominant coral, followed by *Diploria strigosa, Montastrea cavernosa, Colpophyllia* spp., and *Porites astreoides*. Convention should therefore dictate a change in zonal designation for high-diversity reefs at the Flower Garden Banks to *Montastrea-Diploria* Zone to reflect the true order of coral dominance above 36 m. For convenience, however, the older designation is retained.

Crustose coralline algae are abundant on the high-diversity reefs and add substantial amounts of calcium carbonate to the reef substratum. Standing

TABLE 6.1. Flower Garden Reef Builders[a]

Reef Builders	Depths of Collection and/or Observation	Abundance			
		DMP	Mad	SM	AS
Red Calcareous Algae					
Corallinaceae					
Porolithon	23–32	**	*		
Hydrolithon	23–65	****	**	?	*
Archaeolithothamnium	23–72	*	*	?	*
Lithophyllum	23–80	**	**	?	**
Lithoporella	23–85	*	*	?	***
Tenarea	23–90+	**	**	?	***
Lithothamnium	23–90+	*	*	?	****
Mesophyllum	15–86+	*	*	?	**
Fosliella?	23	*		?	*
Squamariaceae					
Peyssonnelia	23–90	**	**	?	***
Green Calcareous Algae					
Codiaceae					
Halimeda	21–91	*	?		**
Halimeda tuna	48–61				*
Udotea	40–64	?			*
Udotea cyathiformis	58				*
Foraminiferans					
Gypsina plana	21–68	**	****	?	***
Corals					
Astrocoenidae					
Stephanocoenia michelini	21–52	*		***	
Pocilloporidae					
Madracis	15–92	***			**
Madracis asperula	48–84				**
Madracis decactis	15–41	***			
‡*Madracis* cf. *formosa*	62				*
Madracis mirabilis	23–40		****		
‡*Madracis myriaster*	113				*
Agariciidae					
‡*Agariciidae* (saucer-shaped)	18–82	*		*	***
Agaricia	15–76	**		**	***
Agaricia agaricites	20–24	*		?	**
‡*Agaricia fragilis?*	20–53	*		?	*
Helioseris cucullata	20–84	*		?	**
Siderastreidae					
Siderastrea siderea	21–50	**		*	
Poritidae					
Porites astreoides	21–40	***		*	
Porites furcata	21	*			
Faviidae					
Colopophyllia	21–47	***		**	
Colpophyllia amaranthus	21–26	*		?	
Colpophyllia natans	21–26	**		?	
Diploria strigosa	15–55	***		**	
Montastrea annularis	18–43	****		*	
Montastrea cavernosa	15–60	***		**	*
Mussidae					
Mussa angulosa	21–54	**		*	
‡*Scolymia*	18–46	*		*	
Scolymia cubensis	21–27	*		?	

TABLE 6.1. (Continued)

Reef Builders	Depths of Collection and/or Observation	Abundance DMP	Mad	SM	AS
Milleporidae					
Millepora alcicornis	15–55	***		***	*
Caryophylliidae					
‡*Oxysmilia?*	82–101				
‡*Paracyathus?*	19–?	*		?	?

[a] When information is available, relative abundance is indicated as follows: **** = very abundant; *** = abundant; ** = common; * = present. DMP = *Diploria-Montrastrea-Porites* Zone; Mad = *Madracis* Zone; SM = *Stephanocoenia-Millepora* Zone; AS = Algal-Sponge Zone. Information concerning occurrences in the *Stephanocoenia-Millepora* Zone is limited. Probable presence is indicated by a question mark.

‡*Notes*

1. *Madracis* cf. *formosa*, *M. myriaster*, *Oxysmilia* sp., and *Paracyathus* sp. are not considered by us to be "hermatypic."
2. "Saucer-shaped" Agariciidae are not recognizable to species *in situ*. Collected specimens are *Helioseris cucullata*, *Agaricia agaricites*, and, if the identification is correct (see 3), *Agaricia fragilis*.
3. Two collected specimens; identification as *Agaricia fragilis* uncertain.
4. Possibly only one species, *S. cubensis*.
5. *In situ* sightings only of *Oxysmilia?* sp. *Oxysmilia rotundifolia* is generally abundant on, and has been collected from, deeper parts of shelf-edge carbonate banks in the northwestern Gulf of Mexico. The authors speculate that Flower Garden sightings are this species.
6. One specimen identified by Dr. S. Cairns as "*Paracyathus* sp. or *Polycyathus* sp." *Paracyathus pulchellus* is common on other banks in the northwestern Gulf of Mexico. The closest collected specimen of *P. pulchellus* came from 28 Fathom Bank (8 NM due east of the East Flower Garden Bank) at 76-m depth.

crops of leafy algae on the high-diversity reefs are consistently low, possibly kept so by the grazing activities of fishes and mobile invertebrates, such as gastropods and the urchin *Diadema antillarum**.

The 253 species of reef invertebrates and 103 reef fishes reported in Bright and Pequegnat (1974) were almost all taken from the *Diploria-Montastrea-Porites* Zone at the West Flower Garden. Subsequent studies imply a nearly identical community structure and diversity for the *Diploria-Montastrea-Porites* Zone at the East Flower Garden.

Among the typically caught sport and commercial fishes that frequent the high-diversity coral reefs are several species of grouper and hind, *Mycteroperca* and *Epinephelus;* amberjacks, *Seriola;* Great barracuda, *Sphyraena barracuda;* Red snapper, *Lutjanus campechanus;* Vermilion snapper, *Rhomboplites aurorubens;* Cottonwick, *Haemulon melanurum;* porgy, *Calamus;* and Creole-fish, *Paranthias furcifer*.

Spiny lobsters, *Panulirus argus*, are known to occur on the high-diversity reefs at both Flower Garden Banks and have been seen by the author on several other banks in the northwestern Gulf of Mexico (Sonnier Bank, 18 Fathom Bank, Bright Bank). *Panulirus guttatus* has been seen on the shallow coral reefs (26 m) at the East Flower Garden and probably occurs also at the West Flower Garden. The Shovel-nosed lobster, *Scyllarides aequinoctialis*, is reported from the high-diversity reef at the West Flower Garden and undoubtedly occurs also at the East Flower Garden. These species of lobster are probably widely distributed on the outer continental shelf banks in the northwestern Gulf of Mexico, but nothing is known of the magnitude and dynamics of their regional populations or whether they could support a commercial lobster fishery.

*Within the last 2 years a disease that causes mass mortalities in *Diadema* populations has spread northward from the southern Caribbean. It apparently reached the Flower Gardens between November 1983 and August 1984, killing large numbers of the previously abundant *Diadema antillarum*.

Stephanocoenia-Millepora Zone. Between 36- and 38-m depth at both banks a transition is apparent from the *Diploria-Montastrea-Porites* assemblage to a reef zone of lower diversity which extends generally down to 46 m, with components to 52 m. Among the 12 varieties of hermatypic corals known from the zone 8 are particularly conspicuous: *Stephanocoenia michelini, Millepora* sp., *Montastrea cavernosa, Colpophyllia* spp., *Diploria* sp., *Agaricia* spp., *Mussa angulosa*, and *Scolymia* sp., probably in that relative order of abundance. Therefore the designation *Stephanocoenia-Millepora* Zone is appropriate (Figures 6.1 to 6.3 and 6.8).

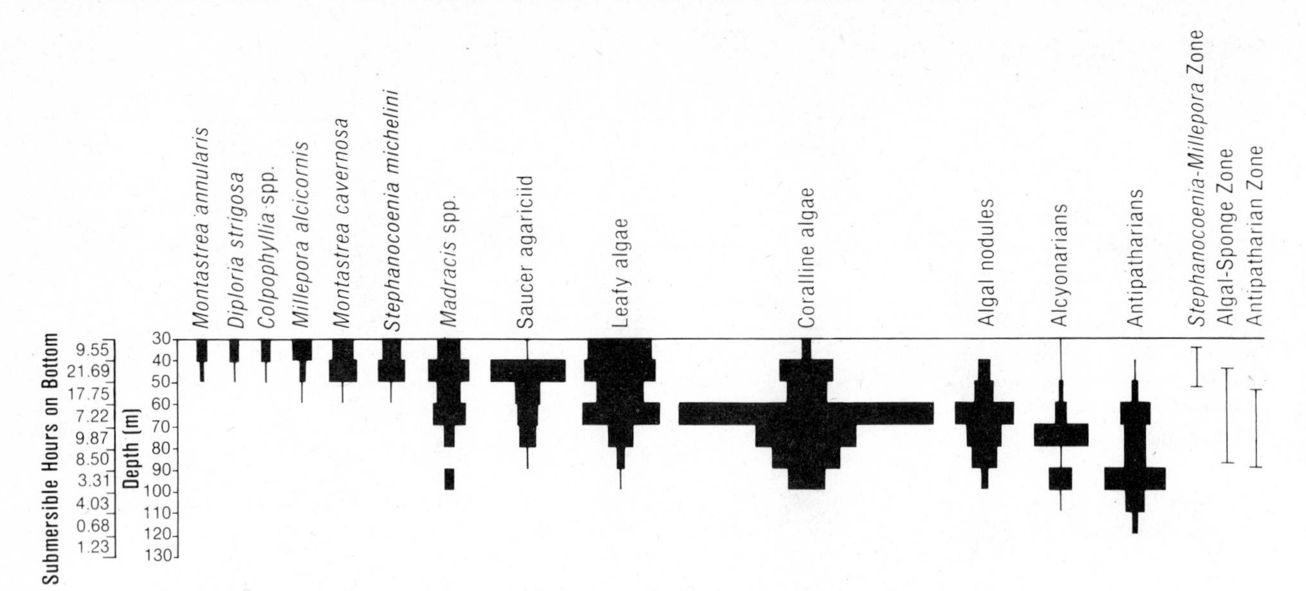

Figure 6.8. Relative abundance of benthic algae, stony corals, alcyonarians, and antipatharians below 30-m depth (high-diversity coral reefs excluded). Based on a quasi-quantitative compilation and ranking of observations made from the Texas A&M Research Submersible DIAPHUS from 1974 through 1980. Raw abundance values were derived from recorded sightings of the various taxa, each sighting was weighted (ranked 1, 5, 10, or 20) according to the observer's (Bright's) assessment of taxon abundance made at the time. Raw abundance values from the East and West Flower Gardens were combined within 10-m depth intervals and then divided by the total number of hours spent on the bottom making the observations within each depth interval (data were standardized to abundance per hour of observation). Results adequately describe the depth-abundance relationships within taxa. Between-taxon relationships are more approximate but representative.

Population levels of these corals have not been quantitatively determined but visual observations indicate considerably lower total live coral cover than in the *Diploria-Montastrea-Porites* Zone and a great deal of variation in percentage of cover and relative abundance from place to place. Crustose coralline algae are substantially more conspicuous in the *Stephanocoenia-Millepora* Zone and apparently are the predominant encrusting forms that occupy dead coral reef rock.

Little is known of the assemblage of organisms inhabiting the *Stephanocoenia-Millepora* Zone. The reef-fish populations appear less diverse than in the *Diploria-Montastrea-Porites* Zone. Population density of the black urchin, *Diadema antillarum,* which is a significant bioeroder of reef rock, may be similar in both zones. Exceptional numbers of the American thorny oyster, *Spondylus americanus,* have been seen in the *Stephanocoenia-Millepora* Zone.

Successional relationships, if any, between the shallower, high-diversity *Diploria-Montastrea-Porites* reefs and deeper, low-diversity *Stephanocoenia-Millepora* reefs at the Flower Garden Banks cannot be determined from existing information. The low-diversity reefs could represent depauperate remnants of high-diversity reefs which have been displaced downward to a habitat too deep to support a majority of the contemporary coral species. In this case much of the dead reef rock on which the corals of the low-diversity reefs now grow would be composed of the remains of coral species now living on the high-diversity reefs. Conversely, contemporary high-diversity reefs may have developed on the tops of low-diversity reefs or other local topographic mounds (possibly accumulations of the coral *Madracis mirabilis)* whose crests achieved a depth above which high coral diversity and rapid coral growth are possible. To help resolve these questions more information is needed concerning the nature of the reef rock in both zones, rates of deposition of carbonate rock and sediment, and vertical movements of the substratum and sea level in relation to each other in the last several thousand years.

Leafy Algae and Madracis Zones

On peripheral parts of the main reefal structure between 28- and 46-m depth at the East Flower Gar-

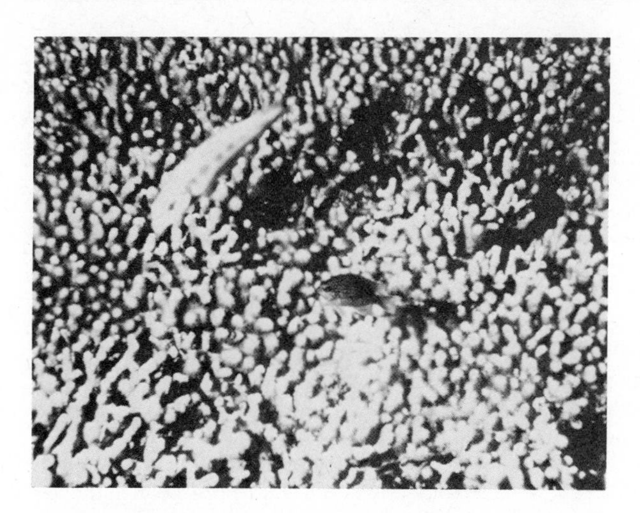

Figure 6.9. *Madracis mirabilis,* dominant coral of *Madracis* zone (28-m depth, East Flower Garden).

Figure 6.10. Leafy algae zone (28-m depth, East Flower Garden).

den, large knolls occur which generally lack significant populations of massive head corals. Certain of these knolls are overwhelmingly dominated by the small branching coral *Madracis mirabilis (Madracis* Zone) (Figure 6.9). Other knolls are covered by lush assemblages of leafy algae (Leafy Algae Zone) including species of *Stypopodium, Caulerpa, Dictyota, Chaetomorpha, Lobophora, Rhodymenia, Valonia,* and *Codium* (Figure 6.10). None of these assemblages resembles the adjacent coral-reef community in structure, although it is likely that most or all of the species found on the knolls also occur on the reefs.

The steeper margins of both main reefs (north, east, and south reef faces) are underlain by thick deposits of coral gravel made up of skeletal remains of *Madracis mirabilis*. Some of these gravel deposits vaguely resemble large, rounded "spurs and grooves." Others are blufflike but many simply appear to be mounds at the reef margins. Living populations of *Madracis mirabilis* occur only at these reef margins, atop the gravel deposits, on which, standing as thickets a few centimeters high, they are the predominant organisms.

Typically, living *Madracis mirabilis* thickets are accompanied by conspicuous populations of leafy algae and sponges. Elsewhere the tops of *Madracis* gravel mounds are overgrown by coralline algae, which form a stabilizing, rindlike surface crust beneath which the loose gravel lies. In the grooves and valleys between spurs, hills, or mounds *Madracis* gravel is frequently shifted by water movements and thrown into large ripple marks, presumably with a net transport of gravel downward and off the reef.

Examples of progressive invasion of the tops of the *Madracis* gravel mounds by hermatypic corals from the high-diversity reefs are common. Some mounds harbor a few small heads of *Porites, Agaricia, Diploria* and other corals superimposed sparsely on the *Madracis* Zone assemblage. Other mounds support larger but scattered heads. Closely spaced heads on the north, east, and south faces of the high-diversity reefs *(Diploria-Montastrea-Porites* Zone) are typically separated by shifting *Madracis* gravel rather than coarse sand like most of the reef top.

Thus it is apparent that, at least in places on the steeper faces of the main reefs, the substratum has been built upward and outward initially by deposition of *Madracis* remains (a massive, roadcut-like anchor scar on the northeastern face of the West Flower Garden revealed a *Madracis* gravel thickness of at least 15 m). The *Madracis* deposits serve as shallow substratum suitable for occupation by the 16 hermatypic coral species of the *Diploria-Montastrea-Porites* Zone. This successional relationship between the *Madracis* Zone and *Diploria-Montastrea-Porites* Zone probably facilitates the slow spreading of the main reefs onto the adjacent Algal-Sponge Zone.

Algal-Sponge Zone

Although most zones overlap or grade into one another to some extent, they are nevetheless recognizable. Their depth ranges correlate consistently with those of known sedimentological facies (Figure 6.11). The Algal-Sponge Zone includes a number of biotope types that occur between 46 and 82 m at the East Flower Grden and 46 and 88 m at the West Flower Garden. Coarse carbonate sand and gravel

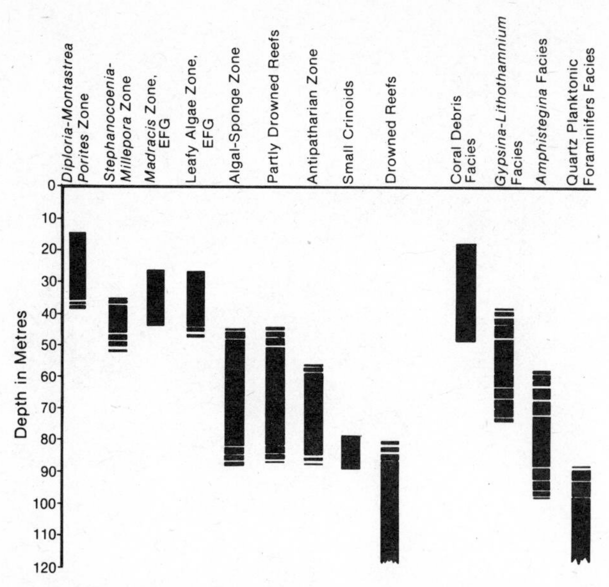

Figure 6.11. Depth ranges of recognizable biotic zones and sedimentary facies. Broken bars indicate varying upper and lower depth extremities for zones, depending on locations.

surrounding the living coral reefs (Coral Debris Facies) mark a geobiological transition between the coral reefs and the surrounding platform, which is largely covered with carbonate sand, gravel, nodules, and partly drowned reefal structures. In general, the Algal-Sponge Zone is spatially coincident with this platform (Figures 6.1, 6.2, and 6.3).

On both banks large areas on which nodules predominate constitute the *Gypsina-Lithothamnium* sedimentological facies. The most important contemporary producers of carbonate nodules and crusts on gravel and reefal structures are the coralline algae. The nodules are typically referred to as rhodoliths or algal nodules (Figure 6.12). Coralline algae are also the overwhelmingly dominant living organisms of the Algal-Sponge Zone.

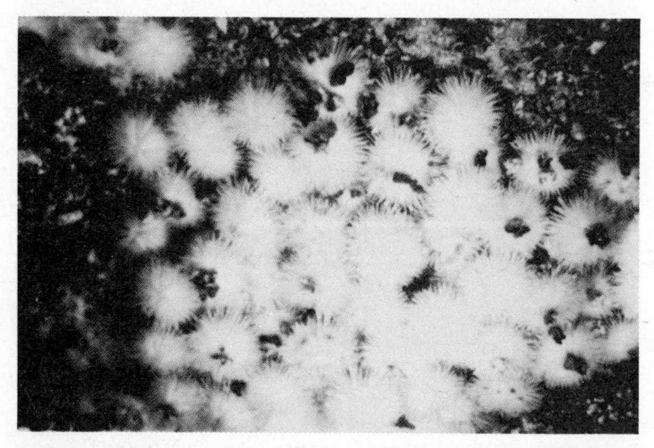

Figure 6.12. Aggregation of white urchins, *Pseudoboletia maculata,* at 59-m depth on bottom covered with small nodules. Note the nodules displayed on spines of urchins.

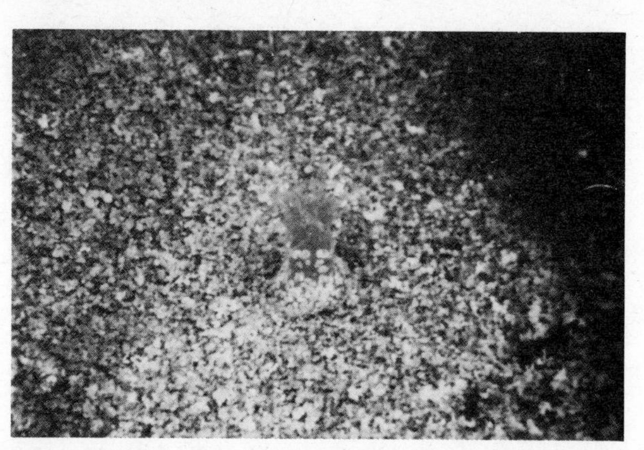

Figure 6.13. Patch of living *Halimeda* sp. at about 55-m depth, East Flower Garden. Large organism in the center is an anemone.

The algal nodules range in size from less than 1 cm to 10 cm or more and in most places cover 50 to 80% of the bottom. They create a biotope that harbors an infaunal and epifaunal community which is probably comparable in diversity to the living coral reefs. In addition to coralline algae and the encrusting foraminifer, *Gypsina,* the nodules themselves house an abundance of boring species and attached epibenthos (Abbott, 1975). Numerous mobile invertebrates and small fishes find shelter under, between, and within the nodules. Beneath the nodules, coarse carbonate sand contains active soft-bottom infaunal populations, as evidenced by the presence of numerous burrows.

Most of the leafy algae at both banks occur among the algal nodules and on reefal structures within the Algal-Sponge Zone (the aforementioned algae-covered knolls, Leafy Algae Zone, in shallower water are comparatively small in area). Leafy algae are pervasive among the nodules and on hard surfaces within the Algal-Sponge Zone but are neither uniformly distributed nor uniformly abundant from place to place. At certain locations lush growths of *Stypopodium, Peyssonnelia,* or *Lobophora* may obscure all else on the bottom. The highly productive and rapidly renewable benthic algae populations must furnish substantial amounts of food to the surrounding communities.

Calcareous green algae, *Halimeda* and *Udotea,* contribute to sediment production within the Algal-Sponge Zone (Figure 6.13). Patches in excess of 10-m diameter composed almost exclusively of *Hali-*

meda have been seen adjacent to algal nodules at both banks. These patches are apparently long-lived, semipermanent features because the platelike remains of dead *Halimeda* extend at least several centimeters into the substratum. *Halimeda* also occurs as individual plants among the nodules and on reefal structures.

Several species of hermatypic corals are abundant enough among the algal nodules to be considered major sediment producers within the Algal-Sponge Zone. Saucerlike colonies of *Helioseris cucullata* and *Agaricia* are pervasive but unevenly distributed, with populations varying from less than 1 to more than 10 colonies per square meter. Several small species of *Madracis* likewise occur with varying abundance among the algal nodules. Colonies of *Montastrea cavernosa* have been encountered occasionally in the Algal-Sponge Zone, where, with increasing depth, they tend toward a flat encrusting growth form, presumably in response to decreased light.

Deep-water alcyonarians, primarily Ellisellidae and Paramuriceidae, are abundant in the lower Algal-Sponge Zone. Large, white, coiled, antipatharian whips of the genus *Cirrhipathes* (= *Cirripathes)* and bushy colonies of *Antipathes* are also present in the deeper parts of the zone.

Among the various species of sponges that are conspicuous and abundant within the Algal-Sponge Zone, *Neofibularia nolitangere* is most distinctive. Crusts of this sponge, a meter or so in diameter, occur on nodules, sand, or rock within the zone. Fishes and mobile invertebrates are attracted to the sponge, swimming or crawling among its chimney-like spires.

Populations of echinoderms within the Algal-Sponge Zone must add significantly to the carbonate substratum. Sizable comatulid crinoids and a number of asteroid species are to be found everywhere on the banks except on the living coral reefs. A particularly large population of the asteroid starfish *Linckia nodosa* and great numbers of the urchins *Pseudoboletia maculata* (Figure 6.12) and *Arbacia punctulata* were seen on the algal nodules and reefal structures of the platform west of the main reef at the East Flower Garden. It is interesting that similar concentrations of these particular asteroids and echinoids have not been seen on other parts of either bank.

Small gastropods and pelecypods are abundant on and among the nodules, and gastropod shells are known to be the nuclei around which some of the nodules are formed. The largest abundant pelecypod in the Algal-Sponge Zone is the American thorny oyster, *Spondylus americanus,* which is attached to nodules as well as reefs. Its distribution, as with many of the conspicuous organisms in the zone, appears to be irregular and locally contagious (patchy) and results in a high degree of lateral variation in population levels. Small clumps (approximately ½ m in diameter) of worm-shell gastropods, *Siliquaria,* occur infrequently among the nodules and on sand bottoms within the Algal-Sponge Zone. Their contribution to the carbonate sediment is probably minor. The coiled tubes of this species sometimes occur in masses embedded in the sponge *Chelotropella.*

Obviously many species of plants and animals associated with algal nodules within the Algal-Sponge Zone are involved to varied extents in the frame-building processes. Their successful effort in this respect is probably the dynamic factor on which the stability of benthic community structure within the zone depends.

Small Yellowtail reeffishes, *Chromis enchrysurus,* are the most abundant of the conspicuous fishes that congregate around irregularities in the Algal–Sponge Zone. Conical burrows 1 m across and ½ m deep, produced by the Sand tilefish, *Malacanthus plumieri,* are scattered about the zone from the base of the main coral reef down to at least 70 m. Other particularly characteristic fishes among the algal nodules are the small Cherubfish, *Centropyge argi,* and Orangeback bass, *Serranus annularis.*

Partly Drowned and Drowned Reefs

The nature of both types of living coral reef has been discussed. As indicated, uncertainty exists concerning the successional relationship, if any, between the high- and low-diversity coral reefs. The developmental history of the deeper, and presumably older, partly drowned and drowned reefs may be even more complex.

The biota now occupying partly drowned and drowned reefs at the Flower Garden Banks probably do not reflect the history of the reefs so much as they do contemporary environmental limitations on depth distributions of the organisms. Accordingly, drowned reefs are defined as those reefal structures now existing at depths too great for hermatypic corals to exist and where crustose coralline algal populations are insignificant (below 82 m at the EFG, below 88 m at the WFG). Partly drowned reefs are those reefal structures covered primarily by living crusts of coralline algae, with occasional crusts and heads of hermatypic corals. They exist now at depths

Figure 6.14. Crusts of coralline algae on hard substratum of "partly drowned reef" at the East Flower Garden (60-m depth).

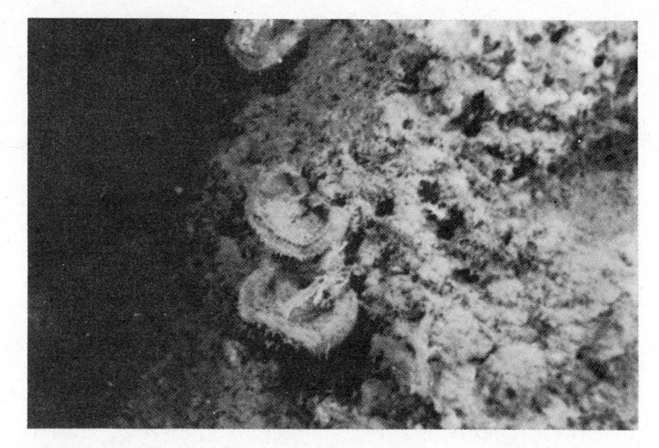

Figure 6.15. Hard sponges attached to carbonate rock coated with veneer of sediment. This is representative of drowned reefs (105 m, Diaphus Bank).

in which hermatypic corals have limited capabilities for growth and the predominance of crustose coralline algae is favored (46 to 82 m at the EFG, 46 to 88 m at the WFG).

Not unexpectedly, partly drowned reefs are generally restricted to the Algal-Sponge Zone, of which they compose a major biotope component. They bear crusts dominated by coralline algae, accompanied by other sessile organisms that are also typical of the algal nodules (Figure 6.14). In addition, they house large anemones such as *Condylactis gigantea* and *Lebrunia danae,* an abundance of large comatulid crinoids, occasional basket stars, limited crusts of the hydrozoan coral *Millepora,* and, infrequently, small colonies of the hermatypic corals *Agaricia, Helioseris cucullata, Montastrea cavernosa,* and *Stephanocoenia michelini.*

The partly drowned reefs attract a number of fish species that also occur on the living coral reefs. Most of these "expatriate" reef fishes are found consistently on similar structures at other banks in the northwestern Gulf that do not support living coral reefs. Therefore they may not necessarily be recruited from the shallower coral reef fish assemblage at the Flower Garden Banks. The most abundant fish to frequent the partly drowned reefs is the small Yellowtail reeffish, *Chromis enchrysurus,* which is not often seen on the high-diversity coral reefs.

Drowned reefs occur below approximately 82 m at the East Flower Garden and 88 m at the West Flower Garden, where coralline algae do not thrive and hermatypic corals are absent. Comatulid crinoids, small deep-water octocoral whips, Ellisellidae; octocoral fans, Paramuriceidae; antipatharians, *Cirrhipathes, Antipathes;* encrusting sponges; and solitary ahermatypic corals are the most conspicuous attached organisms (Figure 6.15).

Among the fishes that frequent drowned reefs are Red snappers, *Lutjanus campechanus;* Spanish flag, *Gonioplectrus hispanus;* Snowy grouper, *Epinephelus niveatus;* Bank butterflyfish, *Chaetodon aya;* scorpionfishes, Scorpaenidae; and, most characteristically, the Roughtongue bass, *Holanthias martinicensis.* The snappers are highly mobile schooling fish that congregate around reefal structures at all depths on the banks but seem to prefer the deeper "drop-offs" and bank-edge features and show little affinity for the high-diversity reef tops. The other fishes listed are commonly found only on the drowned and deepest partly drowned reefs. The drowned-reef ichthyofauna is therefore substantially different in basic species composition and of much lower diversity than that of the partly drowned or coral reefs.

Drowned reefs at the Flower Garden Banks exist in comparatively turbid water and are generally covered with veneers of fine sediment, the veneers being thicker on the deeper reefs. Light penetration, water turbidity, sedimentation, and temperature are probably the most important factors controlling the present distribution of hermatypic corals and coralline algae on the banks. It is suspected that, were it not for the chronically turbid bottom water and sedimentation around the peripheries of the banks, the living algal nodules, partly drowned reefs, and other elements of the Algal-Sponge Zones would extend to slightly greater depths, as they do on certain other banks farther offshore, where the surrounding soft bottoms are deeper (Rezak and Bright, 1981a, Vol. 4).

Transition Zones

White, bedspring-shaped antipatharian whips, *Cirrhipathes,* occur from 52- to more than 90-m depth and, where they are most abundant (generally around 80 to 90 m), mark a transition between biotic assemblages that exhibit distinct shallow-water affinities (leafy algae, abundant coralline algae, hermatypic corals, sizable shallow water reef fish populations) and those that are deep-water oriented. The upper parts of this supposed ''Antipatharian Zone'' blend with the Algal-Sponge Zone and it is impossible to find any sharp demarcation between the two. We might just as well speak of a lower Algal-Sponge Zone with a sizable antipatharian population (Figures 6.8 and 6.11).

Deeper parts of the ''Antipatharian Zone'' (over 80 m) are recognizably less diverse and are characterized by antipatharians, comatulid crinoids, few if any leafy algae, thin to sparse populations of coralline algae, and a distinctly limited fish fauna which includes *Holanthias martinicensis, Bodianus pulchellus, Chromis enchrysurus, Chaetodon sedentarius, Holacanthus bermudensis,* and a few others.

Between 73 and 78 m the nature of the bottom changes, usually rather abruptly, from algal nodules and crusts to a soft, level bottom of mixed coarse calcareous sand, with an abundance of nonliving foraminifera (*Amphistegina)* tests and fine silt- to clay-sized particles which are easily stirred up and remain in suspension for a long while. *Amphistegina* is known to live on the algal nodules at both banks. Nonliving tests of this protozoan account for a large portion of the *Amphistegina* sands that occur on both banks (Figures 6.2 and 6.3). Presumably the remains of spent *Amphistegina* from the algal nodule biotope are transported downslope to become incorporated into the *Amphistegina* sand.

The *Amphistegina* Sand Facies is characterized by the presence of a conspicuous population of echinoderms, particularly the urchin *Clypeaster ravenelii* and the asteroids *Chaetaster* and *Narcissia trigonaria*. Also, patterned burrows (6 to 12 small burrows in circular aggregations with diameters up to approximately ½ m) are numerous. In places between 80 and 90 m a large population of small comatulid crinoids clings to carbonate gravel and other objects on the *Amphistegina* sand. These small crinoids do not occur abundantly elsewhere on the banks and their distribution does not appear to be related to the distribution of larger crinoid species that occupy the lower Algal-Sponge Zone, rocks, and drowned reefs to depths exceeding 120 m. The presence of these crinoids indicates a final transition from shallower, higher diversity, clear-water communities dominated by frame-building corals and coralline algae to subdued, deep-water communities subjected to turbidity, sedimentation, and chronically low light levels. With increasing depth, beyond approximately 90 m, the coarser sediments of the *Amphistegina* Sand Facies are replaced by mud in the Quartz-Planktonic Foraminifers Facies (Figure 6.11).

The deep-water populations, whether on hard or soft bottom, represent an assemblage of organisms that differs substantially from the clear-water assemblages. Few species of fish and invertebrates occur in both the clear, shallow-water and the turbid deep-water environments on the banks.

POPULATION STUDIES OF HIGH-DIVERSITY REEFS

Coral Populations

The Flower Garden Banks harbor 21 species of scleractinian corals and one hydrocoral (Table 6.1), among which 18 are considered to be hermatypic (reef-building). Sixteen of the hermatypes occur in the *Diploria-Montastrea-Porites* Zone. Population levels of the hermatypes identified have been measured by using photographic mosaics (Figure 6.16) of 64 transects, each 8 m in length, taken at three sites: 23 transects on the East Flower Garden at 20-m depth on the central reef top (station EFG 20); 23 transects on the East Flower Garden at 26-m depth on the reef top near the reef edge (station EFG 26); and 18 transects on the West Flower Garden at 24-m depth on the central reef top (station WFG 24).

Transect positions were restricted to hard bottom; sand substratum was avoided. Otherwise no systematic method of transect location was used, and randomness is assumed with substantial confidence. The three samples therefore are considered to be stratified (hard substratum only) and random.

Overlapping photographs were taken along a graduated fiberglass tape transect marker with a Nikonos III camera and electronic strobe mounted to a frame that described an 80 X 55 cm field of view on the reef. The color prints were assembled in the laboratory as transect mosaics. A string was stretched along each transect mosaic and the lengths of coral colonies, by species or genera, directly beneath the line were recorded. Metric scale units from the transect tape pictured in the photographs were used for size reference. Most of the 16 hermatypic coral species known to occur on the high-diversity reef were identifiable in the photographs, but the two species of *Colpophyllia* and the three species of *Agaricia* could not be identified beyond the generic

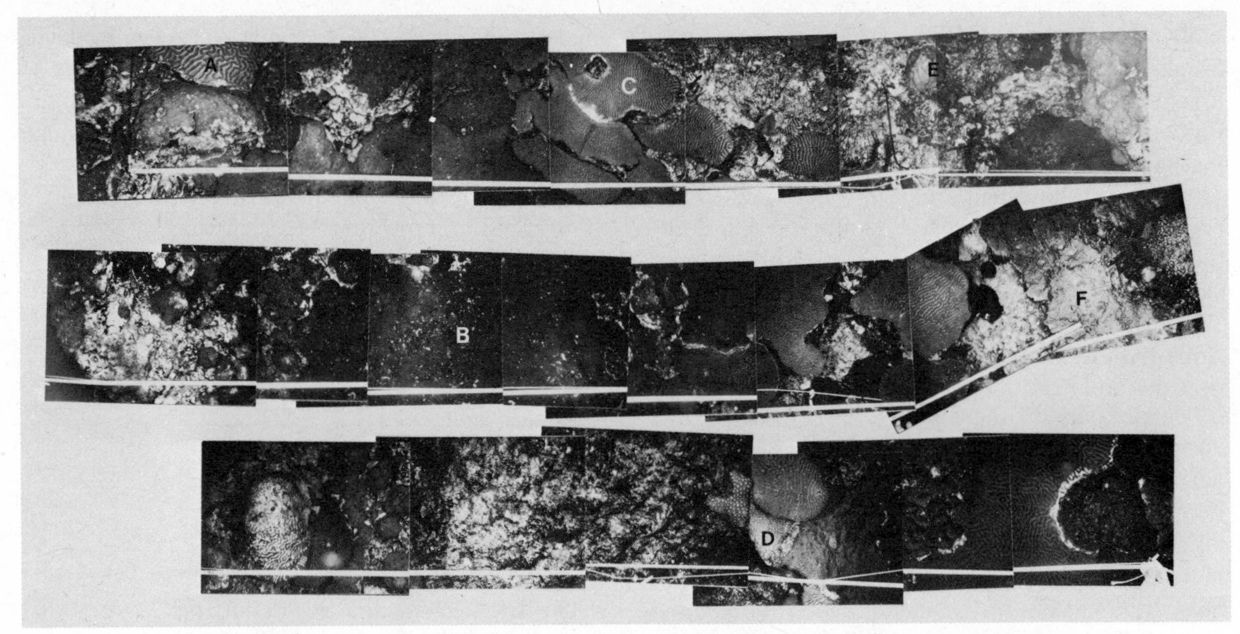

Figure 6.16. Typical photographic mosaic of 10-m-long transect at 24-m depth on the West Flower Garden. *A* = *Colpophyllia; B* = *Montastrea annularis; C* = *Diploria strigosa; D* = *Montastrea cavernosa; E* = *Porites astreoides; F* = *Millepora alcicornis.*

level. These genera were treated as individual taxa in the analyses. *Porites furcata* is so rare that it was never encountered in any of the transects. Failure to detect or resolve these species resulted in 12 measureable taxa (Table 6.2). The determinations are comparable to those achieved using the line transect method described by Loya (1978).

Hard substratum, composed primarily of hermatypic coral reef rock, occupies 85% or more of the bottom within the *Diploria-Montastrea-Porites* Zone. The remainder is carbonate sand and gravel. This was determined by analysis of wire-line underwater television records of transects traversing the zones at both banks (5 hours at the East Flower Garden, 2.5 hours at the West Flower Garden). On both reefs somewhat more than half of the hard substratum is covered with living coral (Table 6.2). One-way analyses of variance (ANOVA, 95% confidence level) and Duncan's multiple range tests (DMR) indicate a statistically significant, slightly higher percentage of coral cover at EFG 26, near the reef edge (65%), than at the other two more central reef-top sites (50 to 55%) (Table 6.2). This is consistent with observations by Goreau and Wells (1967) and Porter (1972) of increased coral abundance on reef margins and steep reef slopes in the Caribbean.

Montastrea annularis, the dominant hermatypic species on the high-diversity coral reefs, covers 20 to 40% of the hard substratum. ANOVA and DMR analyses imply significantly lower *M. annularis* population levels at the shallowest site (EFG 20), compared with the other two sites (EFG 26 and WFG 24). *Colpophyllia* spp. populations were statistically lower at WFG 24. Abundances within each of the other taxa were statistically uniform among the three sites. Therefore, the observed reef-edge increase in total coral cover is apparently a reflection of the lower reef-top populations of *M. annularis* at the East Flower Garden and of *Colpophyllia* spp. at the West Flower Garden.

Corals other than *M. annularis* cover minor amounts of the substratum, none more than 9%, and most much less (Table 6.2). Together, *Diploria strigosa, Colpophyllia* spp., *Montastrea cavernosa, Millepora alcicornis,* and *Porites astreoides,* in that order of abundance, occupy approximately 20 to 25% of the hard substratum. The importance of the two species of *Colpophyllia* in relation to one another is unresolved, but *Colpophyllia amaranthus* population levels are probably considerably lower than those of *Colpophyllia natans* (Tresslar, 1974c).

Millepora alcicornis, The only hydrocoral, tends to encrust; it spreads rapidly as a thin layer over unoccupied reef rock with a lateral encrusting growth rate of approximately 2 mm/mo (Bright et al., 1982). This is by far the fastest rate of encrusting

TABLE 6.2. Dominance (% cover) of Major Hermatypic Corals on the Hard Substratum Within the *Diploria-Montastrea-Porites* Zone (High-Diversity Coral Reef)[a]

Major Hermatypic Corals	Weighted Averages, All Sites Combined	Dominance Expressed as the Mean Percentage of Cover (95% confidence limits of the means)		
		Station EFG 26 (N = 23)	Station EFG 20 (N = 23)	Station WFG 24 (N = 18)
M. annularis	31.80	40.06 (34.88–45.20)	22.88 (17.32–28.51)	32.63 (22.45–42.93)
D. strigosa	6.23	4.43 (2.01–5.03)	8.63 (4.18–13.14)	5.46 (1.93–8.99)
Colpophyllia spp.	5.33	7.33 (3.49–11.12)	6.62 (3.57–9.62)	1.11 (0.17–2.05)
M. cavernosa	3.86	3.68 (1.48–6.13)	3.84 (2.82–5.89)	4.10 (1.28–6.92)
M. alcicornis	3.61	3.21 (2.01–4.38)	3.69 (1.52–5.94)	4.03 (2.04–6.20)
P. astreoides	2.26	2.26 (1.28–3.24)	1.94 (1.17–2.72)	2.68 (1.60–3.76)
M. decactis	1.91	3.02 (0.47–5.61)	0.88 (0.47–1.31)	1.79 (0.15–3.43)
S. siderea	0.90	0.00	1.27 (0–3.32)	1.56 (0–3.85)
Agaricia spp.	0.83	0.76 (0.34–1.19)	0.90 (0.52–1.28)	0.83 (0.26–1.40)
S. michelini	0.30	0.16 (0.02–0.34)	0.21 (0–0.57)	0.59 (0–1.43)
M. angulosa	0.26	0.60 (0.19–1.02)	0.00	0.17 (0.01–0.33)
S. cubensis	0.03	0.03 (0–0.06)	0.03 (0–0.09)	0.03 (0.–0.08)
Total live coral	56.82	64.53 (59.6–69.46)	50.42 (45.1–55.74)	55.15 (23.77–86.53)

[a]Hard substratum constitutes approximately 85% of the bottom within the zone. Determinations are based on 23 transects at each EFG site and 18 at the WFG site. The number of each station is also the depth of collection in meters.

growth among the corals at the Flower Garden Banks (see below). Consequently, the species is abundant (Table 6.2). *M. alcicornis* is an opportunistic encrusting form, and may not be nearly so effective as a substratum builder, in terms of sustained accretionary growth, as the dominant scleractinian corals such as *Montastrea annularis,* which tend to form massive, rounded colonies.

Stephanocoenia michelini, although not abundant on the high-diversity reef tops (Table 6.2), is one of the dominant head-forming species below about 36-m depth on the fringes of the two main reefs and on outlying low-diversity reefs (Figure 6.8).

Although *Mussa angulosa* and *Scolymia cubensis* are minor species on the reef, it is likely that they were seriously undersampled in the transects because of their small size and often cryptic occurrence. For these reasons we discount the results of ANOVA–DMR comparisons which indicate significantly higher population levels for *Mussa angulosa* at the deepest site (EFG 26) than at the other two sites. *M. angulosa,* however, is one of the several species whose distribution persists onto the deeper, low-diversity reefs.

Shannon–Weaver indices of coral diversity were calculated for the three sites using a method described by Pielou (1966) for estimating the diversity of vegetatively reproducing organisms in large, incompletely censused sample areas with cover as the measure of abundance (Table 6.3). Evenness estimates were derived by assuming that a maximum of 13 coral taxa were detectable in the photographic

TABLE 6.3. Shannon-Weaver Indices of Coral Diversity and Evenness[a].

Station	Depth (m)	Diversity [mean (95% CI)]	Evenness[c]	N^d
EFG 26	26	1.41 (1.24–1.58)	0.55	18
EFG 20	20	1.69 (1.43–1.95)	0.66	22
WFG 24	24	1.62 (1.16–2.09)	0.63	14

[a]Calculated by using the Pielou (1966) technique for estimating diversity of vegetatively reproducing organisms in communities that cannot be fully censused. Ninety-five percent confidence limits are given in parentheses. Evenness estimates were calculated by using the mean Shannon–Weaver diversity values and by assuming 13 recognizable coral taxa within the *Diploria-Montastrea-Porites* Zone.

[b]Diversity = H'_c.

[c]Evenness = H'_c/H'_{max}.

[d]N = number of sets of pooled transects used to estimate the Shannon–Weaver Index by Pielou's method.

transect mosaics; *Porites furcata* would have been recognized had it occurred in the photographs. Strongly overlapping 95% confidence intervals and similarity of means between the EFG 20 and WFG 24 stations imply that coral diversity and evenness are similar at these central reef locations. Comparatively lower means for EFG 26 may indicate slightly decreased diversity and evenness at the deeper, reef-edge station. This would be consistent with the observed increase in dominance of *M. annularis* at EFG 26 (Table 6.2), which would tend to decrease both diversity and evenness.

Even when considering the minor variations in abundance of certain corals from place to place on the East and West Flower Garden Banks, there is a strong impression that the high-diversity reef tops of the two banks are essentially similar in coral species composition, cover, dominance hierarchy, diversity, and evenness.

With only 18 species of hermatypic corals (and notably no acroporids or shallow-water alcyonarians) the Flower Garden coral community is less diverse than those of comparable reefs in the southern Gulf of Mexico (34 species), South Florida (51 species), and the Caribbean (55 species). Yet in spite of the qualitative deficiencies in coral species composition, those hermatypes that occur at the Flower Garden Banks constitute an apparently hardy coral community.

Calcareous Algae and Encrusting Foraminifer Populations

At least nine genera of crustose coralline algae (Corallinaceae) occur at the Flower Garden Banks (Table 6.1), among which *Hydrolithon* is particularly abundant on the high diversity coral reefs. In the deeper Algal-Sponge Zone *Lithothamnium, Tenarea,* and *Peyssonnelia* (a squamariacean) predominate.

Coverage of crustose coralline algae on hard substratum within the *Diplora-Montastrea-Porites* Zone was estimated from transect mosaics at the EFG 20 and WFG 24 study sites. As a group, crustose corallines occupy approximately 15 to 20% of the hard bottom. The algae tend to inhabit cryptic as well as exposed substratum and are found abundantly on such things as living gastropod shells, hermit crab shells, attached pelecypods, and shell and coral debris on the sand between coral heads.

Following the mortality of part of a coral colony the newly exposed reef rock is almost always occupied first by filamentous algae and then by crusts of coralline algae, which tend to persist until overgrown in turn by coral, sponge, or some other superior competitor for space. Observations derived from studies of coral growth and mortality leave the impression that the coralline algae on the Flower Garden high-diversity reefs are pervasive opportunists, capable of rapidly colonizing and covering exposed reef rock but incapable of preventing subsequent encroachment by the dominant hermatypic corals. Hence their importance as frame builders is overshadowed by the corals, which have a competitive advantage at the shallowest depths on the two banks.

The coral's advantage deteriorates with increasing depth, leading to increased, but still secondary, dominance by coralline algae on the low diversity coral reefs and virtual dominance by coralline algae within the Algal-Sponge Zone, where rhodolith formation is extensive (Figure 6.12). The Algal-Sponge Zone, in fact, occupies a much greater area on the banks than the coral reefs (Figures 6.2 and 6.3) and the majority of carbonate sediment produced on the banks is of algal origin.

In terms of the banks' overall structure and biology, the coralline algae are extremely important. Like the corals, their patterns of dominance, zonation, and community structure are largely dependent on levels of light intensity, therefore depth (Table 6.1). Below approximately 85 m at the Flower Garden Banks, growth of coralline algae is seriously impaired and reefal development ceases.

As previously indicated, calcareous green algae, *Halimeda* spp. and *Udotea* spp., contribute to sediment production within the Algal-Sponge Zone (Table 6.1). The encrusting foraminifer *Gypsina plana* is an important contributor to rhodolith formation within the Algal-Sponge Zone. It is also abundant on the coral reefs and *Madracis* zone (Table 6.1).

CORAL GROWTH AND MORTALITY

Accretionary Growth

Coring the living heads of *Montastrea annularis* and *Diploria strigosa* with an underwater pneumatic drill (Stearn and Colassin, 1979) and thin-sectioning a collected head of *Stephanocoenia michelini* provided samples for X-radiographic sclerochronological determination of accretionary growth rates (Figure 6.17) by the Hudson et al. (1976) method. Accretionary growth determinations from the cores were

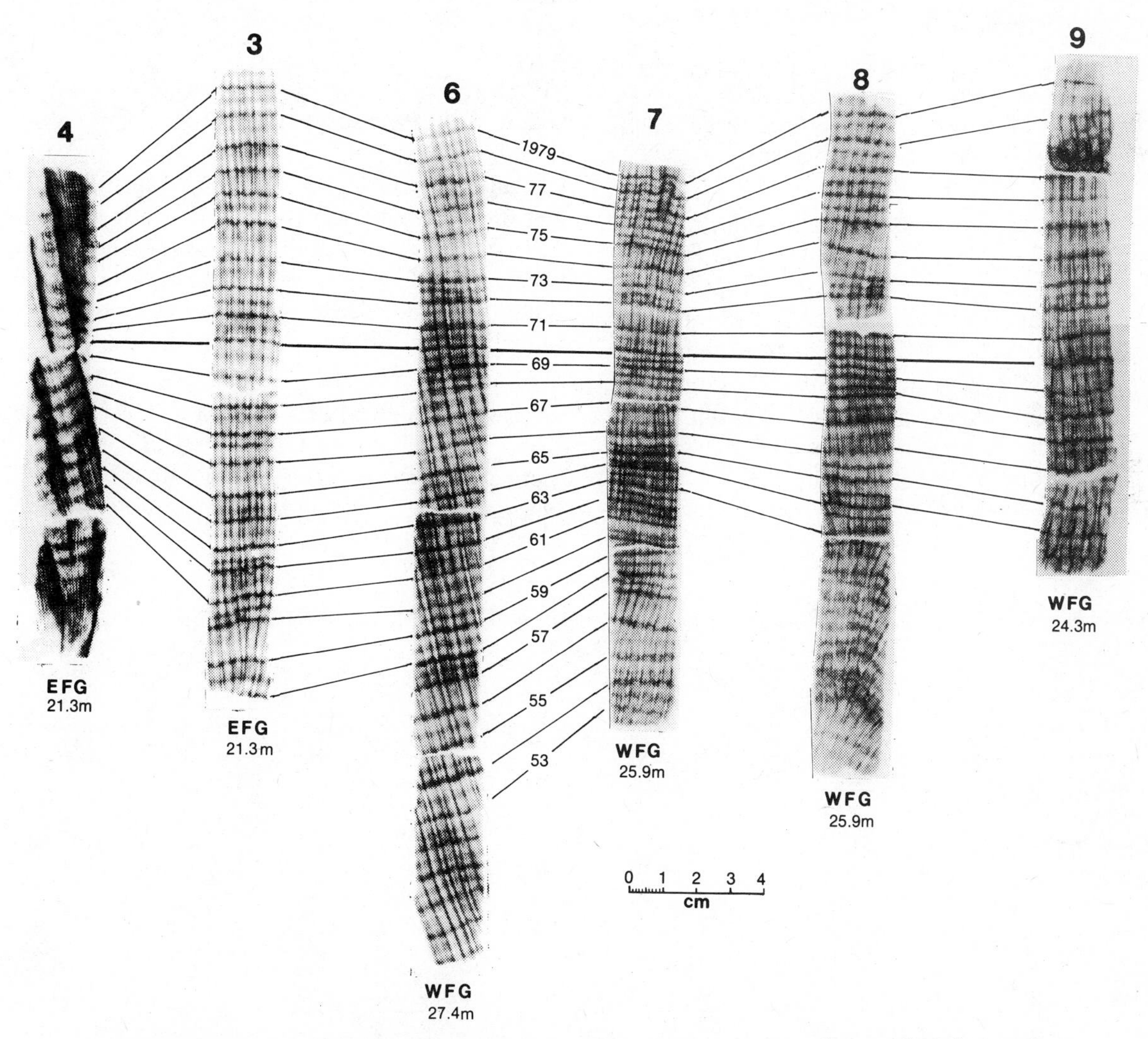

Figure 6.17. X-Radiographs of cores taken from *Montastrea annularis* and *Diploria strigosa* at the East and West Flower Garden Banks, showing annual bands. Locations and depths of collection are indicated beneath each core.

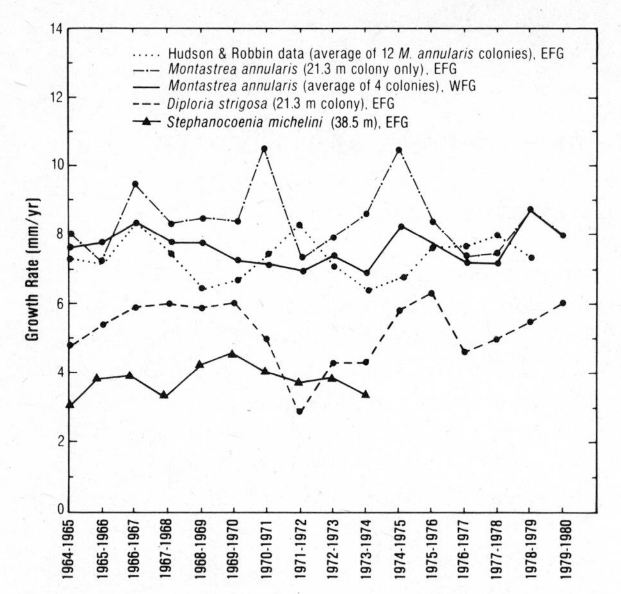

Figure 6.18. Accretionary growth rates for *Montastrea annularis, Diploria strigosa,* and *Stephanocoenia michelini*. From Bright, Kraemer, Minnery, and Viada (1984) by permission of *Bulletin of Marine Science*.

limited to 16 years (1964–1980) because of the core lengths. Average growth rates for four West Flower Garden heads of *M. annularis* collected at 24.3- and 27.4-m depth appear to be essentially the same as average rates for 12 East Flower Garden heads (20-m depth) sampled and analyzed by Hudson and Robbin, 1980 (Figure 6.18). Unfortunately, our short cores do not allow comparison beyond 16 years past. Thus we are unable to say whether the decrease Hudson and Robbin detected in *M. annularis* growth rate at the East Flower Garden (from a 50-year average of 8.9 mm/yr through 1957 to a subsequent average of 7.2 mm/yr through 1979) also occurred at the West Flower Garden.

Our data suggest a decrease in *M. annularis* growth rate with increasing depth (Table 6.4). This would not be surprising, but because of the small number of cores it cannot reliably be supported as a conclusion.

The single core of *D. strigosa* from 21.3 m at the East Flower Garden indicates a substantially lower growth rate for this species (average, 5.0 mm/yr) than for *M. annularis* (range of averages, 5.3 to 8.0 mm/yr) on the reef top.

It is interesting that the apparent growth rate over 44 years (average, 5.8 mm/yr) of our specimen of *S. michelini* (collected in 1974 near the base of the main East Flower Garden reef at 38.5 m depth) is comparable to that of *D. strigosa*. Considering only the years 1964–1974, however, the average accretionary growth rate for *S. michelini* was only 3.8 mm/yr (Figure 6.18). We suspect that the smaller value is more representative and that the 44-year average was inflated by tangential sectioning of growth bands near the central part of the slab that was analyzed. *S. michelini* is a dominant coral on the deeper, low-diversity coral reefs at both banks. It also survives well on other northwestern Gulf of Mexico banks in water that is often, or chronically, turbid.

Encrusting Growth and Retreat

Encrusting growth is the lateral spreading of thin layers of living tissue and newly secreted corallum across reef rock. Its obverse is lateral retreat of col-

TABLE 6.4. Accretionary Growth Rates of Corals from the East and West Flower Gardens[a]

Species	Location and Core Number	Depth (m)	Accretionary Growth: Mean (mm/yr)	Accretionary Growth: Standard Deviation (mm/yr)
D. strigosa	EFG-4	21.3	5.0	0.86
M. annularis	EFG-3	21.3	8.0	1.06
M. annularis	WFG-9	24.4	7.7	1.13
M. annularis	WFG-8	25.9	7.1	0.93
M. annularis	WFG-7	25.9	5.3	0.98
M. annularis	WFG-6	27.4	6.7	0.78
S. michelini	EFG-s	38.5	3.8	0.44

[a]Generated from sclerochronological analyses of sections taken from cores and a collected head.

[b]Core WFG-8 was taken vertically into the top of the head and WFG-7 was taken horizontally into the side of the same head. Results imply that vertically directed accretionary growth on the crest of this head was more rapid than horizontally directed accretionary growth on its flank.

TABLE 6.5. Rates of Encrusting Growth and Retreat of Colony Borders Across Underlying Reef Rock at the East and West Flower Garden Reefs.

	Species	Mean (mm/mo)	95% Confidence Limits of Mean (mm/mo)	Number of Colonies Measured
Encrusting growth	*M. annularis*	0.33	0.16–0.50	13
	D. strigosa	0.46	0.31–0.61	3
Retreat	*M. annularis*	0.20	0.0–0.43	13
	D. strigosa	0.57	0.0–1.15	6

ony borders due to tissue mortality and disintegration. Both encrusting growth and lateral retreat were determined for *M. annularis* and *D. strigosa* by planimetric analysis and comparison of seasonal, close-up (MACRO) photographs taken *in situ* (20–24-m depth) of specific segments of living coral colony borders not in proximity to other competing colony margins (Kraemer, 1982).

Encrusting growth rates measured for *D. strigosa* (average, 0.46 mm/mo) were somewhat greater than those of *M. annularis* (average, 0.33 mm/mo) (Table 6.5). Rates at which disintegrating borders retreated were considerably greater for *D. strigosa* (average, 0.57 mm/mo) than for *M. annularis* (average, 0.20 mm/mo).

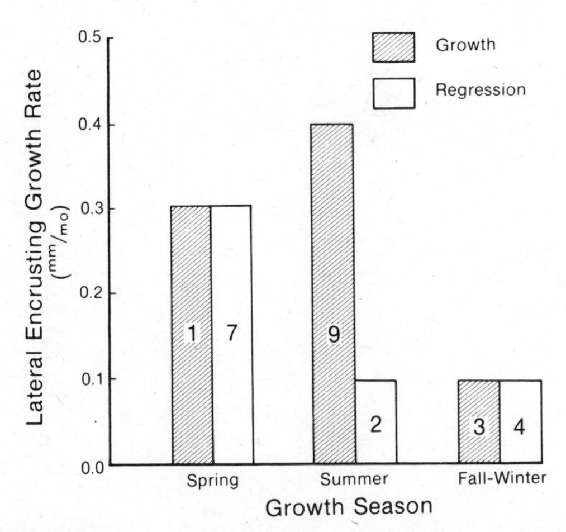

Figure 6.19. Seasonal averages of *Montastrea annularis* encrusting growth and retreat measurements made between September 1978 and September 1980. Spring = mid-February to early June; summer = early June to early September; fall/winter = early September to mid-February. Numbers of coral colonies measured are indicated on the histogram bars. Although the differences are interesting, one-way analysis of variance did not indicate that they were statistically significant.

ANOVA revealed no differences between the East and West Flower Gardens in encrusting growth rates or rates of retreat in *M. annularis,* nor were seasonal differences statistically demonstrated. A simple comparison of the seasonal means (Figure 6.19), however, hints at a decrease in *M. annularis* encrusting growth rates during the fall and winter and a possible increase in rates of retreat in the spring.

Coral dominance, as defined here, is dependent on the percentage of hard bottom covered by living tissue of the various coral species. The rates at which colonies of coral species can expand their borders when not in direct competition with other colonies, compared with their typical mortality rates in terms of border retreat, are undoubtedly important when determining positions of species in the coral dominance (percentage of cover) hierarchy (see also Lang, 1973 for concepts of coral dominance based on physiological competition).

The data collected suggest that *D. strigosa* encrusts at a slightly greater rate than *M. annularis* but *D. strigosa* typically dies back (retreats) substantially faster than *M. annularis*. In addition, *D. strigosa* is particularly susceptible to disease-related mortality at the Flower Garden reefs (Abbott, 1979). Rates of border retreat in apparently diseased *D. strigosa* heads are as great as 6.5 mm/mo, an order of magnitude larger than the average rates of retreat for either species. Thus, hypothetically, the net balance between encrusting growth and retreat at colony margins on the Flower Garden reefs could favor *M. annularis* over *D. strigosa* and, if so, would give *M. annularis* a competitive advantage.

DISCUSSION

Assemblages of stony corals that occupy the Flower Garden reefs are similar to those of submerged reefs

TABLE 6.6. Occurrence of Hermatypic Corals in the Western Atlantic[a]

Hermatypic Corals	Caribbean	Bahamas and South Florida	Southern Gulf of Mexico	Bermuda	Flower Gardens	Florida Middle Ground
Stephanocoenia michelini	***	***	***	***	***	***
Madracis decactis	***	***	***	***	***	***
Madracis asperula[b]	***	***	—	—	***	—
Madracis mirabilis	***	***	—	***	***	—
Acropora palmata	***	***	***	—	—	—
Acropora cervicornis	***	***	***	—	—	—
Acropora prolifera	***	***	***	—	—	—
Agaricia agaricites	***	***	***	—	***	—
Agaricia tenuifolia	***	—	***	—	—	—
Agaricia undata	***	***	—	—	—	—
Agaricia lamarcki	***	***	***	—	—	—
Agaricia grahamae	***	***	—	—	—	—
Agaricia fragilis	***	***	***	***	***	***
Helioseris cucullata	***	***	—	—	***	—
Siderastrea siderea	***	***	***	***	***	—
Siderastrea radians	***	***	***	***	—	***
Porites astreoides	***	***	***	***	***	—
Porites branneri	***	***	***	—	—	***
Porites porites	***	***	***	***	—	—
Porites divaricata	***	***	***	—	—	***
Porites furcata	***	***	***	—	***	—
Favia fragum	***	***	***	***	—	—
Diploria clivosa	***	***	***	***	—	—
Diploria labyrinthiformis	***	***	***	***	—	—
Diploria strigosa	***	***	***	***	***	—
Manicina areolata	***	***	***	***	—	***
Colpophyllia natans	***	***	***	—	***	—
Colpophyllia amaranthus	***	***	***	—	***	—
Colpophyllia breviserialis	***	***	—	—	—	—
Cladocora arbuscula	***	***	***	—	—	***
Montastrea annularis	***	***	***	***	***	—
Montastrea cavernosa	***	***	***	***	***	—
Solenastrea hyades	***	***	—	—	—	—
Solenastrea bournoni	***	***	—	—	—	—
Meandrina meandrites	***	***	***	***	—	***
Meandrina brasiliensis	***	***	—	***	—	—
Dichocoenia stokesi	***	***	—	***	—	***
Dichocoenia stellaris	***	***	—	—	—	***
Dendrogyra cylindrus	***	***	***	—	—	—
Mussa angulosa	***	***	***	—	***	***
Scolymia lacera	***	***	—	—	—	***
Scolymia cubensis	***	***	***	—	***	***
Isophyllia sinuosa	***	***	—	***	—	—
Isophyllia multiflora	***	***	***	***	—	—
Isophyllastrea rigida	***	***	—	***	—	—
Mycetophyllia lamarckiana	***	***	***	***	—	—
Mycetophyllia danaana	***	***	—	—	—	—
Mycetophyllia ferox	***	***	—	—	—	—
Mycetophyllia aliciae	***	***	—	—	—	—
Mycetophyllia reesi	***	—	—	—	—	—
Eusmilia fastigiata	***	***	***	***	—	—

TABLE 6.6. Occurrence of Hermatypic Corals in the Western Atlantic[a]

Hermatypic Corals	Caribbean	Bahamas and South Florida	Southern Gulf of Mexico	Bermuda	Flower Gardens	Florida Middle Ground
Millepora alcicornis	***	***	***	***	***	***
Millepora moniliformis	***	—	—	—	—	—
Millepora complanata	***	***	—	—	—	***
Millepora squarrosa	***	—	—	—	—	—

[a]Present *** (no abundance implied); absent —.

[b]We consider *Madracis asperula* "hermatypic" because of its apparent importance as a substratum producer in the Algal-Sponge Zone at the Flower Gardens and other shelf-edge banks in the northwestern Gulf of Mexico.

Sources. Davis (1982), Dodge et al. (1982), Goreau and Wells (1967), Wells (1973), Porter (1972), Rannefeld (1972), Roos (1964, 1971), Chavez et al. (1970), Chavez (1973), Villalobos (1971), Moore (1958), Fandino (unpublished), Cairns (1977), Jameson (1981), Smith (1971), Hopkins et al. (1977), Grimm and Hopkins (1977), Ginsburg and Stanley (1970), Wilson (1969).

and deep forereefs in the Caribbean and the southern Gulf of Mexico (Goreau, 1959; Bak, 1977; MacIntyre, 1972; Logan, 1969; Kornicker et al., 1959; Rigby and McIntyre, 1966). However, among the approximately 55 species of shallow-water hermatypic scleractinian and hydrozoan corals in the Caribbean, the Bahamas, and South Florida only 18 occur at the Flower Garden Banks (Table 6.6). In comparison, 24 hermatypic species are known from Bermuda, 16 from the northeastern Gulf of Mexico (Florida Middle Ground), and 34 from the southern Gulf of Mexico (Campeche Bank and the Gulf of Campeche).

Bermuda, the Florida Middle Ground, and the Flower Garden Banks represent three separate, northern, biogeographic "end points" in the contemporary distribution of tropical Atlantic coral reefs and coral-dominated assemblages derived from the Caribbean biota. All are depauperate in hermatypic corals, and harbor only 29 to 44% of the known tropical Atlantic species (Table 6.6).

Coral reefs at the Flower Garden Banks and Bermuda are more closely related to one another in terms of community structure and coral dominance patterns than either is to the Florida Middle Ground assemblages, which are not considered true coral reefs. The Florida Middle Ground has only 6 hermatypic coral species in common with the Flower Garden Banks and 8 in common with Bermuda. Ten species co-occur on the Bermudian and Flower Garden reefs. Only 3 scleractinians and 1 hydrocoral occur at all three locations (Table 6.6).

Primary dominants common to the Flower Garden Banks and Bermuda *(Montastrea annularis, Diploria strigosa,* and *Porites astreoides)* are absent from the Florida Middle Ground, where *Madracis decactis, Millepora* spp., and *Dichocoenia* spp. are the important stony corals (Hopkins et al., 1977; Grimm and Hopkins, 1977). On the other hand, shallow water alcyonarians and *Dichocoenia* spp. are abundant at the Florida Middle Ground and Bermuda but are absent from the Flower Garden Banks. In fact, shallow-water alcyonarians are abundant on all Atlantic coral reefs other than the Flower Gardens. This discrepancy alone is enough to distinguish the Flower Garden Banks biogeographically. All three northern localities lack *Acropora palmata* and *Acropora cervicornis,* which are species of material importance on emergent reefs in the Caribbean, South Florida, the Bahamas, and the southern Gulf of Mexico.

A distributional link between the Flower Garden reefs and the more diverse Campeche reefs is implied by the fact that only three hermatypes present on the Flower Garden reefs are not known from the southern Gulf *(Madracis asperula, Madracis mirabilis,* and *Helioseris cucullata).* These species tend to be rare or cryptic on reefs dominated by head corals or *Acropora* and we suspect that they occur in the Gulf of Campeche but have not been reported.

Considering the similarity of the Flower Garden reefs to those in the southern Gulf and their dissimilarity to tropical hard-bottom assemblages in the northeastern Gulf of Mexico, we hypothesize that the Flower Garden biota represent a northward extension of the reefal communities in the southern Gulf of Mexico. The Florida Middle Ground assemblages must be molded by factors inherent in the eastern Gulf, possibly with little direct biological exchange between them and reef systems in the northwestern Gulf. The general patterns of surface currents within the Gulf of Mexico are not inconsistent with this idea (Nowlin, 1972; Merrell and Morrison, 1981). Specifically, the epipelagic currents that pass the Flower Garden Banks are predominantly from the southwest and are tropical and oceanic in character (McGrail et al., 1982a). Oceanic currents that pass the Florida Middle Ground come directly from the Yucatan Channel and are generally restricted to the eastern Gulf of Mexico. Moreover, the Florida Middle Ground, located on an extensive carbonate shelf with numerous exposures of hard substratum occupied by tropical epibenthos, is not so geographically isolated from other components of the Caribbean biota as the Flower Garden Banks, which occur at the edge of a continental shelf of terrigenous clays and sand. Thus probabilities relating to larval transport and recruitment of Caribbean biota may differ drastically in the two regions.

Mid-shelf hydrographic conditions at the Florida Middle Ground are more neritic than the shelf-edge and insular conditions at the Flower Garden Banks and Bermuda. It is likely, therefore, that the similarities between coral populations at the Flower Garden Banks and Bermuda and the dissimilarities between these two localities and the Florida Middle Ground are related to winter water-temperature minima. During February–March surface temperatures drop to about 19°C in Bermuda (Bosellini and Ginsburg, 1971), and to approximaely 18°C at the Flower Garden reef crest (Etter and Cochrane, 1975). Winter temperatures at the Florida Middle Ground dip to 15.7°C (Hopkins and Schroeder, 1981). The accepted minimum for vigorous coral reef development is 18°C (Stoddart, 1969); 16°C is considered a low temperature stress threshold, below which most hermatypic corals lose their ability to capture food (Mayor 1915; Roberts et al., 1982).

Cool winter temperatures at all three locations in relation to South Florida and the Caribbean are likely factors that contribute to the absence of acroporids on the northern reefs. Although the 15- to 23-m water depths at the Flower Garden Banks and the Florida Middle Ground approach the depth limits for *Acropora palmata* and *A. cervicornis,* these same species have been observed by us at similar depths in the warmer waters of Belize and St. Croix in the Caribbean.

Selective exclusion of coral species by unfavorably low water temperatures is generally accepted (Wells, 1957). Bermudian, Flower Garden, and Florida Middle Ground reefs illustrate this well. The Florida Middle Ground, in particular, with the lowest winter temperatures, has no well-developed aggregations of massive coral heads and a dominance structure not typical of Caribbean reef communities. It is probably existing in marginal environmental conditions. The absence of true tropical coral reefs on outcrops and banks closer to shore than the Flower Gardens off Texas and Louisiana is certainly due to the drastic variations in temperature, salinity, and turbidity on the river-influenced inner and middle continental shelf, compared with those of the basically oceanic shelf edge.

Shannon-Weaver diversity index values from the Flower Garden reefs and several other roughly comparable Atlantic reefs were assessed (Table 6.7). Diversities measured on Panamanian (Porter, 1972) and Puerto Rican reefs (Loya, 1976) were, predictably, higher than those from Bermuda (Dodge et al., 1982) and the Flower Garden reefs. The correspondence between Bermudian and Flower Garden values is interesting in view of the qualitative similarity of these isolated, northern reef communities. Considering the number of possible species in the various regions, it is surprising that there was no greater discrepancy between the Caribbean diversity values and the values from Bermuda and the Flower Garden reefs.

Total coral cover at the Flower Garden reefs compares favorably with reefs elsewhere in the western Atlantic (Table 6.7). Based on assessments of reefs in South Florida, the Bahamas, the Virgin Islands, and the Caymans, the Flower Garden reefs rate high in terms of apparent vitality and amount of substratum occupied by living coral.

Accretionary growth rates for corals at the Flower Garden reefs are similar to growth rates for the same species on other Atlantic coral reefs. Hudson (1981), who determined growth rates for *Montastrea annularis* at 12 locations off Key Largo, Florida, found the most rapid growth on shallow (1- to 2 m-depth), mid-shore patch reefs in fairly clear water (average growth 11.2 mm/yr). He found the slowest growth in the forereef zones of offshore barrier-type reefs in water depths of 6 to 12 m (average, 6.3 mm/yr). The latter value corresponds closely to that obtained by Baker and Weber (1975) at 20-m depth in St. Croix, U.S. Virgin Islands (6.5 mm/yr). Our measurements for vertical growth of the same species (6.7 to 7.7 mm/yr at the WFG and 8.0 mm/yr at the EFG) and those of Hudson and Robbin (1980) (7.4 mm/yr on the EFG between 1964 and 1979) indicate rapid growth, considering the depths of sampling at the Flower Garden reefs (20 to 27 m).

For *Diploria strigosa* our estimate of 5.0 mm/yr

TABLE 6.7. Comparison of Coral Abundances and Shannon–Weaver Diversity Values from This and Other Atlantic Reef Studies Based on similar techniques[a].

Atlantic Reef Studies	Depth (m)	Total Coral Cover (%)	Diversity H'_n	Diversity H'_c	Evenness (H'_c/H'_{max})	Number of Hermatypic Corals: In Depth Range	Number of Hermatypic Corals: Known from Area
Panama	17–24			1.5–2.1		28	49
Netherlands Antilles	15–40	30–35				27	50
Puerto Rico	11–17	42–79	2.0–2.2	2.5–2.6(?)	0.81–0.83	20	35
Veracruz	10–17	20				15	34
Bermuda	3–5	14–22		1.3–1.7	0.77–0.82	16	22
Flower Gardens	20–26	43–55[b]	1.5	1.4–1.7	0.55–0.66[c]	16	18

[a] H'_c is the Shannon–Weaver diversity index, based on the amount of bottom covered as the measure of coral abundance. H'_n is the Shannon–Weaver diversity index, based on the number of colonies as the measure of coral abundance. In our opinion only H'_c is appropriate for encrusting or head-forming organisms that exhibit variable colony size, such as the corals. The diversity indices for Panama were converted to natural log format for purposes of comparison. Loya (1976) reported H'_c/H'_{max} but not H'_c; the H'_c values for Puerto Rico were therefore derived from his H'_c/H'_{max} using the maximum number of species (22) he encountered.

[b] Derived by multiplying our values for live coral cover on hard bottom by 0.85, which is the approximate proportion of hard bottom within the *Diploria-Montastrea-Porites* Zone.

[c] Calculated by using 13 as the number of coral taxa recognizable in photographic transects ($H'_{max} = \ln 13$).

Sources. Panama, Porter (1972); Netherlands Antilles, Bak (1977); Puerto Rico, Loya (1976); Veracruz, Rannefeld (1972); Bermuda, Dodge et al. (1982); Flower Gardens, Viada (1980) and herein.

average accretionary growth at 21-m depth on the East Flower Garden compares to estimates made by Vaughan (1916) for the Dry Tortugas (6.9 mm/yr) and Shinn (1975) for Carysfort reef off Key Largo (5.0 mm/yr). Our minimal estimate of *Stephanocoenia michelini* growth at 38.5 m depth on the East Flower Garden (3.8 mm/yr) implies healthy accretion even at considerable depths.

Baker and Weber (1975) recognized a substantial decrease in growth rate of *M. annularis* with depth at St. Croix, U.S. Virgin Islands (6.5 mm/yr at 20-m depth, 2.1 mm/yr at 25-m depth, 1.6 mm/yr at 30-m depth). Our data also indicate a decrease in growth rate with increasing depth for *M. annularis* at the Flower Garden reefs but of a much lesser magnitude (8.0 mm/yr at 21.3-m depth, 7.1 mm/yr at 25.9-m depth, and 6.7 mm/yr at 27.4-m depth).

Accounts of encrusting growth of corals are rare in the literature. Only that of Dustan (1975) seems directly comparable to ours. Using Alizarin staining techniques, Dustan measured encrusting growth of *M. annularis* between 15- and 28-m depth on Jamaican reefs. His values (0.36 to 0.40 mm/mo) are close to ours (average, 0.33 mm/mo) for the same species at the Flower Garden reefs between 20- and 24-m depth.

The exceptional water clarity and consequently high light penetration at the Flower Garden reefs may be the important factors that favor rapid coral growth on the reefs between 20- and 30-m depth. There seems to be no evidence of inhibition of accretionary growth due to the northern location of the Flower Garden Banks.

CONCLUSIONS

The Flower Garden Banks (northwestern Gulf of Mexico), Florida Middle Ground (northeastern Gulf of Mexico), and Bermuda (western Atlantic) represent separate biogeographic extremes in the northward distribution of tropical Atlantic coral reefs and hard-bottom communities dominated by corals. Coral diversity is considerably less in these northern reefal communities than on reefs in the Caribbean, South Florida, the Bahamas, and the southern Gulf of Mexico.

In terms of species composition and dominance

of stony corals, the Flower Garden reefs resemble the Bermuda reefs more than they do the Florida Middle Ground. The Flower Garden reefs seem to be most closely linked biologically to reefs in the southern Gulf of Mexico and are probably the northernmost elements of an arc of reefal communities of common origin that extends from the Campeche Bank to Veracruz, Tuxpan, and the Texas–Louisiana Outer Continental Shelf in the western Gulf.

In spite of their northern location, the abundance and growth rates of hermatypic corals on the Flower Garden reefs are similar to those on Caribbean reefs. The absence of shallow-water alcyonarians at the Flower Gardens, however, sets these reefs apart from all other Atlantic coral-reef communities.

7

CLASSIFICATION AND CHARACTERIZATION OF BANKS

Submarine banks may be classified in a variety of ways, both geologically and biologically. We have settled on a classification based on the general location on the shelf (i.e., mid- or outer shelf) and the nature of the geological structure expressed in them. From this very general beginning we then examine the details of the geology and biology of the banks to characterize them further and to develop groupings of banks with similar attributes. The relation of the geology, biology, and hydrology of the banks is developed in this chapter, with the ultimate aim of providing a basis for predicting the geological and biological environments on less studied banks from limited amounts of remotely sensed data.

SYSTEM OF CLASSIFICATION

Geological categorization is based on structural expression; that is, did the bank or reef develop on relatively undisturbed strata, like the banks off South Texas, or did it grow on a diapiric structure, like the banks off East Texas and Louisiana? A further subdivision may be made on the basis of the nature of the structure that underlies the bank. Is the structure normally associated with salt diapirs or is it inherited from early Jurassic and Triassic tectonic features? The nature of the substrate also is involved in the categorization. Is the substrate made up of bedded Mesozoic and/or Cenozoic sandstones, siltstones, or claystones? Is it a carbonate cap (reef) that totally conceals the original bedrock substrate? Based on these considerations, the geological classification of the banks identifies three categories: mid-shelf bedrock banks, outer shelf bedrock banks with carbonate reef caps, and reefs growing on a relict carbonate shelf.

The most appropriate means of categorizing the banks biologically involves the recognition of a number of distinct benthic biotic zones characteristic of hard banks in the northwestern Gulf of Mexico, with an indication of the banks on which each zone occurs and the depth range of each zone on each bank. Seven characteristic benthic biotic zones have been identified and classified within four general categories, depending on the degree of reef-building activity and primary production:

1. Zones of major reef-building activity and primary production.
   a. *Diploria-Montastrea-Porites* Zone. This zone consists of living, high-diversity coral

TABLE 7.1. Depth Ranges (in meters) of Biotic Zones on Outer Continental Shelf Hard Banks in the Northwestern Gulf of Mexico

Banks	Biotic Zones							
	Millepora Sponge	*Diploria-Montastrea-Porites*	*Madracis*	*Stephano-coenia*	Algal-Sponge	Antipatharian-Transitional	Nepheloid	Soft Bottom
Claypile	40–45						45+	50+
Sonnier	18–52						52+	60+
Stetson	20–52						52+	62–64+
Small Adam						60?	P[a]	64+
Big Adam						60?	P[a]	66+
North Hospital						58–70	70+	68–70+
Aransas						57–70	70+	70–72+
Baker						56–70	70+	70–74+
Blackfish						60?	P[a]	70–74+
Hospital Rock						59–70	70+	70–74+
Mysterious						70?	P[a]	74–86+
Southern						58–70	70+	80+
Dream						62–70	70+	80+
South Baker						59–70	70+	80–84+
32 Fathom						52?	P[a]	55+
Coffee Lump						62–68	68+	70+
Fishnet						66–73	73+	78+
Alderdice					55–67	67–82	82+	84–90+
Ewing					56–72	72–80	80+	85–100+
Bouma					60–75	75–84	84+	90–100+
Parker					60–82	82–?	P[a]	100+
Sackett					67–82[b]	65–85	85+	100+
East Flower Garden		15–36	28–46	36–52	46–82	82–86	86+	100–120+
Applebaum					76?	P[a]	P[a]	100–120+
Bright				37	52–74	74–?	P[a]	110+
West Flower Garden		20–36	P[a]	36–50	46–88	88–89	89+	110–130+
Diaphus						73–98	98+	110–130+
18 Fathom				45–47	45–82	82–?	P[a]	110–130+
28 Fathom					52–92	92–100	100+	110–140+
Jakkula					59–90	90–98	98+	120–140+
Rezak–Sidner					55–93	93–100	100+	120–150+
Sweet					75–80+	P[a]	P[a]	130–200+
Elvers					60–97	97–123?	123+	180+
Geyer	37–52				60–98	98–123?	123+	190–210+
Phleger						?	122+[c]	200+

[a]P = Zone present, but depth range uncertain.
[b]Weakly represented, stressed.
[c]Clear water, but biota typical of nepheloid zone.

reefs. Hermatypic corals are dominant, coralline algae are abundant, leafy algae are limited.

b. *Madracis* Zone and Leafy Algae Zone. The *Madracis* Zone is dominated by the small branching coral *Madracis mirabilis,* which produces large amounts of carbonate sediment. In places large (possibly ephemeral) populations of leafy algae dominate the *Madracis* gravel substratum (Leafy Algae Zone).

c. *Stephanocoenia-Millepora* Zone. A zone that consists of living, low-diversity coral reefs. Hermatypic corals are dominant, coralline algae are abundant, leafy algae are limited. A variation of this zone, also a low-diversity coral reef, occurs at 18 Fathom Bank and has been designated the *Stephanocoenia-Montastrea*-Agaricia zone.

d. Algal-Sponge Zone. A zone dominated by crustose coralline algae that actively produce large quantities of carbonate substratum, including rhodoliths (algal nodules). The zone extends downward, past the depth at which algal nodules diminish in abundance to the greatest depth at which coralline algal crusts are known to cover a substantial percentage of the hard substratum. This is the largest of the reef-building zones in terms of area of sea bottom. Leafy algae are abundant.

2. Zone of minor reef-building activity.

a. *Millepora*-Sponge Zone. A zone in which crusts of the hydrozoan coral *Millepora* share the tops of siltstone, claystone, or sandstone outcrops with sponges and other epifauna. Isolated scleractinian coral heads may be present but rare. Coralline algae are rare.

3. Transitional zones in which reef-building activity may range from minor to negligible.

a. Antipatharian Zone. Limited crusts of coralline algae and several species of coral exist within a zone typified by sizable populations of antipatharians. Banks that support Algal-Sponge Zones (1d) generally possess something comparable to an Antipatharian Zone as a "transition" between the Algal-Sponge Zone and the deeper, turbid-water, Nepheloid Zone of the lower bank.

4. Zone of no reef-building activity.

a. Nepheloid Zone. A zone in which high turbidity, sedimentation, resuspension of sediments and resedimentation dominate. Rocks and drowned reefs are generally covered with veneers of fine sediment. Epifauna are depauperate and variable; deep-water octocorals and solitary stony corals are often conspicuous. This zone occurs in some form on lower parts of all banks below the depths of the Antipatharian or Transitional zones.

This scheme does not represent a final word on benthic zonation on hard banks in the northwestern Gulf of Mexico. The supposed "Antipatharian Zone" and "Nepheloid Zone" are particularly problematic and may not be valid designations in the biological sense. Each surely represents several biotic assemblages of superficial similarity that could all ultimately be given separate zonal designations. No single bank off Texas–Louisiana possesses all of the zones described, although the East and West Flower Garden Banks lack only the *Millepora*-Sponge Zone (Table 7.1 and Figure 7.1). The two Flower Garden Banks harbor the most diverse and thoroughly developed offshore hard-bottom epibenthic communities in the region and differ from other shelf-edge carbonate banks primarily in the degree of development of coral reefs. High-diversity coral reefs (*Diploria-Montastrea-Porites* Zone) are not present on any other northern Gulf banks. Lower diversity coral reefs (*Stephanocoenia-Millepora* Zone) are present at the Flower Gardens and also at two other shelf-edge banks, 18 Fathom and Bright.

The *Millepora*-Sponge Zone, which occupies depths comparable to the *Diploria-Montastrea-Porites* Zone, is characteristic of the Tertiary bedrock substrata of the Texas–Louisiana mid-shelf banks. It is interesting that the zone is present on one shelf-edge carbonate bank (Geyer) but only on a bedrock prominence at the bank's crest.

Upper parts of the relict Pleistocene carbonate reefs of the South Texas mid-shelf banks and certain mid-shelf carbonate banks off North Texas and Louisiana are occupied by benthic assemblages comparable to those of the lower *Antipatharian* Zone found at somewhat greater depths on the North Texas–Louisiana shelf-edge carbonate banks (Table 7.1).

Thus, as further detailed presently, the basic geological categories of the northwestern Gulf outer continental shelf banks are also broadly distinguishable from one another in terms of benthic community structure. The biotic differences between bank types are probably explicable in terms of lateral and depth-related variations in substratum type, water temperature, turbidity, and sedimentation.

SOUTH TEXAS RELICT CARBONATE SHELF REEFS

Geology

A line drawn from Matagorda Bay to the shelf break (Figure 7.1) divides the Texas Continental Shelf into an area of drowned reefs on a relict carbonate shelf

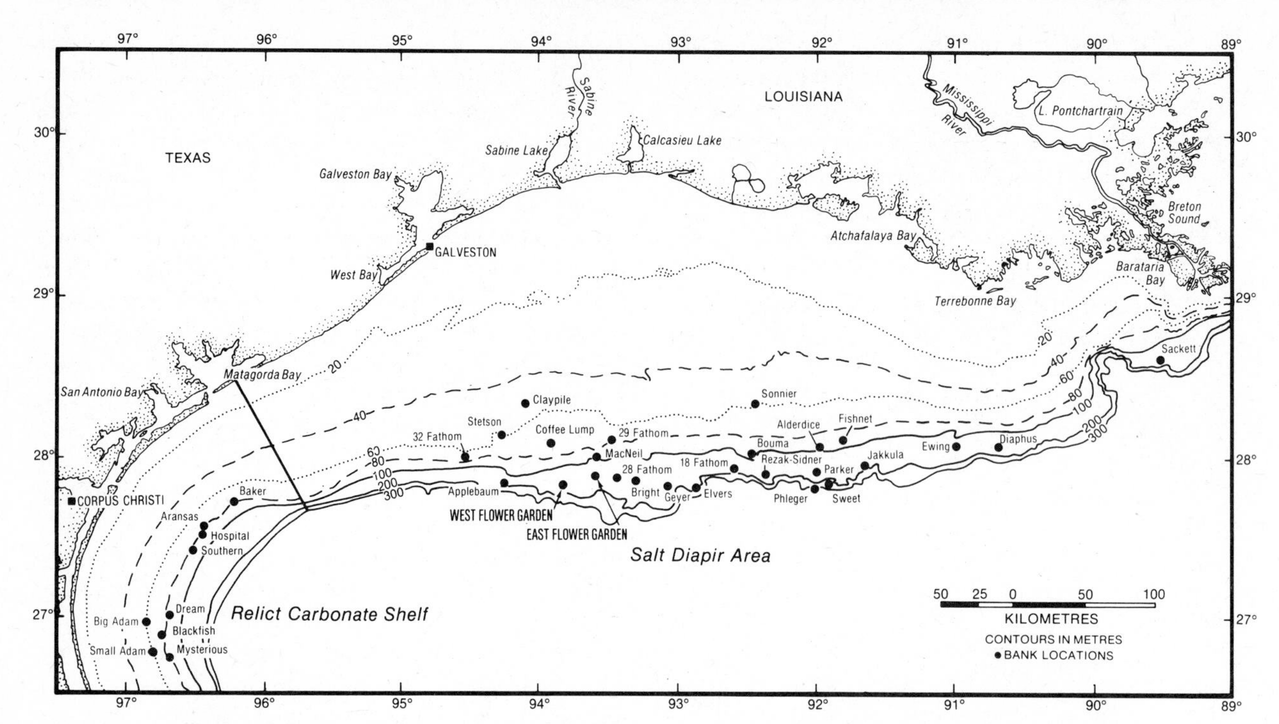

Figure 7.1. Location map of banks discussed in text. Note the boundary between the relict carbonate shelf and the salt diapir area. See Appendix II for exact locations.

and an area of banks situated on salt diapirs. Figure 7.2 illustrates a seismic profile across Baker Bank, about 70 miles east of Corpus Christi, Texas. The only apparent structure involved is normal faulting with little displacement of the shallow reflectors. The shallow reflectors are horizontal and continuous beneath the bank. The keystone fault system displayed in Figure 7.2 is typical of the faulting over shale or salt ridges due to extension in the strata overlying the ridge. Berryhill et al. (1976) illustrate seismic profiles across three other banks on the South Texas Shelf which also show no signs of association with salt diapirs.

There is no doubt that the banks on the South Texas Shelf are drowned coralgal reefs. Rock dredging by the U.S. Geological Survey (Berryhill

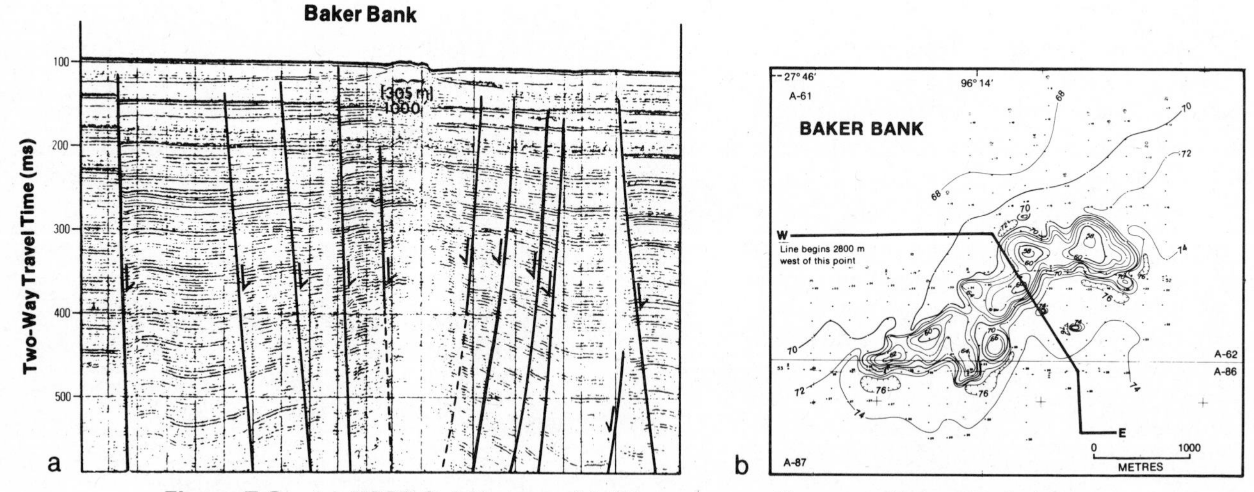

Figure 7.2. (*a*) USGS Del Norte model 561 sparker profile across Baker Bank (after Trippett and Berryhill, 1976, Map I-1287-F); (*b*) bathymetric map of Baker Bank showing location of sparker profile.

et al., 1976) and Texas A&M University (Bright and Rezak, 1976) recovered coralline material from Southern Bank and samples of dead coral from Dream Bank. Radiocarbon (C^{14}) dating yielded ages of 18,000 and 10,580 years BP, respectively. These banks, which are dead reefs that were living near a Late Pleistocene to Early Holocene shoreline, vary in relief from 1 to 22 m. The flanks of all of them are immersed in a layer of turbid water (nepheloid layer) that varies in thickness from 15 to 20 m. Many of the banks are totally immersed in the nepheloid layer. Hospital and Southern Banks have the greatest relief, their crests generally at or slightly higher than the top of the nepheloid layer. Rock samples have been taken from Baker, South Baker, Southern, Dream, Big Adam, and Small Adam Banks. All samples consisted of dead coral or dead coralline algae nodules.

The coral from Dream Bank which was dated at 10,580 ± 155 years BP yields a maximum date for the viability of the reef. Some event that caused the death of corals and most of the coralline algae occurred soon after that time.

Southern Bank is a good example of the relict reefs (Figure 7.3) on the South Texas Shelf. The bank is circular in plan view, approximately 1300 m in diameter, and rises from a depth of 80 m to a crest of 60 m. Four geomorphic features are discernible on the bathymetric map (Figure 7.3):(1) a north–south trending marginal depression up to 4 m deep situated on the east side of the bank; (2) steep slopes extending along the northeast–northwest perimeter of the bank; (3) a nearly flat terrace located between the 68- and 72-m contours; and (4) two peaks rising from the terrace to depths of 60 to 62 m.

Submersible observations allow for a detailed description of the various characteristics of these geomorphic features (Lindquist, 1978). Observations revealed several details. The marginal depression is bounded on its west side by cliffs of coralgal reef rock 1 to 2-m high. The center of the depression contains an accumulation of fine-grained sediment, the surface of which is dotted with numerous patterned burrows 1 to 2 cm in diameter. No observations were made over the depression's east side. The steep slopes on the bank consist of a steplike progression of coralgal cliffs 1 to 2 m high topped by small, irregularly shaped terraces which are coralgal outcrops separated by rubble-filled channels. The terraces become progressively wider as they proceed up the bank from the edge of the marginal depression. Channels on the terraces are approximately 1 m deep and up to several meters wide and

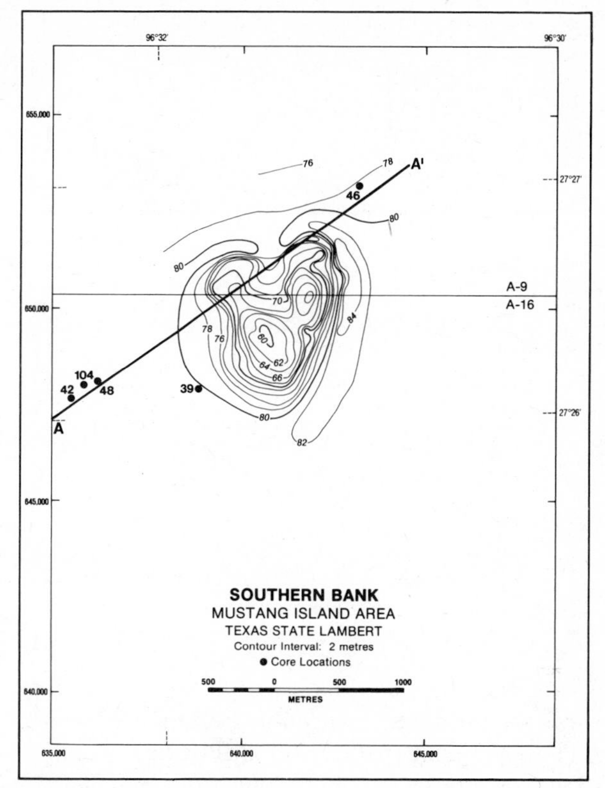

Figure 7.3. Bathymetry of Southern Bank showing locations of cores and cross section in Figure 7.4 (after Lindquist, 1978).

are aligned parallel to the slope of the banks. The major terrace at 66 m consists of relatively flat, rubble-covered coralgal substrate bounded by coralgal cliffs at its outer perimeter. On the substrate, in the sand- and gravel-sized rubble, are ripple beds, with parallel to subparallel symmetrical crests, approximately 5 cm in height and approximately 1 m in wavelength. The crests have a northeast–southwest lineation. Proceeding across the flat portions of the terrace toward the peaks, we find coralgal outcrops approximately 50 to 75 cm high are becoming more common. The peaks are relatively flat-topped and consist of large, rubble-covered coralgal outcrops separated by channels and depressions. The channels decrease in width and depth toward the crests of the peaks. At the crest of the largest peak several groups of small cone-shaped pinnacles approximately 75 cm high and 10 to 50 cm in diameter appear to have been eroded from the coralgal reef rock. Near the crest of this peak several small, intemittent gas seeps were seen emanating from small holes rimmed with a white precipitate.

The surficial sediments consist of varying proportions of three main components: clay, silt, and

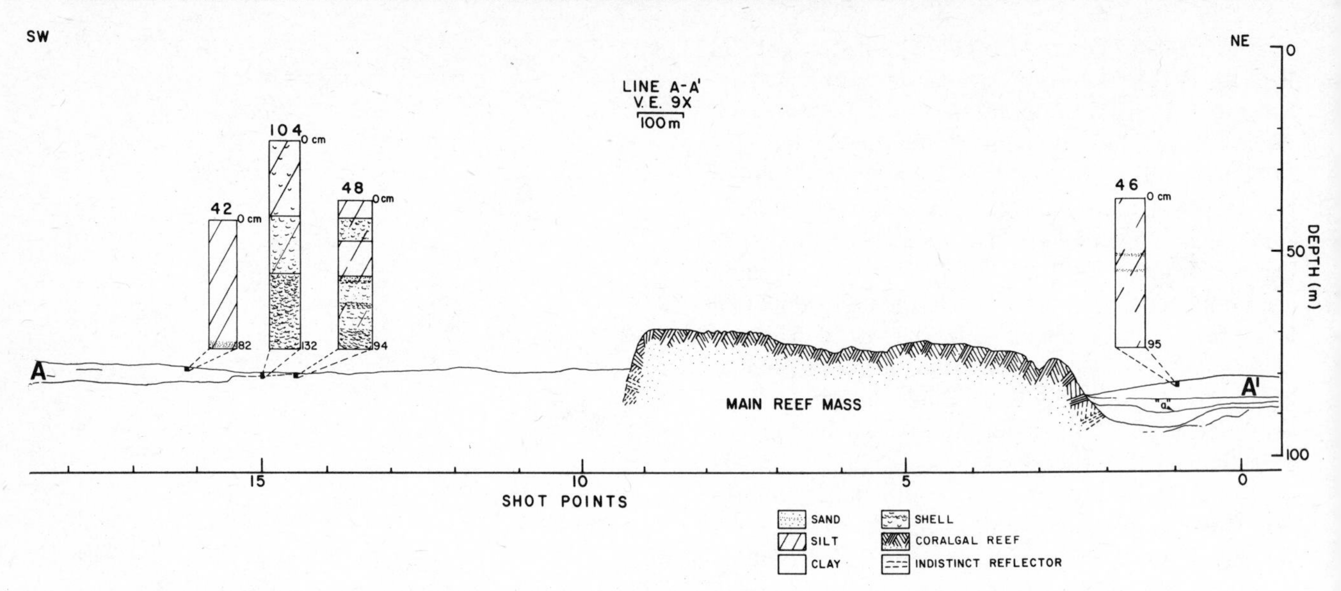

Figure 7.4. Cross section through Souther Bank with locations of four cores, showing sediment types around the bank.

coarse carbonate detritus. Four dominant sediment types can be recognized on and around the bank: (1) a fine-grained, poorly sorted silty clay whose > 62-μm fraction, when present, consists of 20% quartz grains, 53% foraminifer tests, and 24% miscellaneous carbonate grains; (2) a medium-grained, poorly sorted clayey silt whose > 62-μm fraction consists of 75% quartz grains and 15% foraminifer tests; (3) a coarse-grained, extremely poorly sorted shelly mud whose > 62-μm fraction consists of 25% quartz grains, 24% foraminifers, 25% miscellaneous carbonate grains, and 19% shell fragments; and (4) a very-coarse-grained carbonate rubble that was not sampled on Southern Bank because of the hard coralgal substrate beneath it. Samples from a similar facies on other banks were poorly sorted and consisted of 10% quartz grains, 9% foraminifer tests, 30% shell, coral, and algal grains, and 51% miscellaneous carbonate grains. Submersible observations show that it consisted of sand- and gravel-sized grains of shell, coral, and miscellaneous carbonate detritus mixed with cobble-sized algal nodules.

The four dominant sediment types exhibit a distinctive surficial distribution pattern. The surficial sediments of the shelf in the area in which the South Texas Shelf banks are located consist largely of clayey silts (Berryhill et al., 1976) which are predominant in the sediments immediately surrounding the western half of the bank; the exception is a 600-m-wide zone of shelly muds adjacent to the western perimeter of the bank. In a core from station 39 (Figure 7.3) 2.5 m of shelly mud was recovered. The sediments surrounding the eastern half of the bank are silty clays with occasional shell fragments or sand-silt layers. The bank is blanketed by a veneer of fine-grained sediment, which gradually decreases from an approximate thickness of 20 cm on the bank's edge to a trace on its crest. Carbonate rubble is the predominant sediment on the terrace and peaks.

The 3.5-kHz subbottom profiles reveal several reflectors that terminate against the bank. The shallowest can be traced into the surrounding shelf sediments at a depth of about 10 m and is believed to be the Pleistocene–Holocene boundary, termed reflector A by Berryhill et al. (1976). It is the shallowest continuous reflector in Figure 7.4 and is directly overlain by a seismically transparent sediment designated as Unit 4 (Berryhill et al., 1976). Unit 3 (Berryhill et al., 1976) is the shelly mud unit shown in Figure 7.4. Its absence to the east of the bank is simply due to the fact that none of the cores penetrated reflector A, which is the upper surface of Unit 3. On the southwest side of the bank reflector A intersects the seafloor about 600 m to the west of the bank, leaving Unit 3 exposed on the seafloor between that point and the bank.

The coralgal reef rock could not be traced into the subsurface to any significant depth because of the shallow penetration of the 3.5-kHz energy source and the generally poor record quality. Berryhill et al. (1976), however, have illustrated several banks in sparker seismic profiles and the surface of the bank coincides with the surface of Unit 3 in the subsurface (Figure 7.2*a*). We propose that the terrace formed by Unit 3 represents a carbonate shelf that existed during Late Wisconsin time. It seems likely that the event that caused the demise of the reefs was not a rapid rise in sea level but the inception of the nepheloid layer on the South Texas Continental Shelf. The presence of the nepheloid layer has not only caused the death of the once-thriving reefs but also controls the present-day flora and fauna of the South Texas mid-shelf banks.

Biology

The South Texas mid-shelf banks are composed of carbonate substrata overlain by fine sediment veneers of varying thicknesses. Carbonate patch reefs up to 1.5 or 2 m high occur on the banks and are seemingly clustered at particular depths (Figure 7.5). It is on these structures, which provide at least some local relief above the generally sediment-covered carbonate rock bottoms of the banks, that the epifaunal communities are best developed; and it is around them that the greatest number of fishes congregate.

Although there are bank-to-bank variations in population levels, virtually all of the South Texas banks examined are inhabited by fishes and epibenthic invertebrates typical of the Antipatharian Zone (so named because of the conspicuousness and abundance of the large, white, spiraled, sea-whiplike coelenterates *Cirrhipathes*, of the order Antipatharia). A zone similar in biotic composition to the Antipatharian Zone is present in the same depth range on mid-shelf carbonate banks off North Texas and Louisiana (32 Fathom, Coffee Lump, Fishnet Banks) and somewhat deeper on the Texas–Louisiana shelf-edge carbonate banks (Table 7.1). Except for the shelf-edge banks, this zone grades into one of exceptionally high and chronic turbidity and sedimentation near the bases of the banks (Nepheloid Zone). Population levels and diversity of epifauna decrease sharply below approximately 70 m because of chronic turbidity and sedimentation.

Antipatharian Zone assemblages that occupy Southern Bank (Figure 7.5) are representative of those of all the South Texas mid-shelf banks. *Cirrhipathes* is the most conspicuous epifaunal organism on these banks. An almost equally conspicuous macrobenthic organism is the white, somewhat vaselike sponge *Ircinia campana*, which is not particularly noticeable within the Antipatharian Zone on banks off North Texas and Louisiana. Comatulid crinoids are abundant and easily seen everywhere on the upper portion of the bank.

Large, white sea fans, *Thesea*, although apparently less abundant than the organisms mentioned above, were seen frequently because of their size. Other deep-water alcyonarians, mostly paramuriceids, occur on Southern Bank, other South Texas banks, and deep parts of the mid-shelf and shelf-edge carbonate banks farther north. Alcyonarian fans, in particular, seem to be restricted to the Antipatharian Zone and deeper zones in the northwestern Gulf of Mexico.

The only conspicuous stony corals on Southern Bank are saucer-shaped agariciid colonies near the top of the bank in relatively clear water. These small coral patches are not particularly abundant but were encountered several times during submersible dives. Other corals collected included the small, branching *Madracis brueggemanni* and a number of solitary species in moderate abundance. We speculate that there may be a moderate population of the solitary corals, but individually they are quite small. The encrusting coralline algae population on Southern Bank is sparse; it occurs most extensively at the crest of the bank, on which it not only forms isolated patches on the carbonate blocks but also encrusts the tops of pieces of rubble between blocks on the sediment-covered bottom. At best, the current carbonate-producing capacity of these limited populations of corals and coralline algae is feeble and possibly is rivaled by that of the mollusks and echinoderms.

The American thorny oyster, *Spondylus americanus*, appears to thrive in substantial numbers not only in the clearer waters at the crests of the banks but also within the nepheloid layer farther down. It was observed in apparent good health attached to sediment-covered hard bottoms and itself bore a significant veneer of fine sediment. Another small species of calcium-carbonate-secreting organism, the brachiopod *Argyrotheca barrettiana*, is cemented to the rocks but because of its size is certainly not a major substratum builder.

Leafy algae, although present at least at the crest of Southern Bank, are not abundant. A larger population was detected on the crest of Southern Bank in September 1976 than in June 1975. Year-to-year and seasonal variation in leafy algae populations on offshore banks is substantial. We presume that small, variable populations of leafy algae exist similarly on most of the South Texas banks.

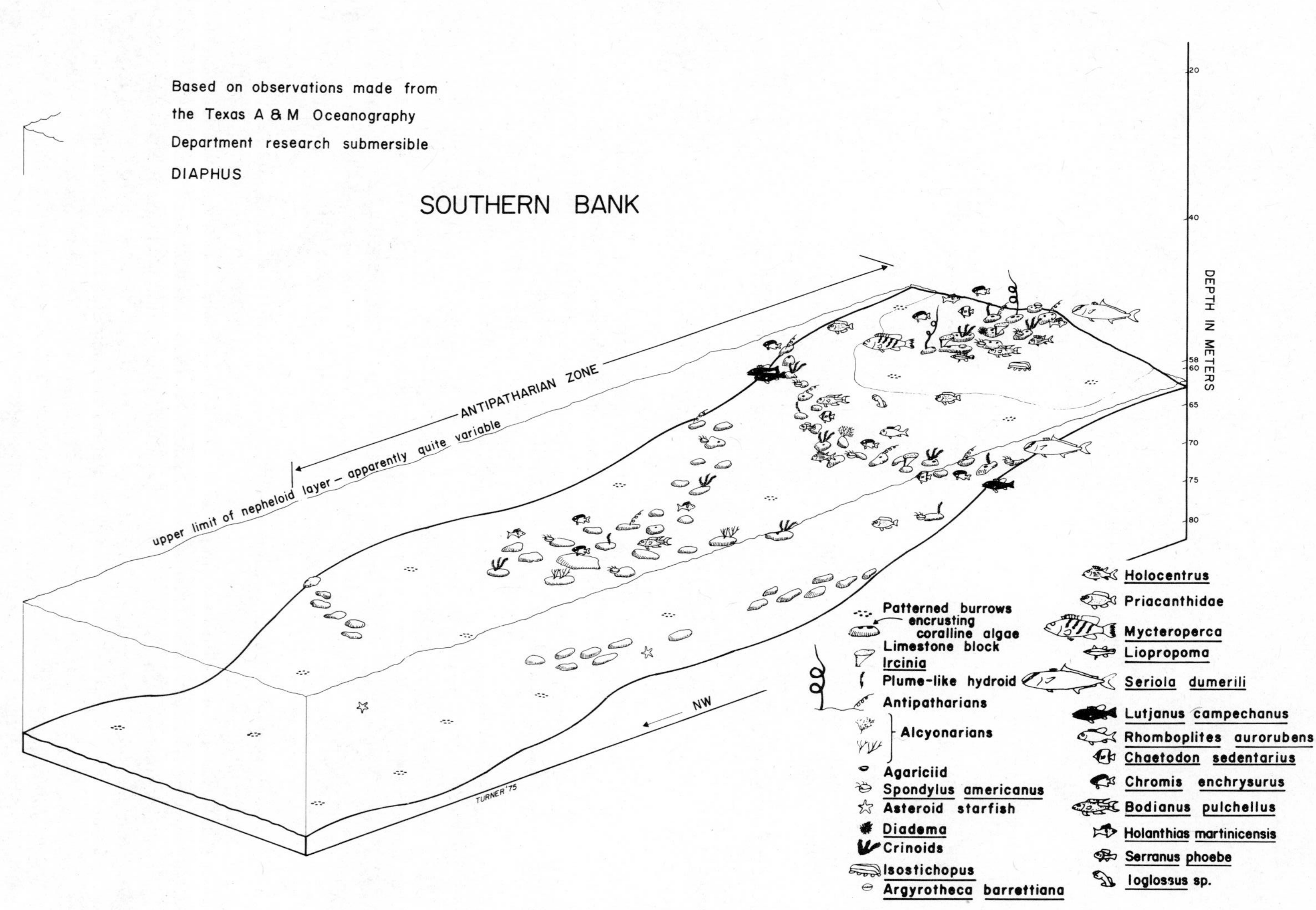

Figure 7.5. Biota of Southern Bank.

Conspicuous among the larger mobile benthic invertebrates are the arrow crab, *Stenorynchus seticornis;* hermit crabs; the black urchin, *Diadema antillarum;* the sea cucumber, *Isostichopus;* and fireworms, *Hermodice*.

Burrows in the sediment on all parts of the bank and adjacent to it are obviously of biological origin, although the nature of the organisms responsible is unknown. The distinctive clusters of holes designated as "patterned burrows" have a form that may indicate involvement of a mobile species, but we have made no observations to confirm it.

Groundfish populations at Southern Bank and all the South Texas banks are strikingly similar in composition and apparent magnitude per unit area to those frequenting the Antipatharian Zones at the Flower Garden Banks and other northwestern Gulf banks. The most characteristic resident species are *Chromis enchrysurus, Holanthias martinicensis, Bodianus pulchellus, Chaetodon sedentarius, Liopropoma eukrines, Priacanthus, Serranus phoebe, Ioglossus,* and *Holacanthus bermudensis,* in apparent order of abundance. *Chromis enchrysurus,* the Yellowtail reeffish, which is particularly abundant, occurs not only in schools of a hundred or more but also singly and in small groups. *Holanthias martinicensis,* the Roughtongue bass, is perhaps most characteristic of the Antipatharian Zone because it is not observed typically in any of the shallower zones found on banks farther north. It seems to be strictly a deep-water variety. All of the other fishes mentioned above, except possibly *Serranus phoebe,* Tattler, are also fairly abundant on the coral reefs and in the Algal-Sponge Zones at the Flower Garden Banks. It is surprising that few large groupers of the genus *Mycteroperca* or hinds of the genus *Epinephelus* have been observed on Southern and the other South Texas mid-shelf banks.

Most of the fishes prefer to congregate around the carbonate reefal structures. The Bigeyes, *Priacanthus arenatus* and *Pristigenys alta,* however, also seem to favor locations on the flatter portions of the bank, particularly over small potholes in the bottom into which they can retreat. The Hovering goby, *Ioglossus calliurus,* likewise favors the level sediment-covered bottom in which it can dig a hole to occupy when not hanging suspended in the water directly above.

Larger migratory fishes, which cannot be considered residents of any particular bank, occur at Southern and include the most important game and commercial fishes; schools of Red snapper and Vermilion snapper, *Lutjanus campechanus* and *Rhomboplites aurorubens;* the Greater amberjack, *Seriola dumerili;* the Great barracuda, *Sphyraena barracuda;* small carcharhinid sharks, and the Cobia, *Rachycentron canadum.*

Benthic communities that occupy Baker, South Baker, Aransas, Hospital, North Hospital, and Dream banks are similar in content and abundance to those of Southern Bank. South Baker Bank may be a bit more profuse biologically. Populations of branching antipatharians, *Antipathes,* and gorgonocephalan basket stars are sizable at Baker Bank, and a spanish lobster, *Scyllarides* sp., was seen. These are typical components of the regional Antipatharian Zone.

Because of their relatively low relief above surrounding mud bottom, the southernmost mid-shelf carbonate banks on the South Texas Shelf (Small Adam, Big Adam, Blackfish, and Mysterious) apparently suffer from chronic high turbidity and sedimentation from crest to base and all rocks are heavily laden with fine sediment. Consequently, the epibenthic communities on these banks, although derived from Antipatharian Zone and upper Nepheloid Zone assemblages, are severely limited in diversity and abundance. *Cirrhipathes* is abundant and frequent basket stars cling to the upper extremities of the *Cirrhipathes* whips. Red and Vermilion snapper, large groupers, amberjack, and barracuda were observed. Little else could be determined visually concerning the communities on these low-relief banks because of the extremely poor visibility.

NORTH TEXAS–LOUISIANA REEFS AND BANKS ON DIAPIRIC STRUCTURES

Mid-shelf Banks

Mid-shelf banks are defined as those rising from depths of 80 m or less and having a relief of about 4 to about 50 m. Banks on the North Texas–Louisiana Shelf that fall into this category are Stetson, Claypile, Coffee Lump, Sonnier, Fishnet, and 32 Fathom. These banks are similar to one another in that all are associated with salt diapirs and are outcrops of relatively bare, bedded Tertiary limestones, sandstones, claystones, and siltstones. The geology of two of these banks is representative of the group.

Geology

Sonnier Banks. Sonnier Banks are located at 28°20′N latitude and 92°27′W longitude (Figures 7.1

and 7.6). In Lease Block 305 of the Vermilion Area, South Addition, several peaks arranged in an arcuate pattern rise from a depth of about 60 m to crest at 20 to 58 m. The individual peaks are apparent fault blocks that have been created by the collapse of the crest of the salt diapir (Figure 7.7). The two highest peaks, near the center of Lease Block 305 (Figure 7.6), have been observed from the submersible. The northern "twin" peak consists of steeply dipping Tertiary sandstones, siltstones, and claystones. The strike of the beds varies from 300 to 330° and the dip varies from 45 to 90°. The side-scan sonar record (Figure 7.8) shows linear patterns of outcrops that converge toward the east and diverge toward the peak just west of the survey line (Figure 7.6). The subbottom profile (Figure 7.8) shows a central area of no subbottom reflectors flanked by beds that dip to the north and south. We know from submersible observations and the side-scan sonar records that the peaks and surrounding areas are composed of steeply dipping bedded rocks. The absence of subbottom reflectors in the central area of Sonnier Banks is due to dips that are too steep to return a bedding plane reflected sound wave to the transducer. The strong surface return in that area is due to reflections from the upturned edges of the beds. The predominant unconformity on the north side of

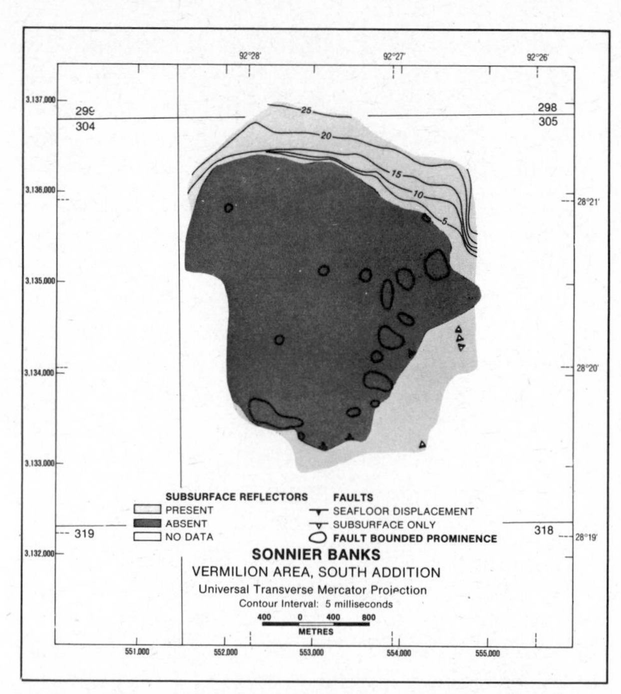

Figure 7.7. Structure/isopach map of Sonnier Banks. Contours are isopachs (sediment thickness) from the seafloor to the unconformity in Figure 7.8.

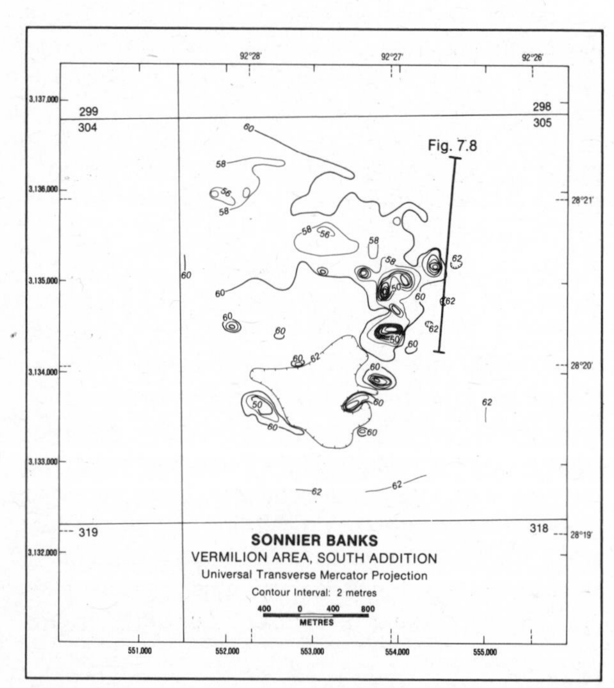

Figure 7.6. Bathymetry at Sonnier Banks, showing location of the subbottom profile and side-scan record in Figure 7.8.

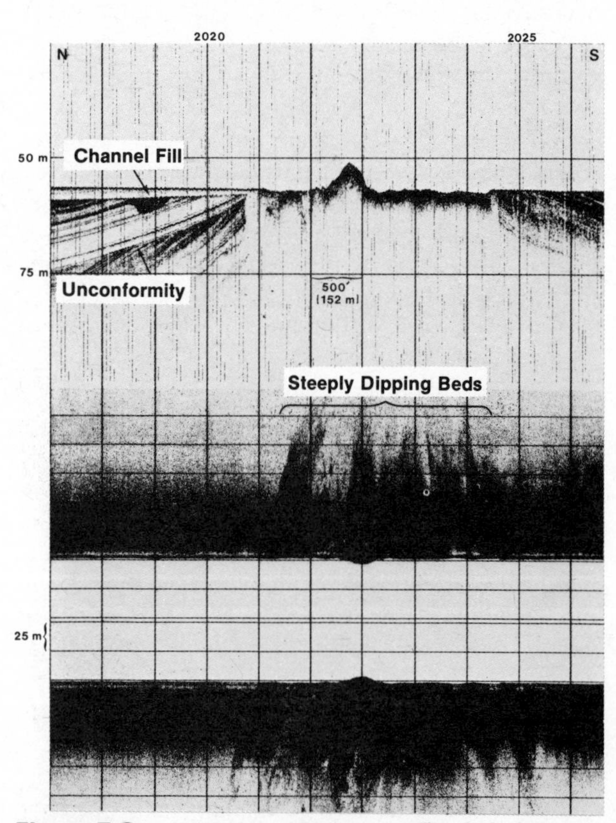

Figure 7.8. A 7-kHz subbottom profile and side-scan sonar record across the east side of the Sonnier Banks area. Location is indexed in Figure 7.6.

the diapir is absent on its south side, possibly because it is below the lowest reflecting horizon on the south side of the diapir.

Fishnet Bank. Fishnet Bank, located at 28°09′N latitude and 91°48′30″W longitude (Figures 7.1 and 7.9), lies in the northeastern quarter of Lease Block 356 in the Eugene Island Area. It is a small, nearly circular bank, covering only 1.9 km^2, with a relatively flat crest that lies at depths of 66 to 70 m. A raised rim along its southeastern and southern margins may be a reef build-up. Three separate peaks on the rim attain minimum depths of just greater than 60 m. Surrounding water depths are about 78 m on all sides of the bank and an east–west channel, 78 to 79 m deep, extends along the base of its north side (Figure 7.9).

Two patterns of local relief are found atop Fishnet Bank: (1) a pattern along its southern and southeastern perimeter, which may be a fringing reef but is more likely an outcrop of a more massive rock unit, and (2) an ellipsoidal pattern of outcrops that result from the truncation of the domal uplift. The fringing reef or outcrop of massive rock occupies the break in slope at the margins of the bank's almost flat top. The ellipsoidal outcrop pattern is caused by steeply dipping, erosionally truncated sedimentary beds in the central part of the bank. These features have less than 1 m of relief.

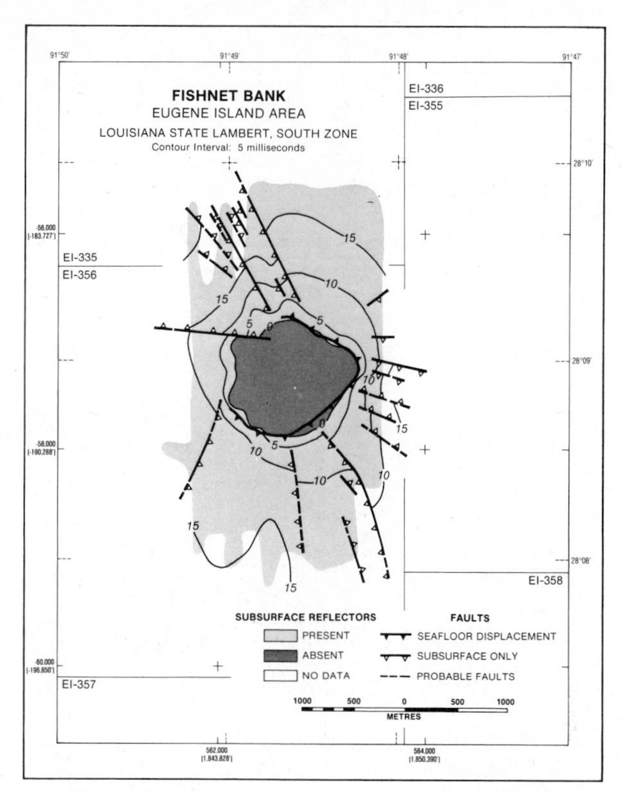

Figure 7.10. Structure/isopach map of Fishnet Bank. Isopachs indicate the thickness of sediments from the sediments from the surface to horizon *H* shown in Figure 7.11.

Two seismic units were mapped on Fishnet Bank: (1) an area in which subsurface reflectors are absent, that is, the main body of the bank, and (2) the surrounding, uppermost sedimentary unit (Figures 7.10 and 7.11). The sediment sequence mapped around its margins onlaps the bank and underlying sediments and appears to be only slightly tilted upward (Figure 7.11). This suggests that the uplift of the bank was nearly completed before the deposition of the last sequence.

The sequence below the mapped unit is truncated by an angular unconformity that appears to have formed at the same time as the truncation of the beds that crop out on top of the bank. This sequence is severely fractured by a pattern of radial faults, many of which extend into the nonreflective unit of the bank proper. These faults were formed by the doming of the strata that overlay the salt diapir as it was

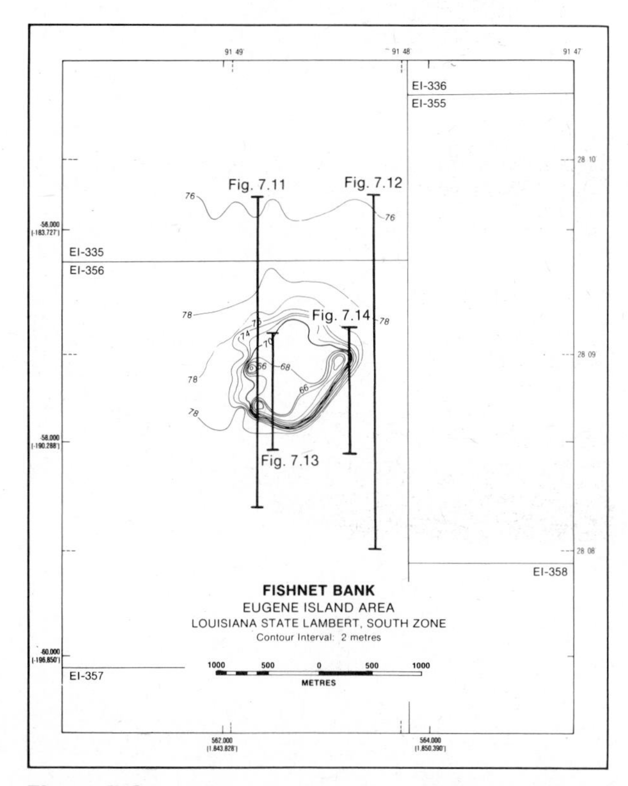

Figure 7.9. Bathymetry of Fishnet Bank, indexing location of profiles shown in Figures 7.11 and 7.12, and side-scan sonar records shown in Figures 7.13 and 7.14.

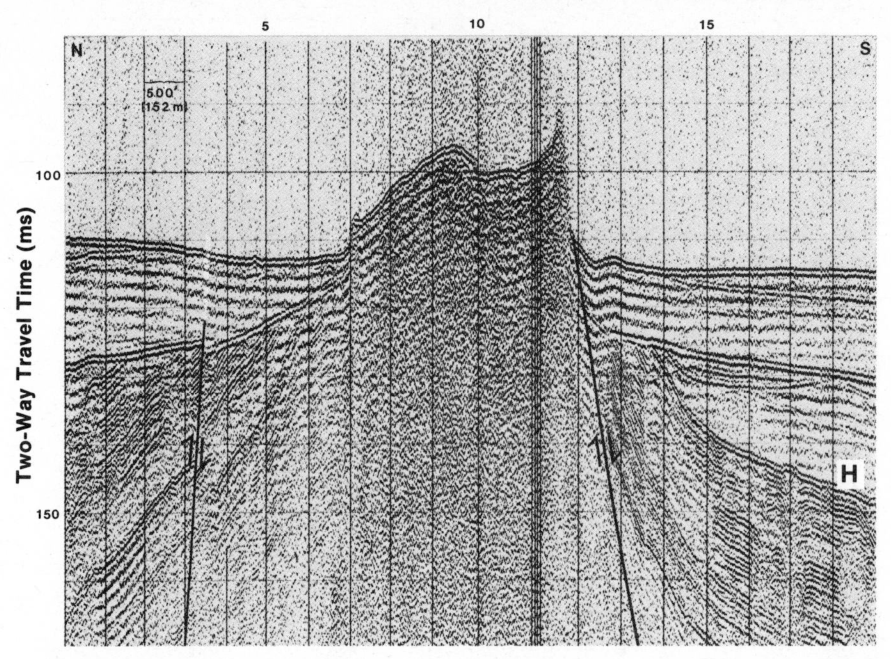

Figure 7.11. A boomer profile across Fishnet Bank. Location is indexed in Figure 7.9.

rising. They and their associated sediments were truncated during the Late Wisconsin low stand of sea level. With the beginning of the Holocene transgression, deposition of the overlying sediments began to bury the unconformity. Renewed movement on the faults in Recent time has displaced the unconformity and created broad, shallow depressions on the seafloor (Figure 7.12).

A diffuse pattern of reflections within the water column over the bank proper is observed in all of the boomer crossings of the bank (Figure 7.11). This is in sharp contrast to the crossings off the margins of the bank, which show considerably less of this pattern. Wipe-out of bedding reflectors by the presence of gas is also displayed on the subbottom profiles. Although there are no obvious vents, it does appear that Fishnet Bank is seeping a considerable amount of gas into the water column. In June 1974 W. E. Sweet observed gas seeps on the bottom (unpublished data).

The 3.5-kHz profiles and side-scan sonar records (Figures 7.13 and 7.14) indicate a lack of sediment

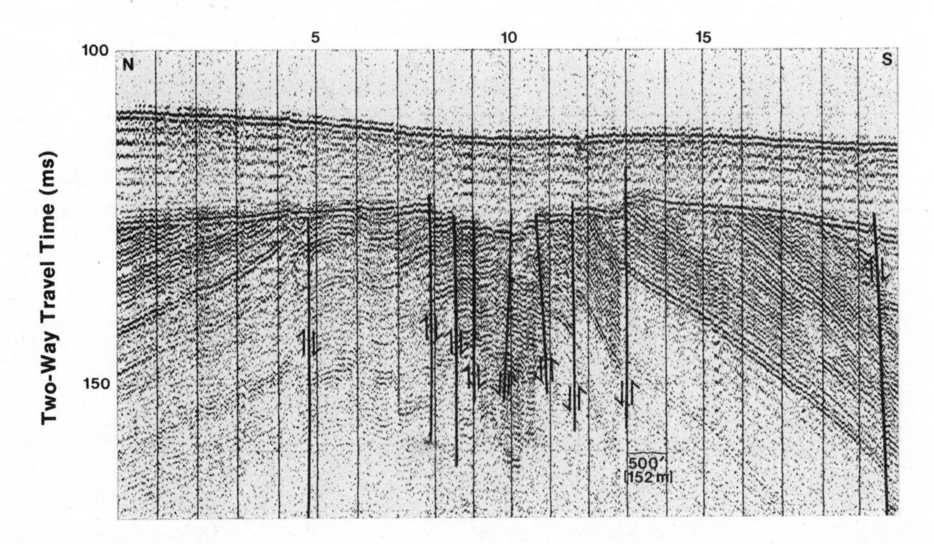

Figure 7.12. A 3.5-kHz subbottom profile across the eastern edge of Fishnet Bank (see Figure 7.9).

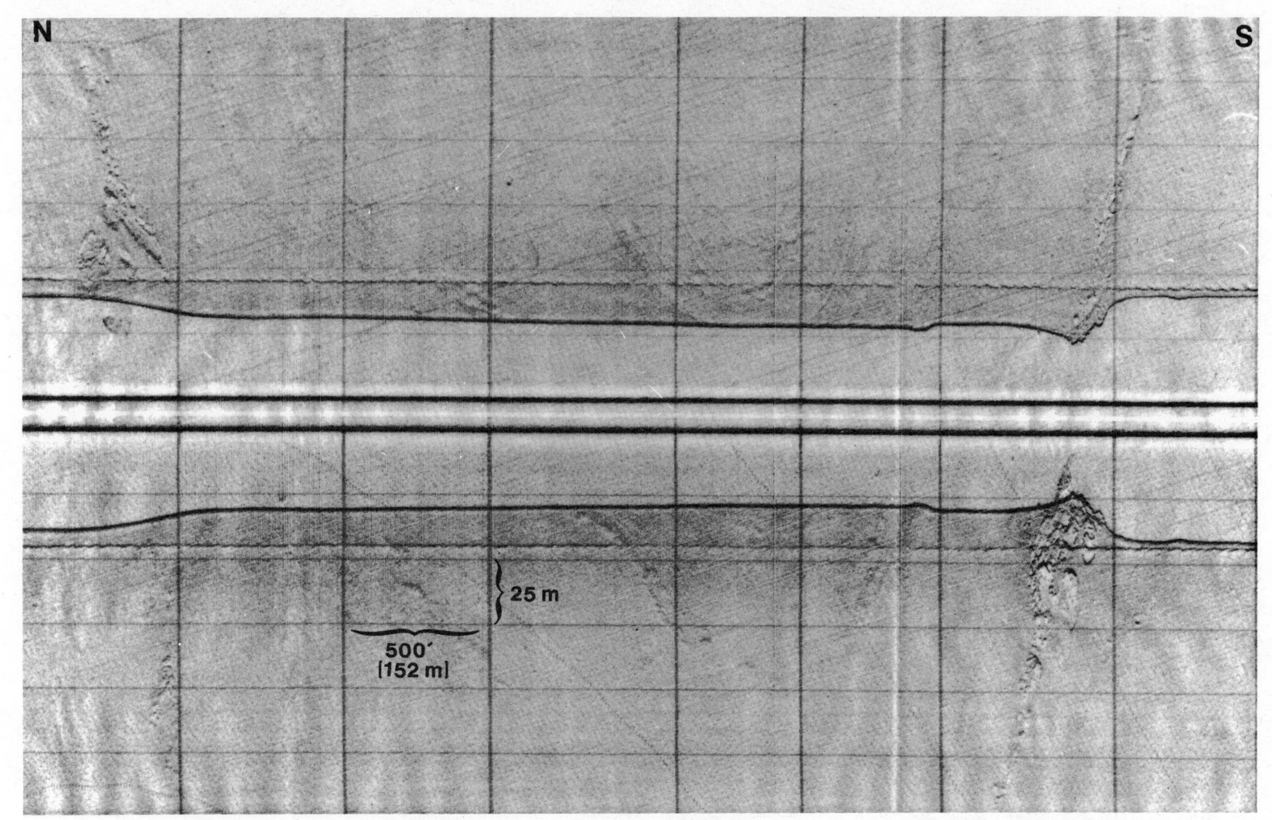

Figure 7.13. Side-scan sonar record, showing the pattern of outcrops at Fishnet Bank. Location is indexed in Figure 7.9.

cover at Fishnet Bank. Sweet's log of a submersible dive on 20 June 1974 indicates that a high blocky ridge lies along the southwest side of the bank, where maximum relief occurs (unpublished data). Numerous large blocks had broken off the ridge and slumped down the flanks of the bank. Other discontinuous ridges were noted. On top of the bank a thin veneer of fine sediment overlies a fairly hard substrate composed of shale. Sweet reported that the top of the nepheloid layer was encountered at a depth of 79 m and that the visibility at 82 m was about 1 m.

During our submersible dive on 13 October 1978 the top of the nepheloid layer was at a depth of 61 m. The bottom at the beginning of the dive was at 78 m and consisted of a very fine, easily suspended mud with shell hash and finger-coral fragments on the surface. On the way up the slope large blocks of rock similar to those described by Sweet were encountered. The top of the talus slope lies at 70 m depth. From that point to the top of the ridge, at a depth of 64 m, the rock forms a vertical cliff. All of the rocks encountered during this dive were Tertiary bedrock. On a second dive that day two peaks were observed that were quite different from those seen on the earlier dive. These peaks consist of cavernous rocks that resemble the drowned reefs observed on other banks.

Biology

The biotic assemblages that occupy North Texas–Louisiana mid-shelf banks, with surrounding depths of 62 m or less (Sonnier, Claypile, and Stetson), are distinct and compose a *Millepora*-Sponge Zone dominated by hydrozoan fire corals (*Millepora*) and various sponges. The biota are best developed on Sonnier Banks.

The crest of Sonnier (18- to 21-m depth) is almost entirely encrusted with fire coral (*Millepora alcicornis*) and the sponges *Neofibularia nolitangere* and *Ircina* (Figure 7.15). The fire coral population extends downward to 40-m depth but is severely diminished below the bank's crest. Dead branches and broken pieces of fire coral occur abundantly in the unconsolidated sediment at the bases of the shallower outcrops, along with siltstone chips and fine silt- and clay-sized particles. The coral must there-

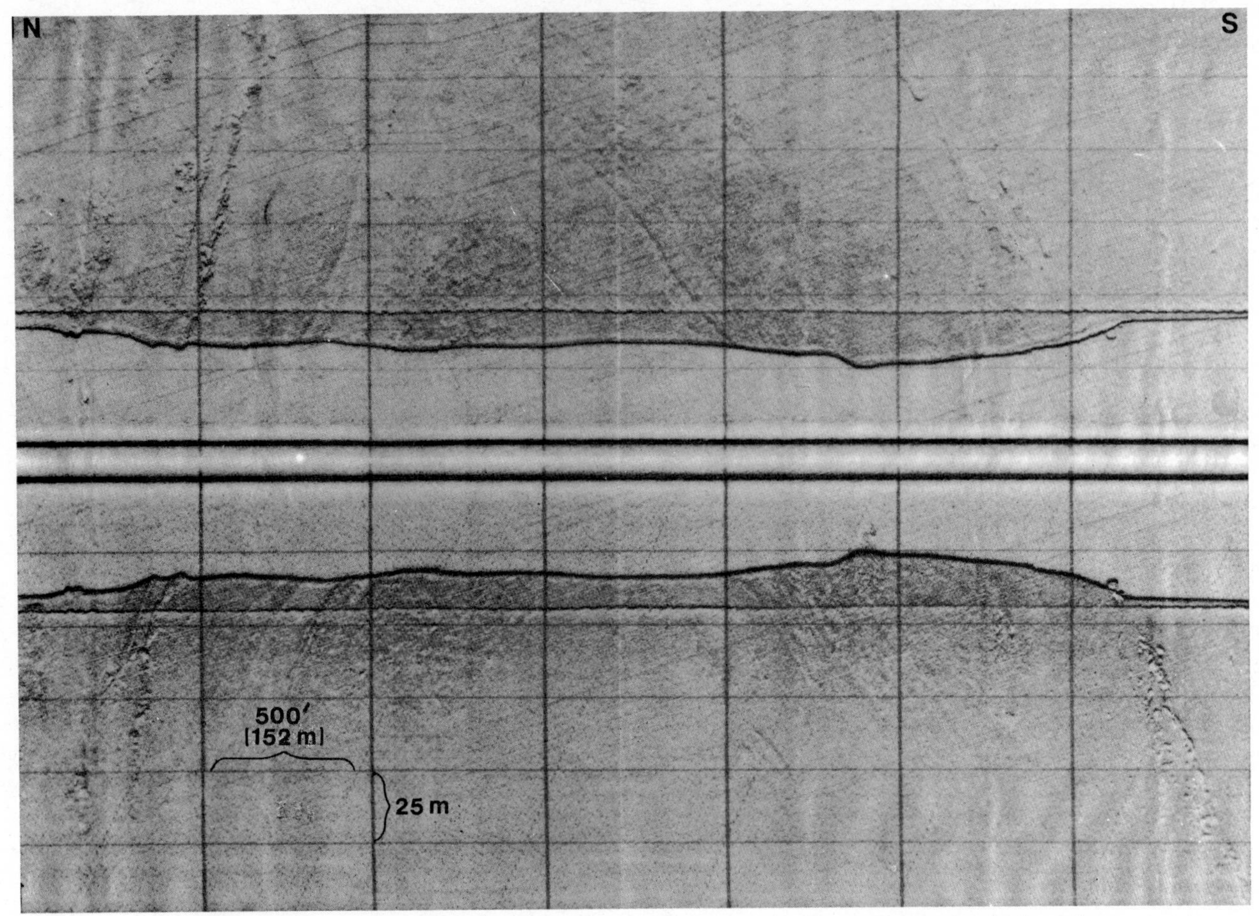

Figure 7.14. Side-scan sonar record, showing outcrop patterns on Fishnet Bank. Location is indexed in Figure 7.9.

fore contribute significantly to the sediment that is produced on the bank and ultimately transported to the surrounding level bottom.

Heads of the hermatypic anthozoan coral *Stephanocoenia michelini,* none of which was more than 1 m in diameter, were seen at 36-, 38-, and 41-m depth. Dead patches on the deepest head indicate possible periodic mortality and regrowth. *Stephanocoenia michelini* is widespread on the mid-shelf and shelf-edge banks off North Texas and Louisiana. It appears to tolerate conditions of low light intensity and moderate turbidity better than other types of reef-building coral in the northwestern Gulf *(Stephanocoenia* also occurs at Stetson Bank as deep as 50 m). The only other stony coral encountered at Sonnier Banks was the saucer-shaped agariciid, which seems to be even more tolerant of such conditions than *Stephanocoenia* and was seen at 52-m depth. Encrusting coralline algae occur on Sonnier Banks down to 47-m depth. Its abundance is moderate above 40 m, becoming sparser as the depth increases.

Populations of fishes and conspicuous, mobile invertebrates at Sonnier Banks are diverse and abundant above 45-m depth and comparable to populations at Stetson Bank. Among the more numerous reef fishes are several species of large angelfishes and butterflyfishes, Chaetodontidae; damselfishes, Pomacentridae; Bluehead, *Thalassoma bifasciatum;* hogfishes, *Bodianus;* Creole-fishes, *Paranthias furcifer;* Rock hind, *Epinephelus adscensionis;* groupers, *Mycteroperca;* and others typical of submerged reefs and banks in the northwestern Gulf. Large schools of Vermilion snapper, *Rhomboplites aurorubens,* were seen above 35-m depth and Red snapper schools, *Lutjanus campechanus,* were encountered near the base of the bank. A species of *Chromis* similar to *Chromis enchrysurus* in shape, size, and habit is extremely abundant above 50-m depth. *Chromis enchrysurus,* which is abundant on most

SONNIER BANK

Figure 7.15. Biota of Sonnier Banks.

of the other banks in the northwestern Gulf, was not observed at Sonnier Banks during the dives made in September 1977.

A sizable population of the large sea cucumber *Isostichopus* is present between 35- and 45-m depth. The molt of a spiny lobster, *Panulirus,* was seen at 27 m depth.

A correlation between the depth distribution and abundance of epibenthic communities and patterns of chronic turbidity at Sonnier Banks is probable. During the September 1977 submersible dive, observers noted that water turbidity at Sonnier Banks was greater below 42 m than above, although a highly turbid nepheloid layer was not encountered above 52-m depth. The abundance of *Millepora,* sponges, coralline algae, and most of the other encrusting epifauna is greatly reduced below 40-m depth.

The transition to soft bottom on Sonnier Banks starts at approximately 52-m depth. As the distance away from the largest bank structures increases, outcrops become smaller and give way to a mud bottom with loose boulders and cobbles of sandstone and siltstone, all covered with veneers of sediment.

At 58-m depth mud is the predominant sediment, with few rocks on the sediment surface.

Although certain species that are abundant on the banks extend some distance out onto the surrounding level bottom (*Neofibularia nolitangere* and *Isostichopus* were seen at a 54-m depth), benthic communities that occupy this turbid-water, rock-strewn, soft bottom differ considerably from those found on the bank. Conspicuous organisms adjacent to the bank, but not on it, include the antipatharians *Cirrhipathes* and *Antipathes,* comatulid crinoids, the sponge *Ircinia campana,* and the Hovering goby, *Ioglossus calliurus.* Abundant burrows, tracks, and trails in the mud attest to a substantial population or large, infaunal, and mobile benthic organisms.

Biotic communities on Sonnier and Stetson Banks are basically the same and above 52-m depth represent a distinct *Millepora*-Sponge Zone. Sonnier populations are somewhat more abundant. A well-developed *Millepora*-Sponge Zone is present between 37- and 52-m depth on a claystone–siltstone pinnacle that protrudes from the carbonate crest of Geyer Bank at the continental shelf edge. This offshore manifestation of an otherwise mid-shelf hard-bank assemblage implies that substratum type is the primary factor controlling the community structure in the *Millepora*-Sponge Zone or that the *Millepora*-Sponge Zone is a pioneer community that populates newly exposed bedrock.

Relatively clear water is probably a necessary requirement for healthy development of this assemblage. Claypile Bank, a Tertiary bedrock structure farther inshore, which has a greater crest depth and less vertical relief than the others, harbors a depauperate *Millepora*-Sponge Zone. This is probably due to conditions of higher average turbidity and sedimentation than occur at Sonnier and Stetson and certainly Geyer Bank, the crest of which is in perpetually clear, oceanic water.

A few sizable heads of *Stephanocoenia michelini* encountered at Claypile Bank in depths to 42 m further indicate the resistance of this species to turbidity, sedimentation, and reduced light. Light intensity at the crest of Claypile Bank nevertheless is adequate to support significant populations of leafy and filamentous algae. In fact, the predominant organisms that occupied the broad top platform of the bank in September 1976 were low-growing mats and clumps of soft algae. Extensive meadows of attached *Sargassum* were observed on the bank in June 1972. These algal populations are certainly seasonal and irregular, as they are on other offshore banks in the northern Gulf.

Farther offshore than Stetson, Sonnier, or Claypile is a series of mid-shelf banks inhabited by transitional assemblages roughly comparable to those of the Antipatharian Zone. These "transitional" banks (Fishnet, Coffee Lump, and 32 Fathom) have surrounding depths in excess of 65 m but less than 80 m. They are most similar in biotic structure to the South Texas mid-shelf banks, which, though of different geologic origin, arise from approximately the same surrounding depths.

General characteristics of Coffee Lump are illustrative of the "transitional" mid-shelf banks. Most of the bottom at Coffee Lump is covered by muddy sand, shell, and gravel (64.0- to 67.7-m-depth), with occasional large boulders 1 m or so across and ½ m high. Major hard-bottom features examined included a reeflike carbonate rock ledge (62.5- to 65.5-m depth) and a large, almost horizontal, slabby outcrop of siltstone (59.5- to 64-m depth).

Where the soft bottom consists of substantial amounts of sand, shell, and gravel there is a predominance of antipatharian whips, *Cirrhipathes;* comatulid crinoids; large asteroids, such as *Narcissia trigonaria* and *Goniaster;* the urchin, *Clypeaster ravenelii;* the sea cucumber, *Isostichopus badionotus;* small branching corals; small benthic fishes; and an enormous population of minute crustaceans, which includes pagurid (hermit), "decorator," and brachyuran crabs and small anomurans.

The extremely large populations of small fishes and crustaceans on the soft bottom above 68 m are significant. These organisms may be the most abundant on Coffee Lump. They are closely associated with the innumerable small burrows on the bank and were seen to produce tracks and trails in the sediment. These animals must be a major source of food for larger fishes and other predators that occupy the bank.

The assemblage of organisms that inhabits hard bottoms above 68-m depth at Coffee Lump is similar in composition and structure to assemblages encountered on South Texas mid-shelf banks such as Southern and South Baker Banks. The most conspicuous, predominant, and abundant organisms are antipatharian whips, *Cirrhipathes;* comatulid crinoids; encrusting coralline algae; sponges, including a large population of *Ircinia campana;* large hydroids; and fragile white "bushes" of serpulid worms, *Filograna.*

Coralline algae cover up to 30% of the rocks on outcrops or ledges. Cobble-sized rocks that lie on the soft bottom frequently bear small patches of coralline algae on their upper surfaces.

A significant population of hermatypic agariciid corals was detected on the shallower parts of the larger hard-rock structures (59.5- to 64-m depth). Other small branching corals were seen on the hard bottom and solitary corals almost certainly occur there. Those parts of the rocks not covered by corals, coralline algae, or encrusting sponges are generally laden with sediment veneers or a sediment-epifauna mat.

Most of the larger fishes seen were associated with the hard bottom. Predictably, the Yellowtail reef fish, *Chromis enchrysurus,* was the most abundant and schools of Vermilion snapper, *Rhomboplites aurorubens,* were present. Other fishes encountered were typical of hard bottom at these depths throughout the northwestern Gulf of Mexico.

Shelf-Edge Carbonate Banks and Reefs

The shelf-edge carbonate banks and reefs are located on complex diapiric structures. Several are described in the following pages to illustrate the complexity of the structures and the similarity of the biotic zonation. Although all of the shelf-edge banks have well developed carbonate caps, local areas of bare bedded rocks have been exposed by recent faulting.

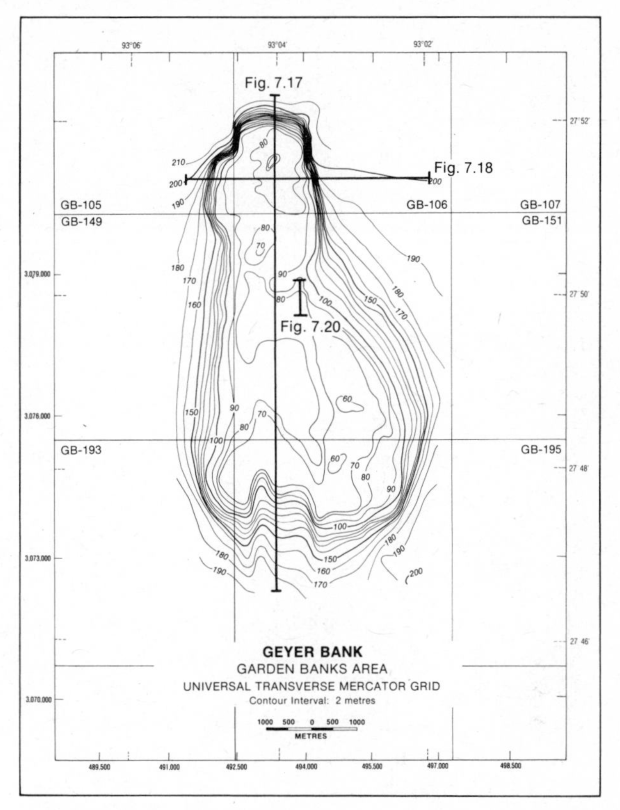

Figure 7.16. Bathymetry of Geyer Bank, showing the location of boomer profiles and the side-scan sonar records shown in Figures 7.17, 7.18, and 7.20.

Geology

Geyer Bank. Geyer Bank, located at 27°51′17″N latitude and 93°04′09″W longitude (Figure 7.1), lies in the Garden Banks Area, Lease Blocks 105, 106, 149, 150, 193, and 194 (Figure 7.16).

Situated just south of the shelf break on the upper continental slope, the bank rises from depths of 210 m on the north and 190 m on the south. It is a north–south elongated structure that covers an area of about 55 km^2. The steepest slopes are found at the northern end of the bank, gentle slopes on the east and west sides, and moderately steep slopes on the southeast and southwest. The bank is "ham-shaped," the "shank" end to the north. The top of the bank is broad and relatively flat, with prominences on the north and south ends that rise to depths of less than 60 m and are separated by a saddle around 90-m depth.

The unusual distribution of depths surrounding the bank is attributed to the position of Geyer Bank in the northeastern part of an arcuate salt diapir complex. The southern boundary of Geyer Bank is on a saddle between the bank and the next diapir to the southwest.

The internal structure of the core on Geyer Bank is difficult to interpret because the surface of the bank reflects a large part of the seismic energy. This difficulty is illustrated on the profile in Figure 7.17 by the strong surface return and the presence of a strong multiple. However, the surface configuration of the bank, as shown on Figures 7.17 and 7.18, provides some clues to the structure that are substantiated by direct observations of the bottom from the submersible during a dive made 21 October 1978.

Two structural units were mapped on the basis of the seismic profiles: (1) the nonreflective area, which underlies most of the bank proper, and (2) the surrounding stratified and tilted sedimentary sequence. The nonreflective area displays weak internal reflective patterns due to the loss of signal strength by reflection from the hard surface of the bank and the nearly vertical bedding surfaces within the bank proper. Outcrops of nearly vertical beds were observed from the submersible on 21 October 1978. The overlying stratified unit, seen on the flanks of the bank, is bounded above by the seafloor and below by a more steeply dipping sequence on which

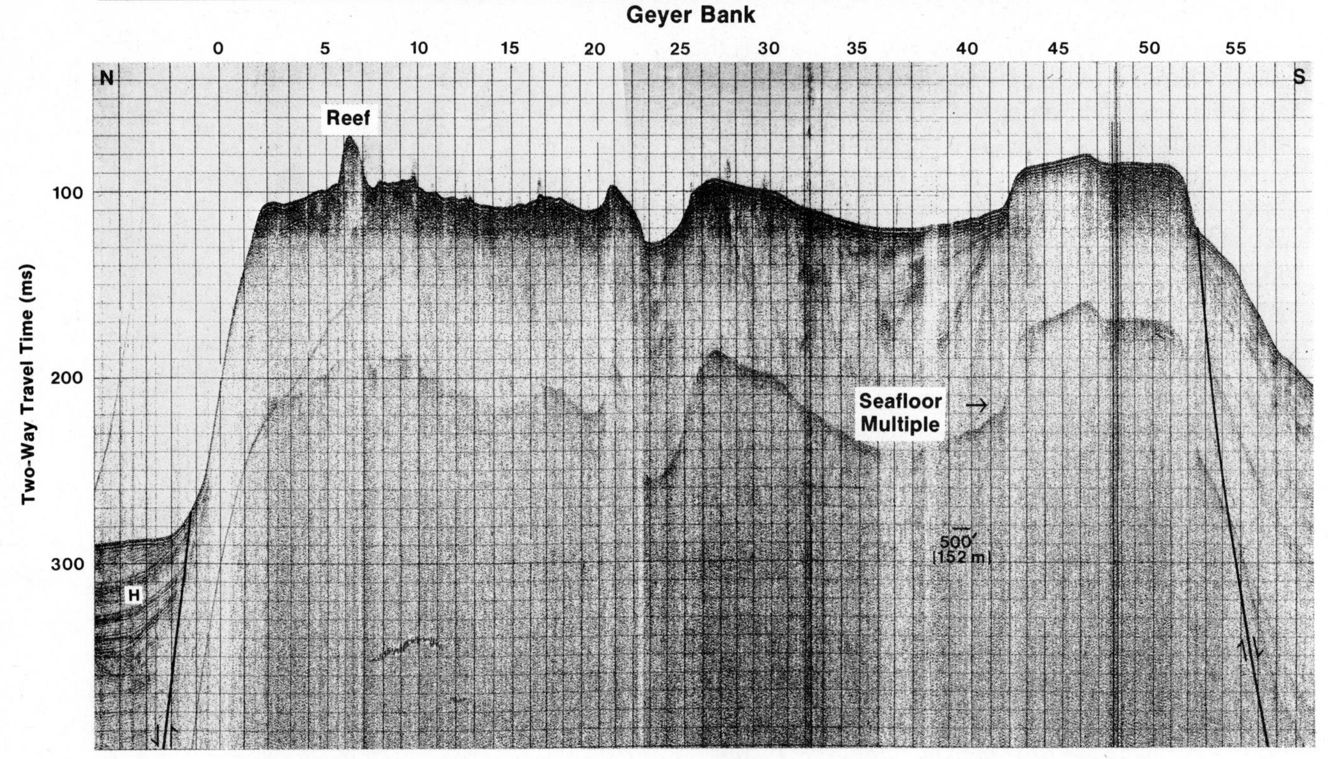

Figure 7.17. North–south boomer profile across Geyer Bank. Location is indexed in Figure 7.16.

the mapped unit overlaps (Figure 7.17 and 7.18). Contours of this sequence are isopachs that represent the thickness of the unit or depth below the seafloor to its boundary.

The steep slopes on the margins of the bank are most likely fault scarps. Only one major and a few minor faults have been mapped on the basis of subbottom data. However, the numerous nearly vertical slopes on the upper surface of the bank are probably fault scarps, which indicate that the upper part of the nonreflective unit is more extensively block faulted than appears on the structural map (Figure 7.19). The areas mapped as slumps on that figure may also be fault blocks.

The entire bank is essentially fault-bounded. The zero isopach or limit of the nonreflective unit may be considered a series of interconnected faults caused by the upward thrust of the salt diapir. The faults on the top of the bank, those inferred from topographic expression and direct observation of nearly vertically oriented beds, are most probably normal faults caused by the collapse of the crest of the bank due to the removal of salt by dissolution.

It should be noted that the darker subsurface area, just beneath the surface at the top of the bank on each of the boomer profiles, is an artifact created by a malfunction of the recorder signal gain control. It does not represent real structure.

The seismic profiles show little evidence of gas being vented or trapped in the sediments at Geyer Bank. During a submersible dive 12 October 1978, however, an area of gas seeps was encountered at a depth of 82 m. Recent movement along faults at the top of the bank is suspected because of the presence of bare rock outcrops and outcrops that are relatively heavily encrusted by coralline algae and other organisms. If all outcrops on the bank had been exposed at the same time, all would now be encrusted with nearly the same amount of reef-building organisms.

Although no sediment samples were taken on Geyer Bank, seismic and side-scan sonar data and direct observations of the bottom from the submersible give a general idea of the nature of the sediments on the bank. There is little or no evidence on the boomer and 3.5-kHz records of recent sediment cover on top of the bank. The side-scan sonar records, however, show large areas of sand waves in the central and southern parts of the bank (Figure 7.20). The wave patterns toward the perimeter of the bank appear to have a general wave structure, regularly spaced with nearly uniform amplitude and with the trend of the crest at right angles to the isobaths.

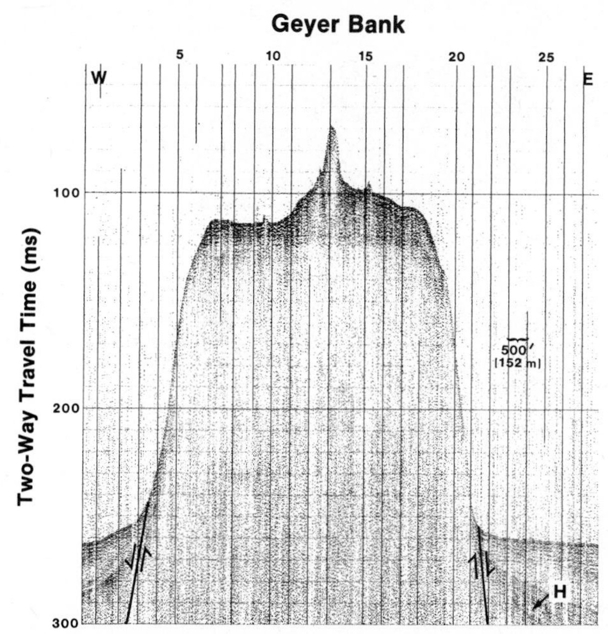

Figure 7.18. East–west boomer profile across Geyer Bank. Location is indexed on Figure 7.16.

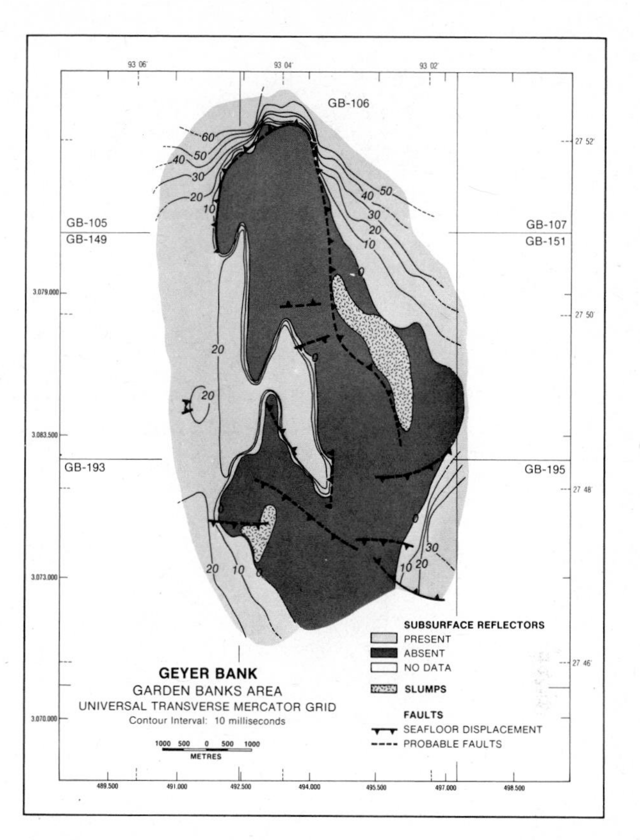

Figure 7.19. Structure/isopach map of Geyer Bank. Isopachs indicate the thickness of sediments from the surface to horizon *H* (Figure 7.18).

The sediment descriptions on the videotapes of the 12 October 1978 dive indicate large areas of algal nodules and carbonate sand between depths of 60 and 75 m. Large-scale sand waves are probably composed of algal nodules, but the small-scale sand waves (Figure 7.20) are probably equivalent to the *Amphistegina* Sand Facies at the Flower Garden Banks.

The sediment at 213-m depth on the north side of the bank is a cohesive clay. Large angular fragments were thrown up into a pile ahead of the battery pod of the submersible as it plowed into the bottom. At a depth of 187 m coralline algal nodules which must have rolled from the top were observed on the muddy bottom. Drowned reefs were observed at 98- to 94-m, 87-, 78-, 76-, and 70-m depths.

Diaphus Bank. Diaphus Bank is located at 28°05′18″N latitude and 90°42′26″W longitude (Figure 7.1 and 7.21) in Lease Blocks 314 to 317 of the South Timbalier Area. It lies close to the shelf edge and is about 50 miles west of the Mississippi Trough. The bank is rectangular and covers an area of about 33 km^2. Superimposed on this rectangle are two ridges that intersect at nearly right angles to form a rough cross. The surrounding water depths range from 110 m on the north to 130 m on the south, with increasing depths to the south, down the upper continental slope. The bank stands about 40 m above the surrounding shelf, with the shallowest depth at a peak in the center of the bank lying at 73 m (Figure 7.21).

The most prominent feature of the bank is an east–west ridge which has an extremely steep (locally about 90 m/km) and linear south side. The slope on the north side is much gentler at only 20 m/km. A smaller ridge extends to the north (about 2.5 km) and to the south (about 1.7 km) from the center of the east–west structure.

Local relief on the bank is concentrated along the crests of the intersecting ridges. This relief is related primarily to the normal faulting that is so evident in the north–south seismic profiles (Figure 7.22).

The seismic reflection records reveal that Diaphus Bank is a domal diapiric structure that has been breeched by a major down-to-the-sea, normal fault, which creates the massive, south-facing scarp that

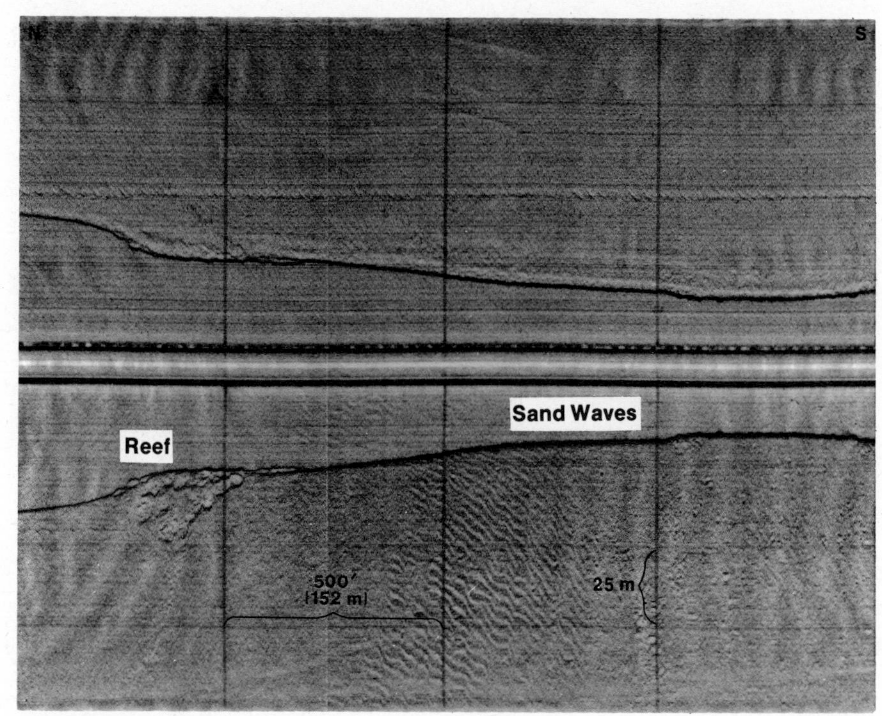

Figure 7.20. Side-scan sonar record, showing sand waves and reef. Location on Geyer Bank is indexed in Figure 7.16.

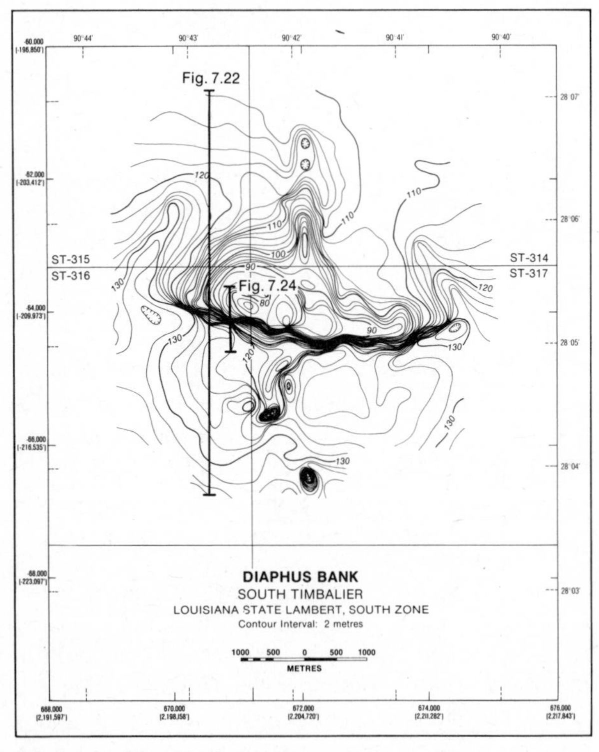

Figure 7.21. Bathymetry at Diaphus Bank, indexing location of boomer profile, and the side-scan sonar record, shown in Figures 7.22 and 7.24.

is so prominent on the bank. Radial faults have created the less prominent north–south ridge.

Two seismic sequences were mapped (Figure 7.23): (1) the exposed acoustic basement unit, which is highly reflective and consists of well stratified sedimentary rock, and (2) a poorly reflective sedimentary unit that surrounds the bank and can be seen to overlie and unconformably onlap the basement unit (Figure 7.22). The distribution and thickness of the upper unit are shown by isopach contours in Figure 7.23. The upper boundary of this unit is the seafloor and the lower boundary is the unconformity. Unlike many of the banks, the overlying sediments are not steeply tilted upward where they onlap the basement unit. This fact suggests that the doming, which produced the primary uplift and tilting of the basement unit, took place before deposition of the overlying sediments.

The dominant sediment at the crest of the western peak (76-m depth) is a coarse, carbonate sand with scattered algal nodules on its surface. This sediment occurs between large coralline algae-encrusted reef masses 4 to 6 m high (Figure 7.24). The dominant particle types in this sediment are coralline algae and *Amphistegina*. The depth at that peak is close to the lower limit of the *Gypsina-Lithothamnium* Facies at

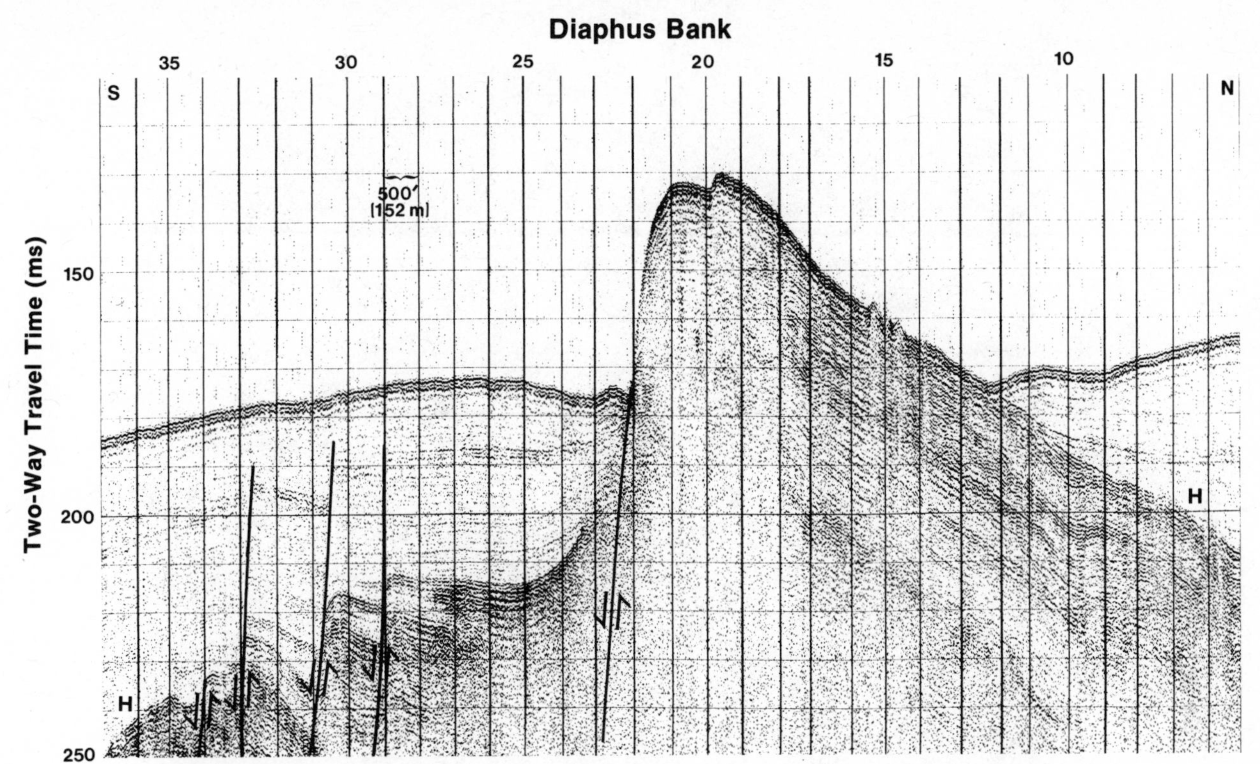

Figure 7.22. North–south boomer profile across Diaphus Bank. Location is indexed in Figure 7.21.

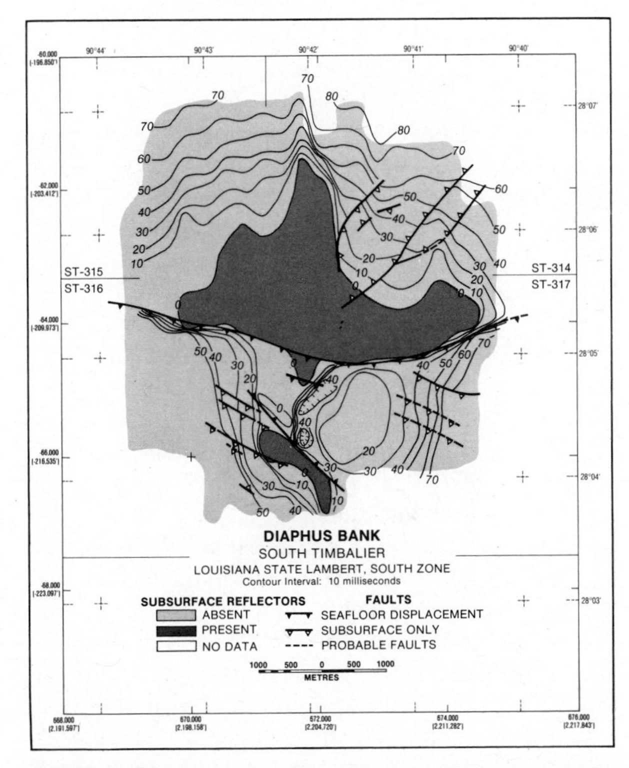

Figure 7.23. Structure/isopach map of Diaphus Bank. Isopachs indicate the thickness of sediments from the surface to horizon *H* (see Figure 7.22).

the West Flower Garden Bank. At Diaphus Bank the *Gypsina-Lithothamnium* Facies and the *Amphistegina* Sand Facies occur together, in a combination typical of the transition zone between the two facies at the West Flower Garden Bank.

Alderdice Bank. Alderdice Bank is located at 28°04′40″N latitude and 91°59′36″W longitude (Figure 7.1) in Lease Blocks 170, 171, 178, and 179 of the South Marsh Island Area (Figure 7.25). The bank is an oval, elongate in an east–west direction, and covers an area of about 16 km^2. The top of the bank is fairly flat, with depths ranging from 78 to 82 m. Superimposed on this broad surface is a smaller scale relief formed by ridges and peaks. The shallowest bank depths (59 m) are two of these peaks. Depth of the seafloor surrounding the bank on the south, west, and northwest sides is about 92 to 94 m, whereas on the northeast and east sides it is about 84 m. Although the relief is not great along the margins of the bank (generally less than 10 m), the margins are rather steep around its western half. The features on top of the western part of the bank have a rather random distribution. The eastern part, however, displays a dominant north–south ridge. There is also a gentle north–south oriented swell on

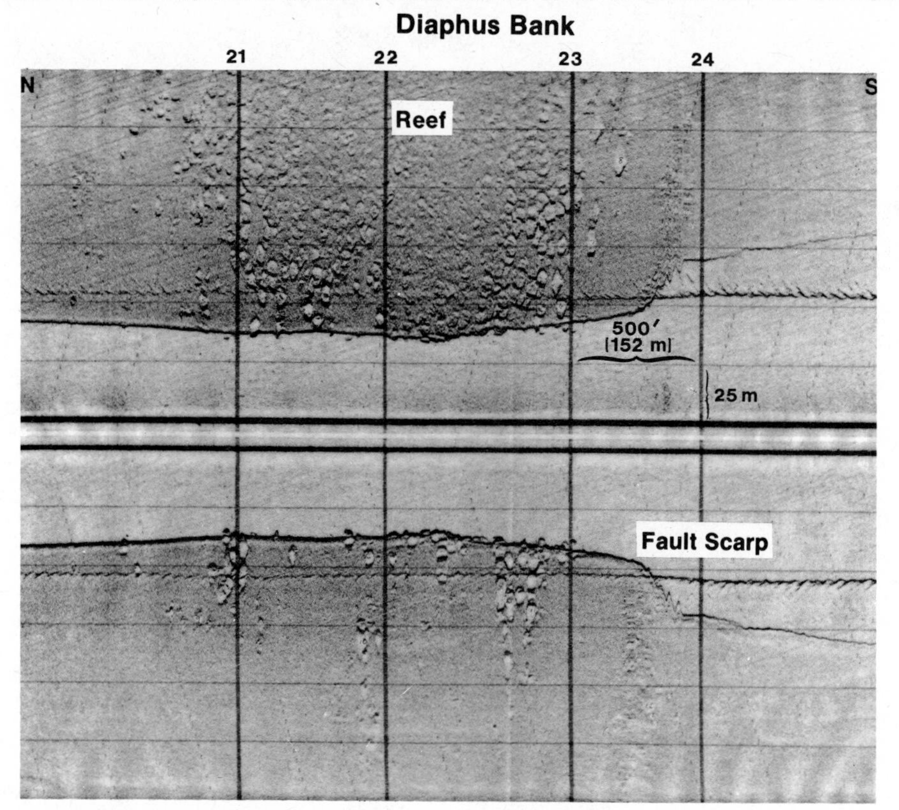

Figure 7.24. Side-scan sonar record across Diaphus Bank. Location is indexed in Figure 7.21.

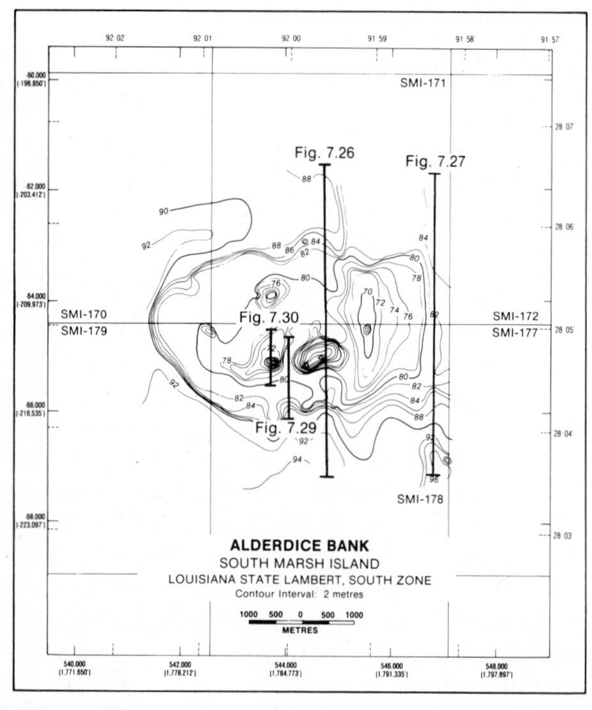

Figure 7.25. Bathymetry of Alderdice Bank, indexing boomer seismic reflection profiles shown in Figures 7.26 and 7.27, and side-scan sonar records shown in Figures 7.29 and 7.30.

the seafloor that extends northward from the eastern part (Figure 7.25). The gentle depression on the northwest margin is the head of a north–south oriented valley that curves around the western margin of the bank. This valley was probably eroded during a lower stand of sea level in Late Pleistocene or Early Holocene time.

Structurally, Alderdice Bank is a uniformly uplifted salt dome with a nonreflective core and surrounded by onlapping sediments that are tilted upward along the margins of the bank and are progressively more steeply tilted with depth (Figures 7.26 and 7.27). The main body of the bank has been mapped (Figure 7.28) as nonreflective core, which may be cap rock and salt.

The sequences mapped in Figure 7.28 are (1) nonreflective core, (2) a reflector on the east end of the bank which has been uplifted with the nonreflective core and may represent the top of the salt, and (3) the onlapping, more recent sediment around the rest of the bank. Contours in Figure 7.28 represent the thickness of the sedimentary sequences down to the deepest reflector that can be interpreted as a sequence boundary. The lower boundaries are marked near the margins of the boomer profiles shown in Figures 7.26 and 7.27.

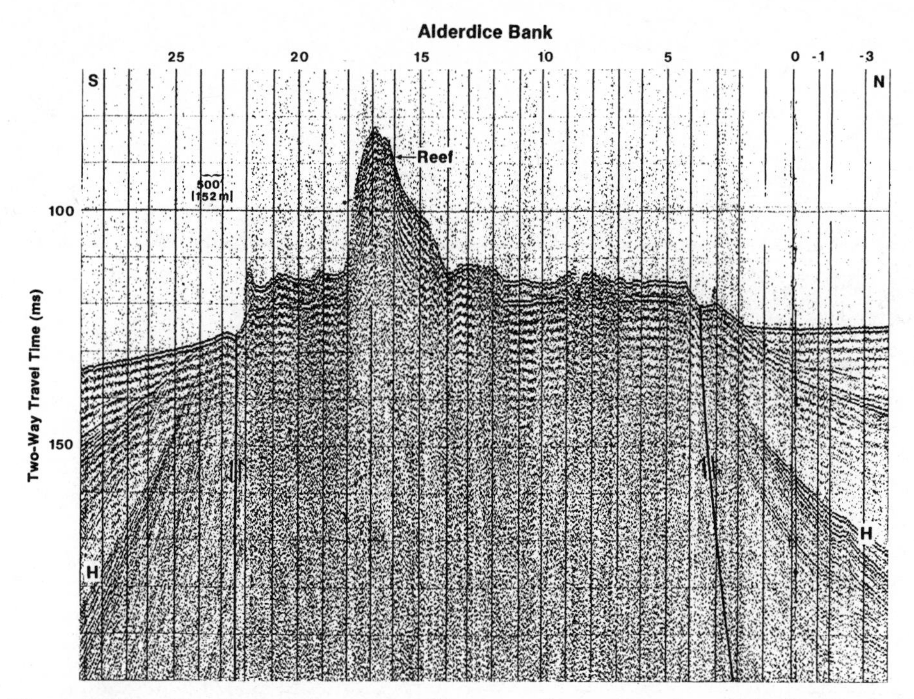

Figure 7.26. North–south boomer profile across the main peak at Alderdice Bank (see Figure 7.25 for location).

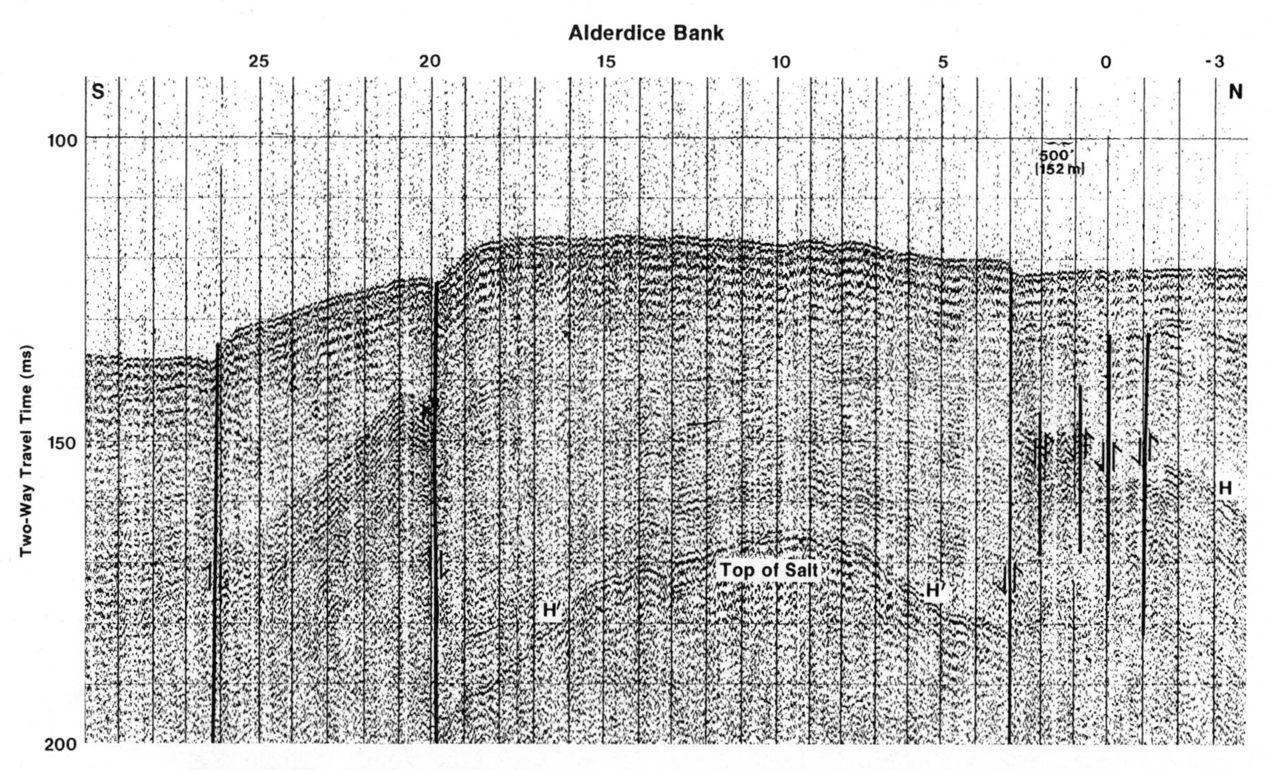

Figure 7.27. North–south profile across the east side of Alderdice Bank, showing renewed uplift of salt; *H* represents the base of the upper sedimentary sequence; *H′* is the base of the lower sedimentary sequence.

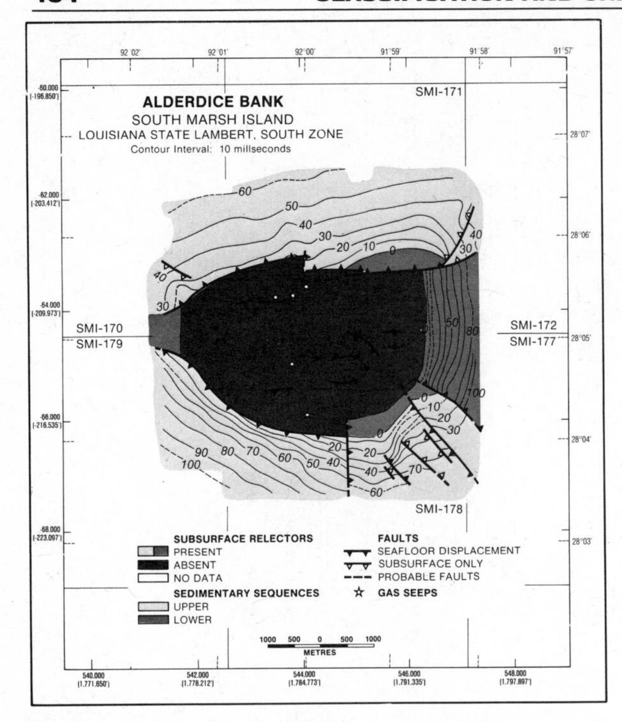

Figure 7.28. Structure/isopach map of Alderdice Bank. Isopachs are the thickness of sediment from the surface to horizons *H* and *H′* (Figure 7.27).

The surrounding sedimentary sequences, mapped in Figure 7.28, contain two prominent erosional unconformities with clear truncation of bedding reflectors on which the overlying sediments onlap toward the bank. The base of the mapped sequence is also an unconformity. The presence of the angular unconformities, as well as the increasing dips with depth, indicates that the bank has been in the process of uplift over a long period of time and includes several periods of erosion and subsequent deposition.

Two patterns of faults are present on the bank: (1) the annular fault that encircles the bank and (2) the radial faults. All show evidence of Recent activity by displacement of the seafloor (Figure 7.27). Along the eastern margin of the bank it appears that the seafloor and subbottom structure have opposite displacements along the same faults—and that is actually the case. The central block is a graben that was formed during the last regression of sea level. With renewed sedimentation following the subsequent transgression, the surface relief on this part of the bank was buried. In very recent time upward movement of salt has reversed the relative movement along these faults and the Recent sediments have been bowed upward over the graben. The directional sense of this movement can be seen where each of the faults intersects the seafloor.

Several seismic profiles and side-scan sonar records show peaks that were suspected of being outcrops of bedrock covered by some thickness of carbonate reef growth (Figures 7.26, 7.29, and 7.30). One ridge (Figure 7.30) on the southwestern peak of the bank was examined by the submersible DRV DIAPHUS. The ridge is about 100 m long, 24 m high, and 5 m wide at the base and lies at a depth of about 77 m. It is a massive ledge of nearly bare basalt that strikes 055° and dips about 80° to the south–southeast. Petrographic analysis of the rock indicates that it is an alkalic basalt (Rezak and Tieh, 1980). K-AR age determination of the bulk rock by Geochron Laboratories yielded an age of 76.8 ± 3.3 my (Senonian). This is the oldest known rock exposed on the continental shelf off Louisiana and Texas.

The feature is interpreted as a dike or sill rafted to the surface by the salt diapir. It has been exposed at the seafloor due to dissolution of the surrounding salt and the subsequent collapse of the adjacent cap rock on both sides of the feature. This implies a sizable root zone still embedded in the salt. Similar features have been observed on Red Sea salt domes in East Africa and in the Zechstein region of Germany. Mounting evidence of Late Mesozoic igneous activity and published multichannel seismic data (Martin, 1978; Humphris, 1978) strongly indicate a rifted origin for the Gulf of Mexico.

Faulting occurs over the entire bank and surrounding seafloor, as revealed by the discontinuous outcrop patterns on the side-scan sonar record (Figure 7.29 and 7.30) and by the displacement of reflectors in the boomer records (Figure 7.27). Evidence of recent movement along faults may be found in the outcrop of basalt and on boomer records. The basalt outcrop is covered by a rather thin crust of coralline algae, sponges, and bryozoans. If this rock had been exposed at the seafloor since Late Pleistocene time we would expect more massive encrustations over the bedrock outcrop like those on the peak just to the east and on other banks such as the Flower Gardens. The presence of such thin crusts suggests a relatively short time span for colonization by encrusting organisms.

A diffuse pattern of reflections in the water column over most of the bank, particularly the western half, is probably due to general gas seepage from nearly vertical beds seen on the side-scan records. Specific vents are also evident over the western part of the bank (Figure 7.28), one of which lies just to the east of the basalt outcrop.

Submersible observations indicate that below 82 m the sediment is primarily fine mud. The bank

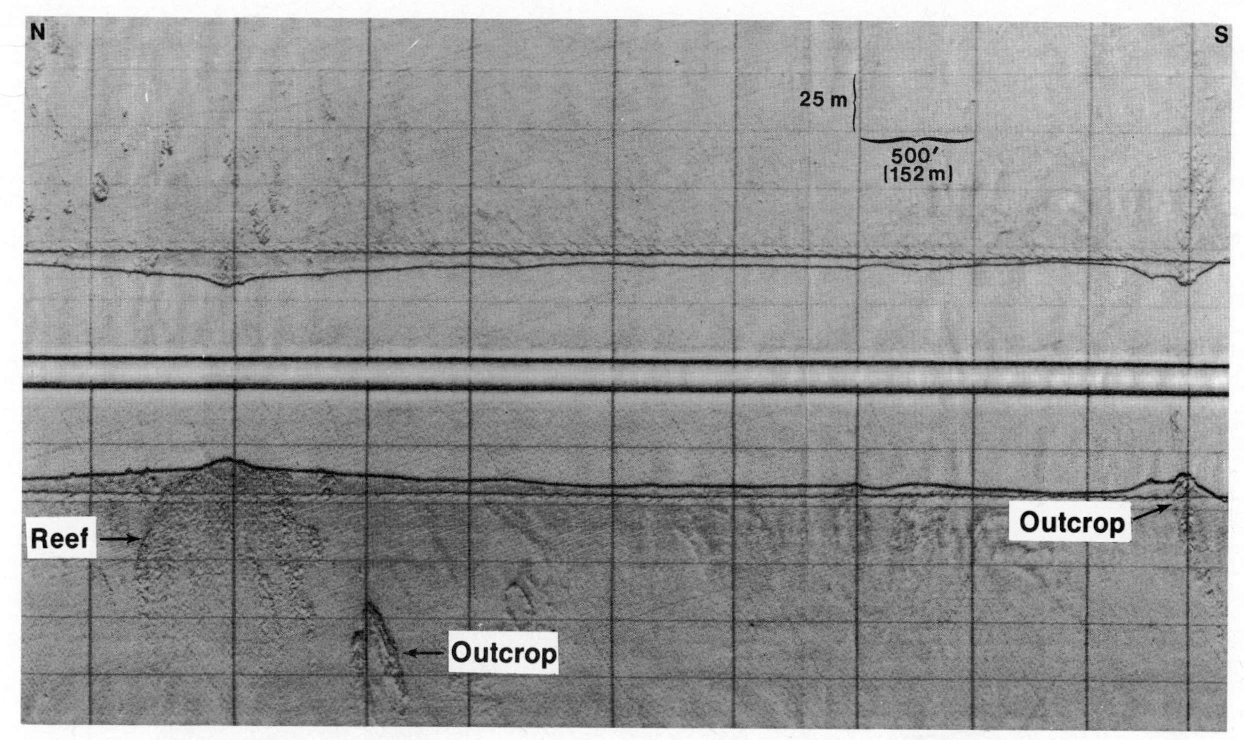

Figure 7.29. Side-scan sonar record across outcrops and reefs at Alderdice Bank (see Figure 7.25 for location).

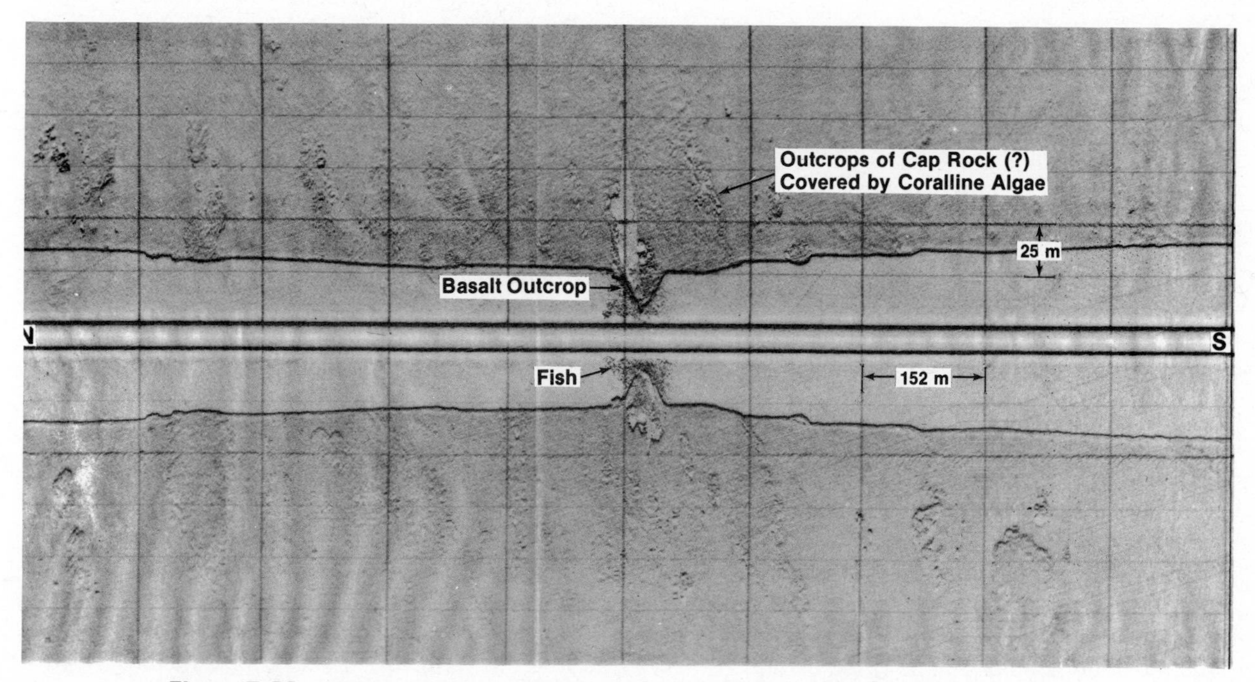

Figure 7.30. Side-scan sonar record across basalt outcrop at Alderdice Bank (see Figure 7.25 for location).

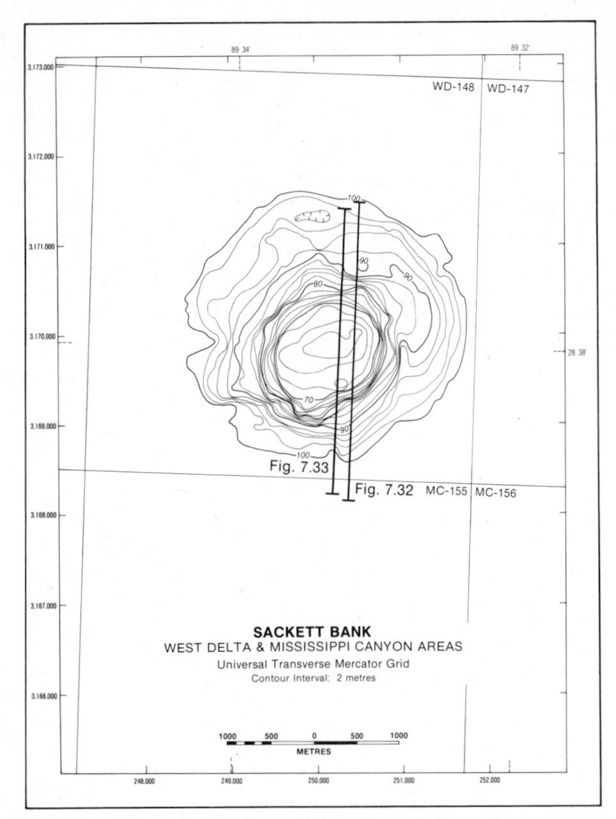

Figure 7.31. Bathymetry of Sackett Bank, indexing location of boomer profiles, and the side-scan sonar records shown in Figures 7.32 and 7.33.

sediments above 75 m consist of carbonate sands and gravels that are representative of the *Gypsina-Lithothamnium* Facies and the *Amphistegina* Sand Facies. Below that depth and down to 82 m the sands and gravels are mixed with silt- and clay-sized sediment.

Sackett Bank. Sackett Bank is located at 28°38′01″N latitude and 89°33′22″W longitude (Figure 7.1) in Lease Block 148 of the West Delta Area (Figure 7.31), close to the shelf edge and about 12 miles east of the Mississippi Canyon. The bank is nearly circular and covers an area of 7.07 km^2. Its crest is broad and relatively flat at a depth of 63 m and its base lies at a depth of about 100 m.

Structurally, the bank is a nearly symmetrical diapir with a nonreflective core and well stratified and faulted upturned beds on its flanks (Figure 7.32 and 7.33). The boomer profile (Figure 7.33) shows hazy reflectors in the subsurface as evidence of gas-charged sediments. Figure 7.33 also shows two active gas seeps in the core of the bank. Figure 7.32 shows gassy zones on the north side of the bank and a gas seep on the south side of its core. The fault patterns on Sackett are mostly oriented in a near east–west direction except for the southwest flank of the bank, where they are oriented northwest–southeast (Figure 7.34). Most of the faults do not displace the seafloor, an indication that Sackett is relatively stable.

Above 65-m depth the bottom sediment is predominantly sandy, with scattered algal nodules. Scattered about on this upper terrace are drowned patch reefs and pinnacles up to 3 m high and 12 m across. Limited amounts of living coralline algae occur on the patch reefs, tops of nodules, and pieces of rubble. The substrate of the 67- to 73-m terrace is basically a carbonate rubble-strewn sandy bottom with significant amounts of silt and clay.

At about 72-m depth a small outcrop of siltstone or claystone, similar in appearance to the rocks at Stetson and Sonnier Banks, was encountered. The presence of bored and relatively unencrusted bedrocks at Sackett does not imply recent exposure because the carbonate productivity at Sackett has been severely retarded by the turbid conditions in this area induced by the outflow of the Mississippi River. Below a depth of 82 m the sediment is primarily silt and clay. In September 1977, when Sackett was explored with the submersible, the top of the nepheloid layer was roughly coincident with the break in slope at 72- to 74-m depth. This accounts for the increased mud content of the sediment with increasing depth.

Rezak–Sidner Bank. Rezak and Sidner Banks are described as a single unit because they belong to a single geological structure, the center of which forms two banks located at 27°57′N latitude and 92°23′W longitude (Figure 7.1) in Lease Blocks 404, 405, 411, and 412 of the Vermilion Area (Figure 7.35). The Rezak–Sidner structure is rectangular in shape and covers an area of 78 km^2. It is bounded by steep slopes on the north, east, and south sides and a gentler slope to the west (Figure 7.35). Local depressions are abundant at the base of the northern, eastern, and southern slopes, but the eastern slope has irregular and complex relief. Although it is a single structural unit, the bank can be divided into northern and southern halves on the basis of bathymetry. The shallowest portion of the northern half (Rezak Bank) has a minimum depth of 60 m on a peak at the northeast corner of the structure. Depths of the adjacent seafloor around the northern half are 120 to 140 m on the north and east sides and 98 to 110 m on the west. The northern half is about twice the width of the southern half and has a gentle slope

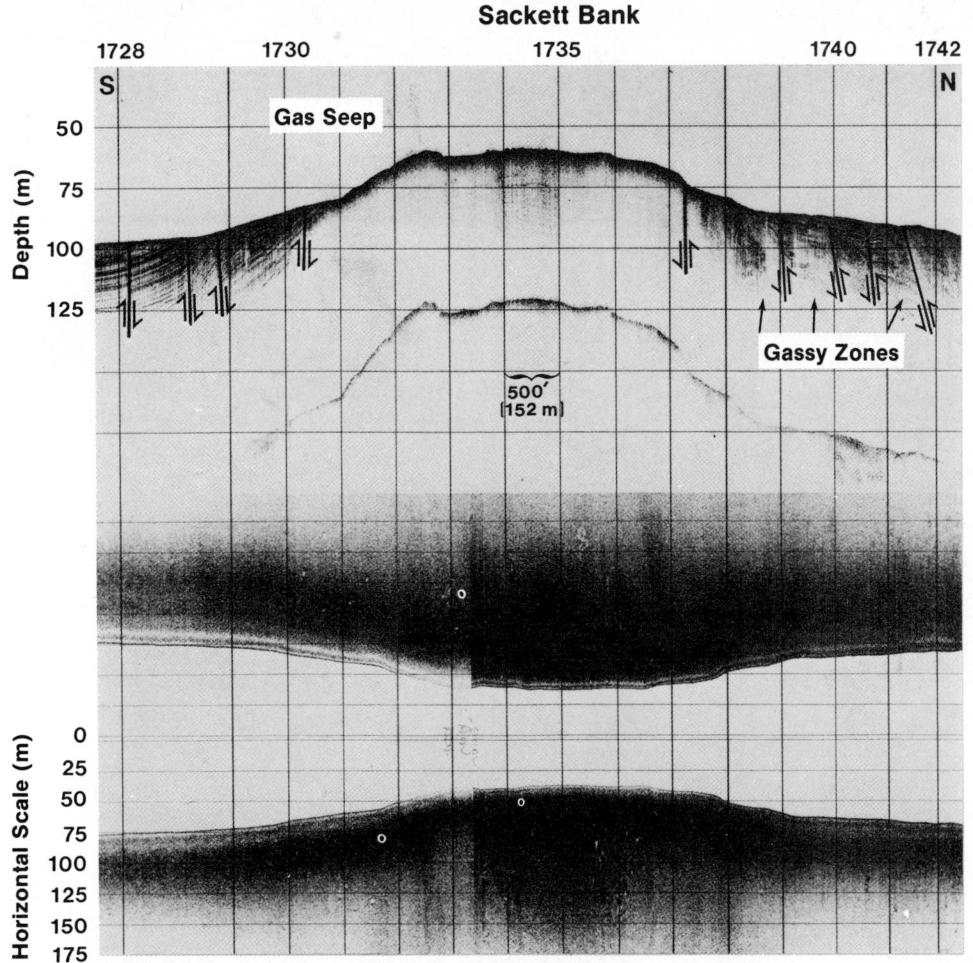

Figure 7.32. A 7-kHz seismic profile and side-scan sonar record across Sackett Bank (see Figure 7.31 for location).

to the west (about 10 m/km). The southern half has a minimum depth of 55 m on the ridge that forms the eastern margin and surrounding depths of 140 to 180 m.

Structurally, the bank is a tilted fault block of well-stratified sedimentary rock that has been uplifted on the east and dips to the west (Figure 7.37). It is bounded by steeply dipping normal faults on its east, north, and south sides. As seen in the north–south boomer profiles in Figure 7.36, the shallow structure of the bank consists of several seismic sequences separated by strong reflectors interpreted as unconformities. The stratification in each sequence is represented by weak yet recognizable reflections. Care must be taken not to confuse the bedding reflections with the strong ringing of the seismic source. Despite the troublesome ringing, it is possible to determine reflection termination patterns that define erosional truncation, depositional onlap, and truncation by faulting (Figures 7.36 and 7.37).

Seafloor relief on the top of the bank is due to faulting, erosion, and, to a lesser extent, carbonate reef growth. The reflectors within the fault block are truncated by faulting on the north and south sides of the bank. The saddle in the middle of the bank appears to be a structural and erosional feature. A large part of the upper seismic sequence has been removed by erosion (Figures 7.36 and 7.38). The erosional unconformities, the surface erosion, and the progressively greater tilt of the individual seismic sequences with depth suggest that this bank has experienced several stages of uplift and tilting during a period of several rises and falls in sea level.

Faulting has been active in Recent time as revealed by the numerous faults that displace the seafloor along the margins of the bank. Faults are not common within the main body of the bank, but two normal faults control the topography of the saddle in the middle.

One apparent gas vent exists in the southwest corner of the bank (Figure 7.38) and another was

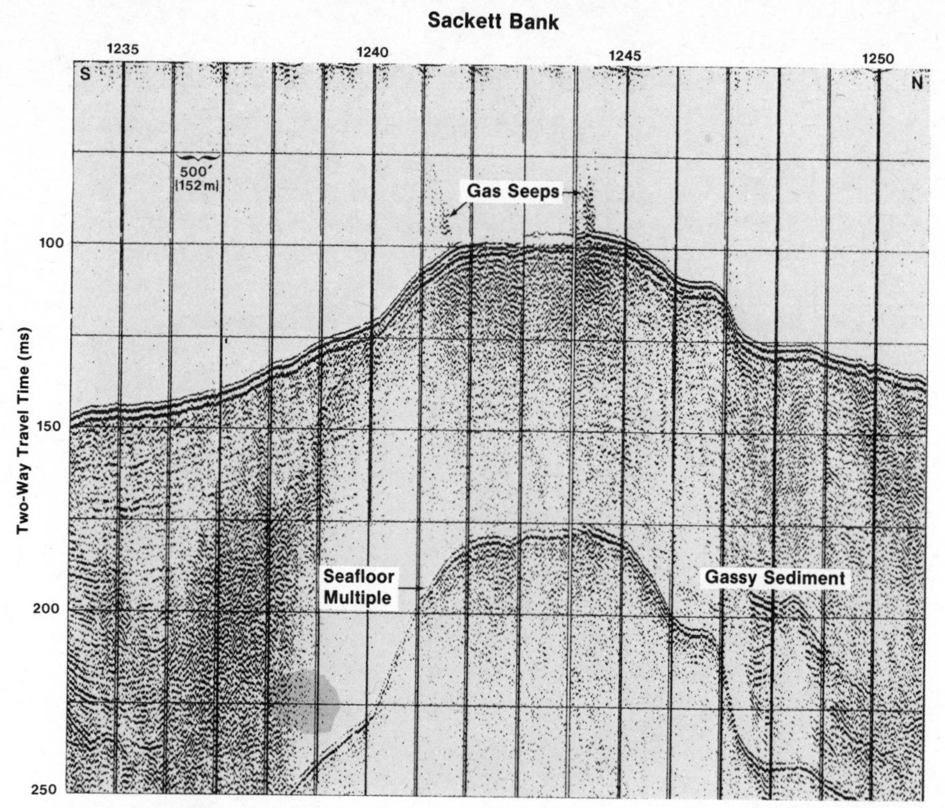

Figure 7.33. A boomer profile across Sackett Bank (see Figure 7.31 for location).

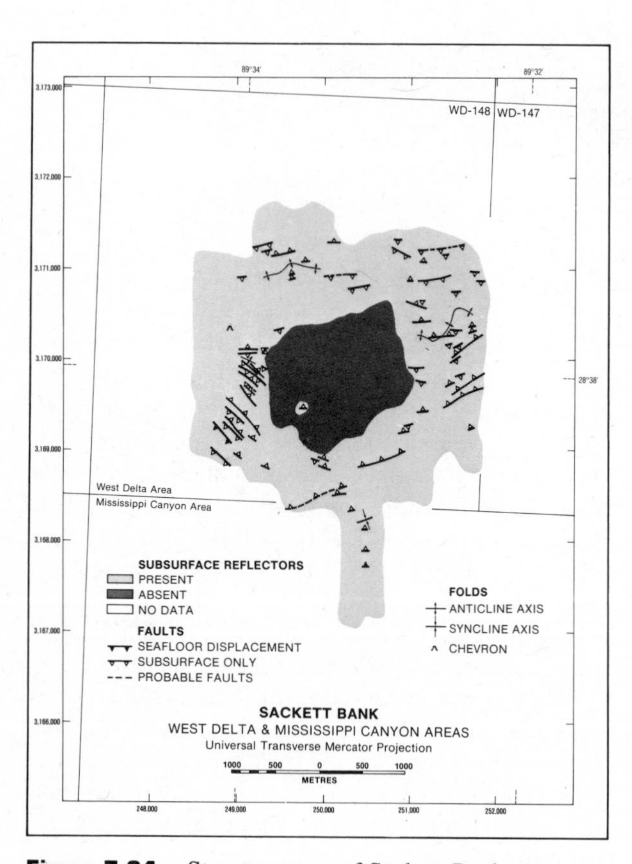

Figure 7.34. Structure map of Sackett Bank.

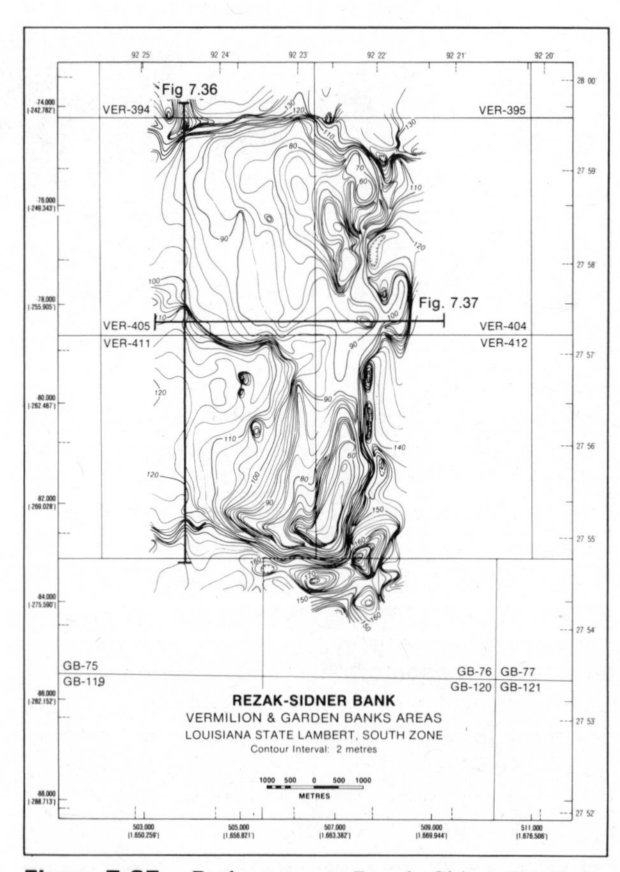

Figure 7.35. Bathymetry at Rezak–Sidner Bank; indexing location of boomer profiles shown in Figures 7.36 and 7.37.

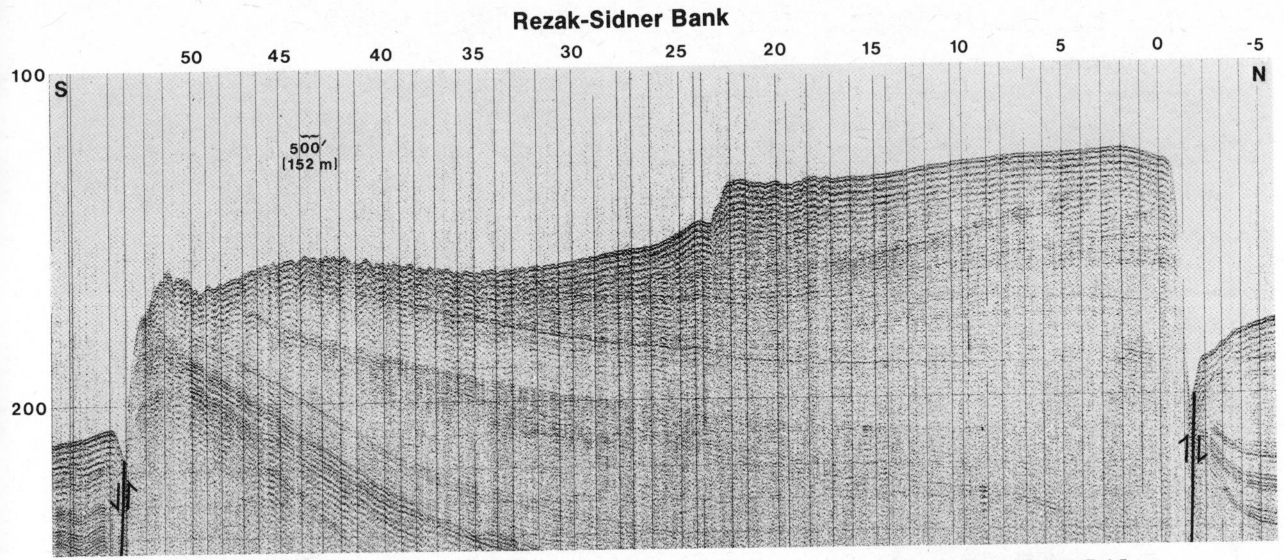

Figure 7.36. North–south boomer profile across Rezak–Sidner Bank (see Figure 7.35 for location).

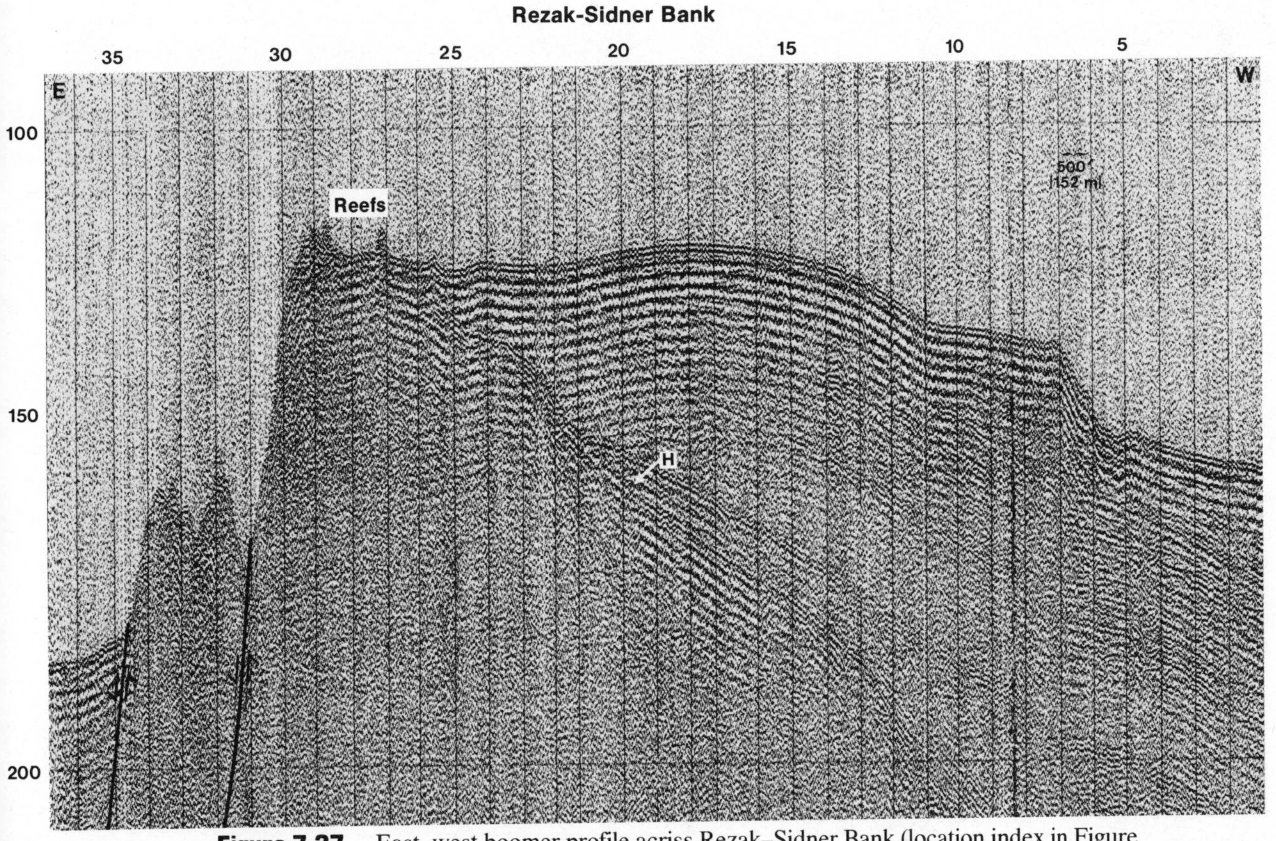

Figure 7.37. East–west boomer profile acriss Rezak–Sidner Bank (location index in Figure 7.35).

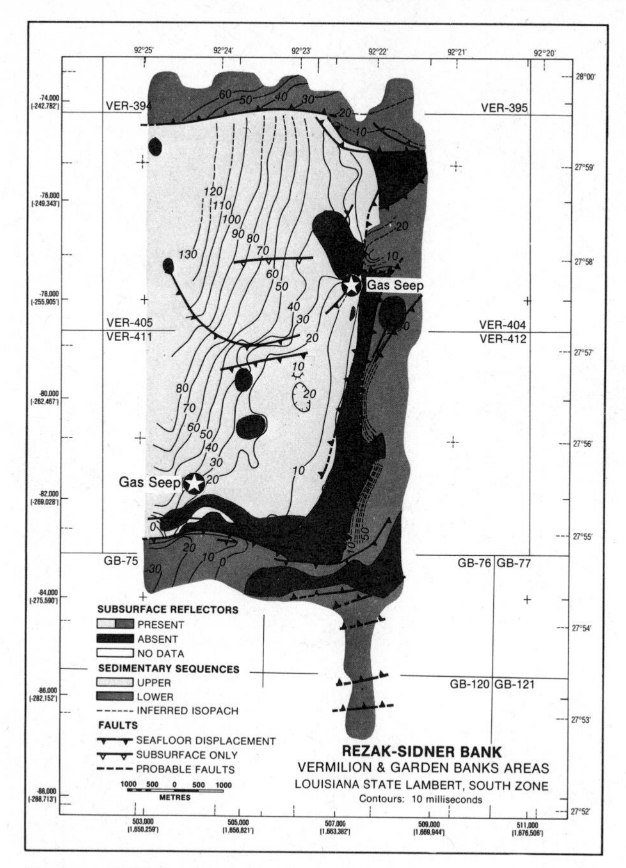

Figure 7.38. Structure/isopach map of Rezak–Sidner Bank. Isopachs indicate thickness of sediment from the surface to horizon *H* shown in Figure 7.37.

observed during a submersible dive at a depth of 69 m just south of the 62-m peak in the southwest part of Lease Block 404.

Submersible observations indicate that the dominant sediments at the top of the bank are coralline algae nodules and a coarse carbonate sand which are equivalent to the *Gypsina-Lithothamnium* Facies and *Amphistegina* Sand Facies at the Flower Garden Banks.

Summary of Structural Geology

The complexity and variety of structural style in the banks described here and in Chapter 4 are due to a number of factors, the most important of which are (1) the regional stress field, (2) the shape of the dome (circular or elliptical), and (3) structures inherited from Late Triassic to Early Jurassic tectonics.

The influences of outline shape and regional strain on salt dome fault patterns have been amply demonstrated by Withjack and Scheiner (1982), who used experimental and analytical models. They found that circular domes in the absence of regional stresses developed radial fault patterns like those we observed at Fishnet Bank (Figure 7.10) and possibly Sonnier and Sackett Banks. With regional extension most normal faults on the crests of circular domes trend in a direction perpendicular to the direction of regional extension. Diaphus Bank (Figures 7.22 and 7.23) is a good example of this kind of fracture pattern. Diaphus lies close to the shelf edge. The major fault on the bank parallels the shelf edge.

With regional compression during doming, most normal faults on the crests and flanks of circular domes strike parallel with the regional compression direction. Strike slip faults that trend 30° from the regional compression direction may form on the flanks. Thrust faults that strike approximately perpendicular to the regional compression may develop on the peripheries. None of the salt domes that we have mapped shows evidence of regional compression. This is most likely because of the location of our study area on the middle and outer continental shelves, where regional compressive stresses are absent. Regional compressive stresses are more likely to be found on the middle to lower continental slope.

With elliptical domes that are not involved with external stresses the pattern of normal faults roughly parallels the long axis of the ellipse but the faults splay outward toward the ends of the long axis. With regional extension the normal faults trend perpendicular to the regional extension direction. For regional compression the faulting is similar in its orientation to that found in circular banks.

Unfortunately, few domes are perfectly circular or elliptical, as revealed by the illustrations in this chapter and Chapter 4. Dome shapes may be strongly controlled by preexisting regional structures. The strongest evidence of the control of shape by preexisting structures is Rezak–Sidner Bank, the structure of which is a rectangular block bounded by normal faults on three sides. The east-facing fault scarp of the bank has a displacement of at least 130 m. Salt domes, by definition, are roughly circular in outline and the overlying beds dip in all directions away from a point. Therefore, according to the definition, Rezak–Sidner Bank is not a salt dome.

If it is not a salt dome, then what is it? The only tectonic processes known to be active on the outer continental shelf are salt diapirism and gravity faulting. Salt diapirs, as we have described them, should be circular or elliptical in plan view. Gravity faults are linear features that parallel the shelf break. Rezak–Sidner Bank is neither of these. The major fault at Rezak–Sidner Bank trends north–south and is approximately perpendicular to the shelf break. Trippett and Berryhill (1982) show that Rezak–Sidner Bank is part of a northwest–southeast trending series of banks formed by a ridge of salt at depth. Many of the banks on the shelf break and the upper slope on this map (Trippett and Berryhill, 1982) appear to belong to complex salt ridges with arcuate patterns. Figure 4.4 demonstrates that the East Flower Garden is located at the intersection of two salt ridges, and Figure 4.10 shows that the nonreflective area at the West Flower Garden Bank is linear rather than circular. Geyer Bank (Figure 7.16) and Elvers Bank (Rezak and Bright, 1981a) are situated on arcuate bathymetric prominences. Their nonreflective cores are also linear rather than circular. The linearity of these cores must be due to the presence of preintrusion zones of weakness along which the salt was intruded. These zones of weakness may be joints or fault systems inherited from tectonic features occurring before salt was intruded. Many of the banks on the upper slope and outer shelf are complicated by regional extension. Parallel normal faulting, oriented at right angles to the regional extension direction, is common on the East Flower Garden, West Flower Garden, and Geyer Banks. At the East Flower Garden the faulting does not parallel the crests of the intersecting salt ridges but is nearly parallel to the shelf break. At the West Flower Garden and Geyer Banks the faulting is more complex. Some faults parallel the shelf break and some, the salt ridge crest.

Consequently, the patterns of faulting will vary, depending on the developmental history of a salt diapir. Those diapirs that are not associated with preinjection tectonic features will be circular or elliptical in plan view. Those that are associated with preinjection tectonic features will assume the pattern of those features. These two forms are extreme end members in a spectrum of structural styles that lies between them. They are controlled by the history of changes in the regional stress field at a given location on the shelf or slope.

Biology

Fundamentals of biotic zonation on shelf-edge carbonate banks and reefs were presented in Chapter 6, which described the East and West Flower Garden Banks. The Flower Gardens differ from the other banks because they alone possess relatively shallow (15- to 36-m) high-diversity coral reefs (*Diploria-Montastrea-Porites* Zone). Lower diversity coral reefs occur at the Flower Garden Banks (*Stephanocoenia-Millepora* Zone) and on 18 Fathom Bank (*Stephanocoenia-Montastrea-Agaricia* Zone) and Bright Bank between 36- and 52-m depth (Table 7.1). The heartier corals of these low-diversity reef zones (*Stephanocoenia michelini* and *Millepora alcicornis*) have been observed on many of the other banks and probably occur on most.

In terms of amount of substratum occupied, however, the Algal-Sponge Zone assemblage is the most important clear-water community on shelf-edge banks. Its presence is indicative of year-round tropical–subtropical oceanic conditions. Its lateral and vertical distribution and apparent diversity and abundance provide important clues to the degree to which the warm, clear oceanic water mass which facilitates its development is modified by more turbid, less saline, and seasonally cooler neritic waters. All of the shelf-edge banks possess transitional zones (roughly equivalent to the Antipatharian Zone of the South Texas mid-shelf banks) which grade downward from the Algal-Sponge Zone (where present) into turbid deep-water communities (Nepheloid Zone) at the base of the banks.

To supplement the preceding description of the Flower Garden zonal pattern, the biota of four additional banks are illustrated. Geyer Bank represents a system located as far offshore as possible on the continental shelf. Therefore it is subject to minimal neritic influence and has maximal surrounding depths (approximately 200 m). The high relief of this bank, and of its neighbor, Elvers Banks, above the deep surrounding mud bottom prevents chronically turbid waters from rising much shallower than 100 m on the bank, where they could interfere with the development of the Algal-Sponge Zone.

Alderdice Bank represents the other extreme by existing as far inshore as a bank may and still support an Algal-Sponge Zone. Unlike Geyer Bank, it is more exposed to the influence of neritic water masses, and, because of the shallow surrounding depths (85 to 90 m), turbid bottom waters frequently approach its crest, to which the Algal-Sponge Zone is restricted.

Sackett Bank is a special case because it is closest to the mouth of the Mississippi River. Algal-Sponge Zone assemblages, although present, are weakly developed on this bank. Diaphus Bank, the next shelf-edge bank west of Sackett, apparently is subject to some degree of coastal influence. Consequently, reef building is diminished and biotic communities are less well developed compared with those of the shelf-edge banks farther west. These factors and others are assessed in addition to the bank descriptions in an effort to explain regional patterns of biotic community structure and abundance on outer continental shelf hard banks in the northwestern Gulf of Mexico.

Geyer Bank. Zonation of benthic biota on Geyer Bank is correlated with substratum type and depth. The community established on rock outcrops at the crest of the bank bears a substantial resemblance to those occupying outcrops at similar depths on Stetson and Sonnier Banks and is recognizable as a *Millepora*-Sponge community (Figure 7.39). The hydrocoral *Millepora* is particularly conspicuous near the crests of bedrock peaks and outcrops (the peak crests examined were at 37 and 49 to 52 m). Sponges, including *Neofibularia nolitangere, Agelas,* and various massive demosponges, are possibly more evenly distributed and predominate on the steeper slopes or cliffs. Coralline algae occur as crusts and patches on the rock but are not significant reef builders in the *Millepora*-Sponge Zone. Leafy algae are abundant locally on the outcrops, particularly on their crests. Only two small colonies of hermatypic corals were seen—one a platelike crust of an agariciid and the other a monocentric variety with small polyps. The large anemone *Condylactis gigantea* is conspicuous and abundant on the bedrock (in places it is the prevalent invertebrate) and *Diadema antillarum* and *Spondylus americanus* populations are substantial.

The most abundant fish to frequent the outcrops are Yellowtail reeffish, *Chromis enchrysurus,* and Creole-fish, *Paranthias furcifer.* Other conspicuous species include the Marbled grouper, *Epinephelus inermis;* Rock beauty, *Holacanthus tricolor;* and, at the 37-m peak, the Brown chromis, *Chromis multilineatus.* At least 28 additional species were encountered.

Much of the rock is not occupied by epifauna, possibly because it is poorly cemented and apparently disintegrates easily. The rock is riddled by holes, probably produced by rock-boring pelecypods. Unconsolidated sediments on the peak-tops are composed primarily of rock chips, *Millepora* rubble, and shell fragments. Finer erosional products occur in "channels" on the less steep slopes of the bedrock peaks. Near the bases of the slopes talus aprons of cobbles and chips occur, the lower portions of which (64 m or so) are mixed with carbonate gravel and algal nodules derived from the surrounding terrace.

Between 64 and 83 m the biotic communities and surficial sediments are dominated by calcium carbonate-producing organisms, primarily coralline algae with populations of small branching corals of the genus *Madracis*. Coralline algal nodules and gravel underlain by coarse carbonate sand constitute most of the substratum between 64- and 76-m depth. The nodules in this range are generally large and their exposed surfaces are almost totally covered with growing coralline algae engaged in the production of carbonate substratum.

Patch reeflike structures occur on Geyer Bank from 64 to at least 98-m depth. The largest were encountered at 64 to 67 m near the bedrock outcrops and at 87 to 95 m at the edge of the upper platform of the bank. Some of these structures were 3 to 5 m in height. Smaller patches ⅓ to 1 m high were seen between 70 and 98 m. All were covered with healthy populations of living coralline algae, but the abundance of coralline algae decreased substantially below 90 m.

Although coralline algae are certainly the most important substratum producers and reef builders on Geyer Bank, immense populations of the small hermatypic branching coral *Madracis* were encountered between 76- and 82-m depth. On the reconnaissance transect live *Madracis,* and *Madracis* remains, composed almost all of the substratum from 79- to 82-m depth. Wherever such populations occur *Madracis* must be the dominant frame-building organism.

Associated with the *Madracis* populations are extensive covers of a flattish, maroon, leafy, calcareous alga, *Peyssonnelia,* which in places occupies 50 to 90% of the bottom on top of *Madracis* remains. This alga is an effective competitor for bottom space that otherwise would probably be occupied by living *Madracis* and consequently may locally retard substratum production by the corals. Other leafy algae occur on the bank at all depths down to 107 m. Clusters and individual stalks of the calcareous green alga *Halimeda* were seen at 61 m on the bedrock outcrops and down to 76 m on the nodule terrace.

Other conspicuous and abundant invertebrates on the part of the bank dominated by coralline algae or *Madracis* include small saucer-shaped agariciid coral colonies (down to 82 m), ellisellid sea whips (72 to 77 m), *Cirrhipathes* (throughout the zone, as well as shallower and deeper), clusters of "worm-shell" gastropods embedded in the sponge *Chelotropella*

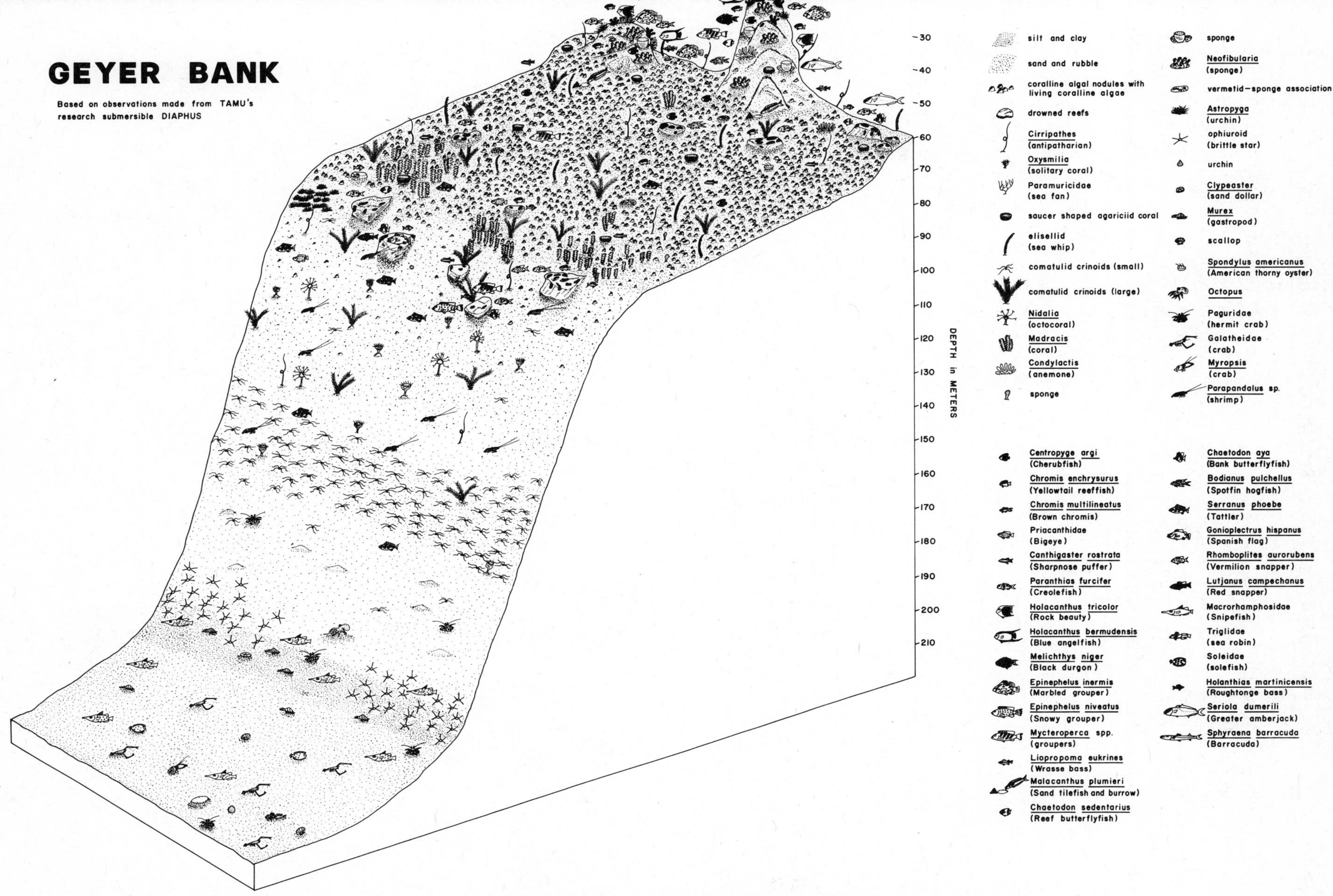

Figure 7.39. Biota of Geyer Bank.

(abundant from 72 to 79 m, attaining the size of a bushel basket), large crinoids (72 m and deeper, tremendous populations between 75 and 81 m), and small yellow sponges of unknown identity, which are extremely abundant between 76 and 78 m depth. The diversity and abundance of invertebrates are probably greater among the algal nodules and *Madracis* (67 to 82 m) than elsewhere on the bank.

The fishes most frequently seen on the nodule and *Madracis* bottom were small: Sharpnose puffer, *Canthigaster rostrata;* Orangeback bass, *Serranus annularis;* Cherubfish, *Centropyge argi;* and Yellowtail reeffish, *Chromis enchrysurus.* Many very small fishes were seen darting about among the algal nodules but none could be identified. Evidence of burrowing by the Sand tilefish, *Malacanthus plumieri,* was found above 72 m.

Where reefal structures occur, more and larger fishes congregate: numerous Creole-fish, *Paranthias furcifer;* Reef butterflyfish, *Chaetodon sedentarius;* Routhtongue bass, *Holanthias martinicensis;* and 25 additional species, including snappers *(Lutjanus* and *Rhomboplites aurorubens),* groupers of the genus *Mycteroperca,* and, at 97 m, Snowy groupers, *Epinephelus niveatus.*

At approximately the lower limit of the profuse growth of small *Madracis,* 82 m, the slope increases to 5° or so. An additional, and greater, slope increase occurs at 91 m. These breaks in slope mark the edge of the upper bank platform and the upper part of an interesting zone of sedimentological and biological transition on the bank slope which extends downward to the mud bottom near 189 m. Scattered algal nodules and sizable carbonate gravel were found as deep at 197 m, but it is felt that most of this material encountered below 91 m has been carried downslope from the upper platform. Generally, below 82 m, the unconsolidated bottom contains less and less coarse carbonate gravel and sand. At 104 m the sediment is a combination of sand and finer grained material with a 30 to 40% cover of carbonate gravel. At 128 m it has graded to a mixture of silt, clay, sand, and gravel, and at 177 m it is mostly muddy, with silt- and clay-sized particles and some remaining coarser material. At 198 m the bottom is sticky, coherent mud and the slope is slight compared to that existing between 91 and 189 m. Although the few sizable algal nodules found on this slope appear to be washed down from above, they were observed in places to be spaced as closely as 1 or 2 m, even at 128-m depth. No living coralline algae were seen on these nodules below 113 m.

A diverse and abundant assemblage of attached epifauna is established on the parts of these nodules not covered by coralline algae. This richness and diversity extend downward to more than 140 m depth and substantial epifauna were seen on nodules as deep as 162 m. The organisms found on the nodules of the bank slope are not the types that would occur on hard substrata on the upper bank platform. They are typical of hard substrata on the deeper flanks of other banks in the northwestern Gulf at similar depths. The most conspicuous attached invertebrates associated with these nodules are small sponges (generally more abundant above 110 m); solitary corals, probably *Oxysmilia* (100 to 160 m); branching corals with the appearance of *Oculina* (123 to 152 m); the octocoral, *Nidalia* (105 to 131 m); and small octocoral fans. The fans are generally oriented parallel to the slope of the bank, indicating a predominance of currents running horizontally.

The bank slope harbors impressive populations of echinoderms. At 91 m an aggregation of several hundred (possibly 700 or more) large black urchins, *Astropyga magnifica,* was seen on the sand and gravel bottom. These aggregations occur on other banks in the northwestern Gulf at similar depths on similar bottoms. They continually cycle surficial sediment through their guts and egest small spherical fecal pellets through the dorsal aboral pore. During examination of an aggregation of this urchin at least two were observed to emit apparent reproductive products from five pores that surround the aboral gut opening. The white reproductive fluid was shed by one urchin after being touched by the submarine's manipulator arm. Subsequently, an adjacent undisturbed urchin also emitted the white fluid, as if some cue from the first urchin had triggered a similar response in the second. As at other banks, several young Marbled grouper *(Epinephelus inermis)* accompanied the *Astropyga* aggregation, swimming above and between the urchins and often contacting their spines.

Another colorful, large, spiny urchin, *Coelopleurus floridanus,* was seen on the slope between 143 and 178 m. Its population must be significant because four sightings were recorded within that narrow depth range. Two, however, were of the remains of urchins, apparently recently eaten. Only one *Stylocidaris* urchin was seen (151 m) on the transect, even though at other banks these organisms are often abundant on such bottoms.

Small crinoids capable of graceful swimming actions occur on the slope from 91 to 175 m, clinging to gravel or any other object on the bottom. Between 136 and 160 m the population of these organisms is phenomenal, peaking at about 137 to 146 m.

The most frequently encountered fish on the steeper bank slope was the Tattler, *Serranus phoebe,* seen down to 175 m but more abundant around 107 m. Numbers of Bank butterflyfish, *Chaetodon aya,* Spanish flag, *Gonioplectrus hispanus,* and Snowy grouper, *Epinephelus niveatus,* were seen on the large reefal structures between 91 and 98 m at the top of the slope but not elsewhere. It is suspected that the range of the Snowy grouper extends considerably deeper.

The slope of the bank decreases below approximately 189 m, but even at that depth a substantial amount of carbonate gravel is mixed with the basically muddy sediment. The gravel content of the sediment decreases below 193 m, leaving a sticky, coherent mud bottom with a thin veneer of loose, easily stirred fine material that probably undergoes repeated disturbance by an active community of mobile benthic animals. Tracks, trails, burrows, depressions, large holes, and small mounds, which are the recognizable features on this bottom, reflect the movements or excavating activities of the limited number of species of echinoderms, mollusks, crustaceans, and fishes that appear to be the main components of the soft-bottom biota.

The most numerous organisms seen between 189 and 197 m were small ophiuroids, 5 cm or so in length, which covered the bottom in places almost arm tip to arm tip (several hundred per square meter). Hermit crabs, mostly occupying *Murex* shells, are frequent below 165 m. Several other types of crab occur on the mud bottom, and rather abundant, small galatheid crustaceans, frequently observed in burrows and small holes between 201 and 213 m, were particularly apparent in the openings of numerous small burrows in the sticky mud rims of larger (⅓ m in diameter) holes, some of which were occupied by groupers.

An octopus was seen in its hole at 186 m, and swimming scallops were seen at 207 and 213 m. A sizable population of *Murex* occurs around 197 m. One was observed in the apparent act of preying upon a *Clypeaster* sand dollar at 201 m. Remains of dead urchins were encountered a number of times between 151 and 213 m. They must be prime forage for more active benthic predators on the bank slope and deeper mud bottom.

The urchins, crabs, and gastropods are certainly responsible for many of the fresh tracks and trails seen on the soft bottom. Some of the more distinctive tracks are produced by flatfishes, Pleuronectiformes, which drag their bodies along the bottom by undulations of the dorsal and anal fins and leave a series of small indentations on both sides of a linear track.

The distinctiveness, variability, abundance, and freshness of the tracks, trails, and burrows are indicative of an active assemblage of organisms on the deep, soft bottom. Some are probably deposit or detritus feeders (primarily the urchins and ophiuroids, possibly some of the crabs and galatheids). Four suspension or filter-feeding types were seen, a pen shell (dead but in place) at 206 m; stalked, colonial coelenterates (sea pens) at 204 m; worms (apparently sabellids) with tentacles extended above the bottom between 193 and 204 m; and the aforementioned scallops. Benthic predators (fishes, crabs, gastropods, octopods, and asteroids) are probably the most active group and may be responsible for most of the tracks and trails.

At least nine different types of fishes were seen on the mud bottom below 189 m. Some lurk in the bottoms of large, steep-sided holes and one was seen inside a dead pen shell. Flatfish and sea robins rested on the sediment. The most interesting and abundant fish, however, was the Snipefish, *Macrorhamphosus scolopax,* which hovered above the bottom, nose down, at a steep angle (also observed at Elvers Bank). These fish, seen from 189 to 213 m, were most abundant between 189 and 197 m and numerous at least down to 204 m. Plankton abundance adjacent to the bottom appeared greater than in the water some distance above. It is possible that the Snipefish makes a good living in the narrow benthopelagic realm near the bottom by feeding on planktonic organisms.

Sackett Bank. Sackett Bank is a shelf-edge, topographic feature capped by carbonate sediments, which include sand, debris, algal nodules, rock ledges and drowned algal reefs (Figure 7.40). Biotic communities here, although composed of species found commonly on the other shelf-edge banks in the northwestern Gulf, are less diverse, and less abundant than are those of the other banks. In terms of community structure, the epibenthic biota of Sackett Bank seem to occupy a position somewhere between those of the South Texas fishing banks (such as Southern Bank and South Baker Bank) and those of the other shelf-edge carbonate banks in the

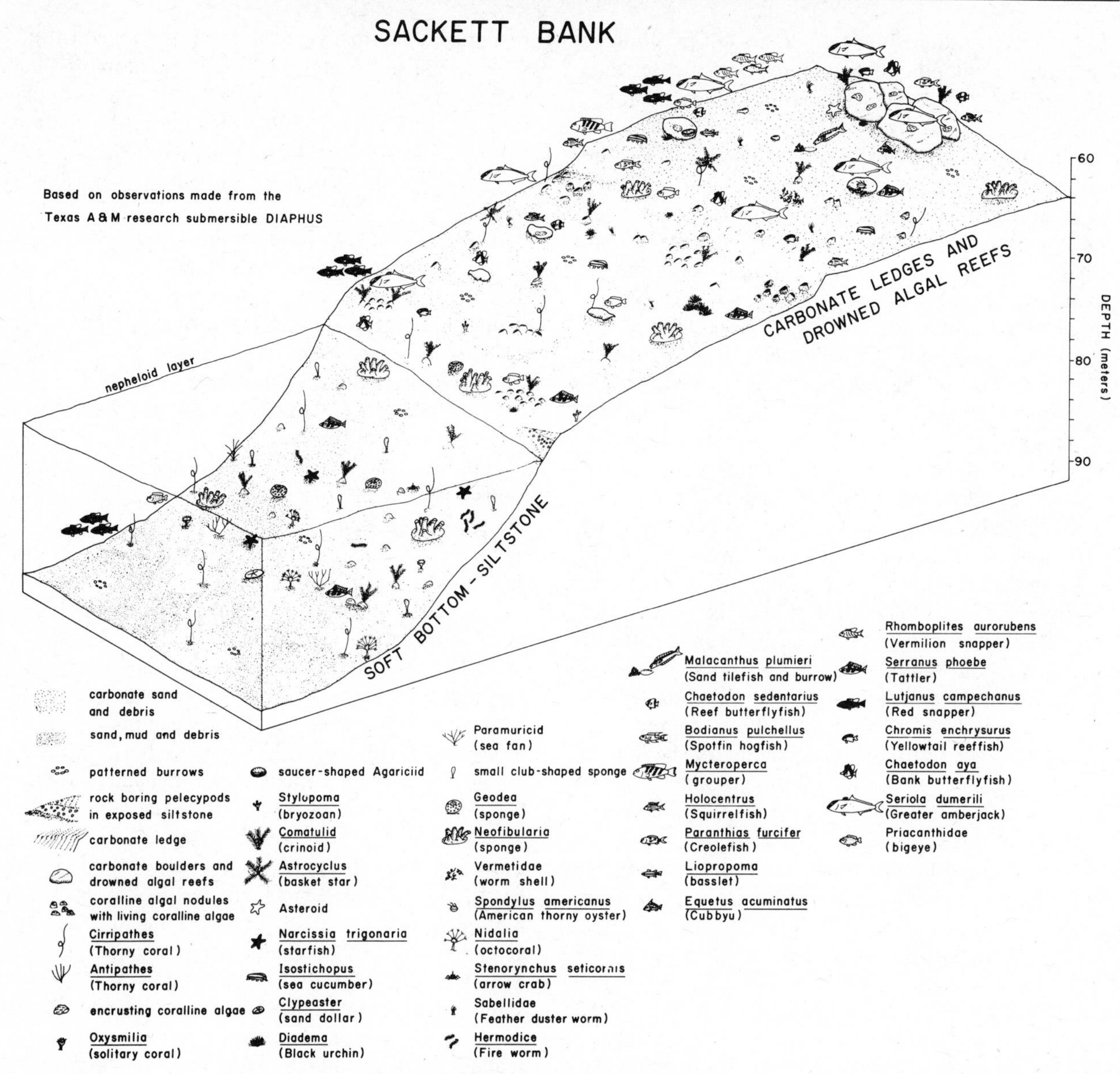

Figure 7.40. Biota of Sackett Bank.

northwestern Gulf of Mexico (such as 18 Fathom Bank and Flower Garden Banks).

The topmost part of Sackett Bank (64- to 65-m depth) is rather flat and sandy, with carbonate gravel, a few scattered coralline algal nodules, and drowned coralline algal reef patches. On this upper terrace limited amounts of live coralline algae occur on the reef patches, nodules, and tops of pieces of gravel. However, the present degree of carbonate production by coralline algal populations on Sackett Bank does not appear to be substantial.

Some of the drowned reef patches are quite large, up to 3 m high and 12 m across (the largest we have seen crest at 61-m depth). All are covered with thin veneers of fine, easily stirred sediment and harbor surprisingly sparse populations of epibenthos and fishes. The primary invertebrates clinging to or encrusting the drowned reefs are comatulid crinoids, encrusting sponges, *Diadema, Cirrhipathes, Spondylus americanus,* the saucer-shaped agariciid, and small patches of coralline algae. Fish are not particularly abundant on the drowned reefs, although the following types are frequent: *Chromis enchrysurus,* Yellowtail reeffish; *Bodianus pulchellus,* Spotfin hogfish; *Equetus umbrosus,* Cubbyu; *Seriola dumerili,* Greater amberjack; *Priacanthus,* bigeye; *Serranus phoebe,* Tattler; *Apogon,* cardinalfish; *Chaetodon sedentarius,* Reef butterflyfish; *Chae-*

todon aya, Bank butterflyfish; *Paranthias furcifer,* Creole-fish; and *Holanthias martinicensis,* Roughtongue bass, all of which are generally the same species encountered over the entire upper portion of the bank above 73 m depth. The population of *Chaetodon aya* seemed large in comparison to other banks examined. Schools of Vermilion snapper, *Rhomboplites aurorubens,* were seen near the drowned reefs.

The predominantly sandy bottom, which occupies most of the upper platform of the bank above 65-m depth, houses a relatively depauperate assemblage of macroepifauna compared to those that occur at similar depths on other shelf-edge banks in the northwestern Gulf. Clusters of burrowing sabellid polychaete worms are abundant however. Comatulid crinoids are frequent on rubble and nodules, and the population of the sponge *Neofibularia nolitangere* is significant. Diademid urchins and the large sea cucumber *Isostichopus* are conspicuous.

Small mounds of "tangled" worm-shell gastropod tubes (probably *Siliquaria)* were seen on the upper terrace (64-m depth). These organisms have also been observed on several other shelf-edge banks and may be important contributors to carbonate sediment on Sackett Bank.

Burrows produced by the Sand tilefish, *Malacanthus plumieri,* are interesting in terms of what they may indicate concerning the nature of the surficial sediments that cover the top of Sackett Bank. These conical burrows, about 1 m in diameter, were seen in what appeared to be a basically sandy bottom with some gravel and a few nodules. Adjacent to the burrows, however, were piles of algal nodules, presumably removed from the burrows by the fishes during construction. This would imply that, even though the visible sediment surface is mostly sand, there are significant amounts of coarse material and dead algal nodules buried just beneath. Considering these facts and that fair-sized drowned reefs are present on the uppermost part of the bank, we hypothesize that in the past Sackett Bank must have supported an active reef-building community dominated by coralline algae.

Actually, living coralline algal nodules are more abundant near the bases of small carbonate ledges and on low mounds on a 67- to 73-m depth terrace bordering the upper platform (Figure 7.40). The carbonate ledges, 1 to 2 m high, separate the upper platform from the terrace in places. The ledges are generally of gentle slope with algal nodules scattered on them. Near their bases, nodules extend sparsely for some distance out onto the terrace. Populations of epibenthos and fishes are somewhat more diverse and abundant on and adjacent to these ledges than on the shallower drowned reefs. Crinoids, sponges, diademid urchins, sabellid worms, and the same types of fishes listed above for drowned reefs occur on the ledges and among the nodules. Large groupers appear to prefer the ledges.

The substratum of the 67- to 73-m depth terrace is basically a carbonate gravel-strewn sandy bottom with significant amounts of silt and clay. Few algal nodules were found on the central part of the terrace, but clusters of them occur more frequently on small mounds toward the break in slope around 70- to 72-m depth. Sabellid worms are abundant on the terrace, and, in general, the conspicuous soft-bottom epifaunal populations are similar to those on the upper platform except for some significant additions. The antipatharian, *Cirrhipathes,* becomes abundant near the outer margin of the terrace. Basket stars are associated with *Cirrhipathes.* Substantial populations of the large urchin *Astropyga magnifica* were observed on the central part of the terrace.

At approximately 72-m depth the slope increases considerably. At this depth, also, a small outcrop of claystone bedrock, similar in appearance to the rocks at Stetson and Sonnier Banks and to outcrops in a basin at Bouma Bank and at the crest of Geyer Bank, was encountered. These rocks are generally soft and probably disintegrate rapidly when exposed directly to water. Chips and fine sediment obviously derived from the bedrock were seen around the base of the outcrop. Small bore holes in the exposed claystone rock at Sackett Bank are identical in appearance to holes in the outcrops at Stetson Bank, where they are considered to be produced by rock-boring pelecypods. The claystone–siltstone outcrops at mid-shelf banks, such as Stetson and Sonnier, and the less conspicuous but similar outcrops on deeper parts of the shelf-edge carbonate banks are interesting. We might speculate that the substantial carbonate reefal communities on the shelf-edge banks have developed on and almost totally overgrown or buried a basic substratum of upthrust claystone and siltstone. In all cases the unconsolidated sediments on the flanks of the shelf-edge carbonate banks grade outward from coarse carbonate sand to mixtures of sand, silt, and clay and finally to a fine mud. A grouper hole 1-m deep in the soft mud bottom at 90-m depth at Sackett Bank was observed to have claystone or siltstone at its bottom.

When Sackett Bank was explored with the submersible, the top of the nepheloid layer was roughly coincident with the break in slope at 72- to 74-m depth. A substantial change in the composition of the benthic epifaunal community is apparent between 73- and 76-m depth. The comatulid crinoid population becomes tremendous; *Neofibularia* sponges are abundant and, surprisingly, white rather than the normal reddish brown; fire worms, *Hermodice,* are exceedingly numerous; Asteroid starfishes, particularly *Narcissia trigonaria* and *Astropecten*-like forms, are conspicuous; *Cirrhipathes* become somewhat more abundant; and the globular, white sponge *Geodea* and small, yellow, club-shaped sponges appear as major components of the soft-bottom community. These conditions persist to at least 85-m depth. In addition, at around 80-m depth *Antipathes* (a branching antipatharian), the club-shaped octocoral *Nidalia occidentalis,* small paramuriceid sea fans, and large solitary stony corals, probably *Oxysmilia,* are found. The paramuriceids, *Nidalia,* and the large solitary corals are abundant down to around 88-m depth. The large polyps of *Nidalia* were contracted and inconspicuous above 83-m depth but well expanded below 85-m depth (daytime observation).

Below 90-m depth almost nothing was visible on the mud surface, the conspicuous epifaunal organisms having nearly disappeared. Holes 20 cm to 1 m wide and deep were observed, however. One contained a large eel, and a grouper occupied another. A school of Red snapper, *Lutjanus campechanus,* was cruising near the bottom.

The fact that the epibenthic communities on the upper part of Sackett Bank are poorly developed, compared with those on shelf-edge banks to the west, makes Sackett Bank interesting ecologically. Its topography, depth, and location would seem to favor development of clear-water carbonate reefal communities, were it not for the proximity of the Mississippi River. Indeed, the presence of large, drowned, coralline algal reefs and large numbers of dead algal nodules buried under carbonate sand indicates that a substantial, active reef-building community, dominated by coralline algae, existed on the bank some time in the past.

The small population of living coralline algae that exists on the bank suggests that environmental conditions do not now altogether preclude limited carbonate substratum production. It is hypothesized, however, that the conditions of water quality, hydrography, and turbidity associated with Mississippi River outflow are responsible for limiting the contemporary development and growth of substantial carbonate reefal communities at Sackett Bank.

These communities, which are basically dependent on photosynthetic organisms, require adequate levels of light throughout the year. The water overlying Sackett Bank was much more turbid when observed than the water above any of the other shelf-edge banks. The upper 10 m of the water column (above the thermocline) was very green and contained an enormous amount of organic matter in the form of plankton and seston (floating, nonliving organic material, often mucuslike). At the thermocline (approximately 10-m depth) large, horizontal "sheets" of mucuslike white organic matter were seen. Mucus "strands," accompanied by large zooplankton populations, extended down past 30-m depth. All this suspended material, substantially the result of high phytoplankton and zooplankton production, effectively blocks out much of the downwelling light, even though the water between the organically turbid surface layers and the bottom nepheloid layers may be quite clear. It is possible, therefore, that the benthic reefal communities of Sackett Bank are held in check partly by the tremendous productivity of nutrient-rich, near-surface marine waters in the vicinity of the Mississippi Delta. Long-term observations of light penetration, salinity, temperature, nutrients, suspended organics, and water-column productivity at Sackett Bank and shelf-edge banks that support substantial Algal-Sponge Zones could reveal much concerning the environmental factors that limit development of clear-water, carbonate, reef-building communities in the northwestern Gulf of Mexico.

The structure and abundance of biotic communities at Sackett Bank are intermediate between those described for shelf-edge carbonate banks (such as Geyer Bank and the Flower Garden Banks) and South Texas fishing banks (such as Southern Bank). Living coralline algae populations at Sackett Bank are more substantial than those found on the South Texas banks but much less developed than those of most other shelf-edge banks. Sackett Bank's closest shelf-edge neighbor, Diaphus Bank, exhibits similar limitations in the degree of development of clear-water, reef-building zones.

Diaphus Bank. Although scleractinian corals and coralline algae live on Diaphus Bank, no substantial populations were encountered during the survey. Reef building appears to be arrested at present, but drowned reef patches occur at least down to 107-m depth. These currently inactive (nongrowing) reefs are covered with fine sediment or sedi-

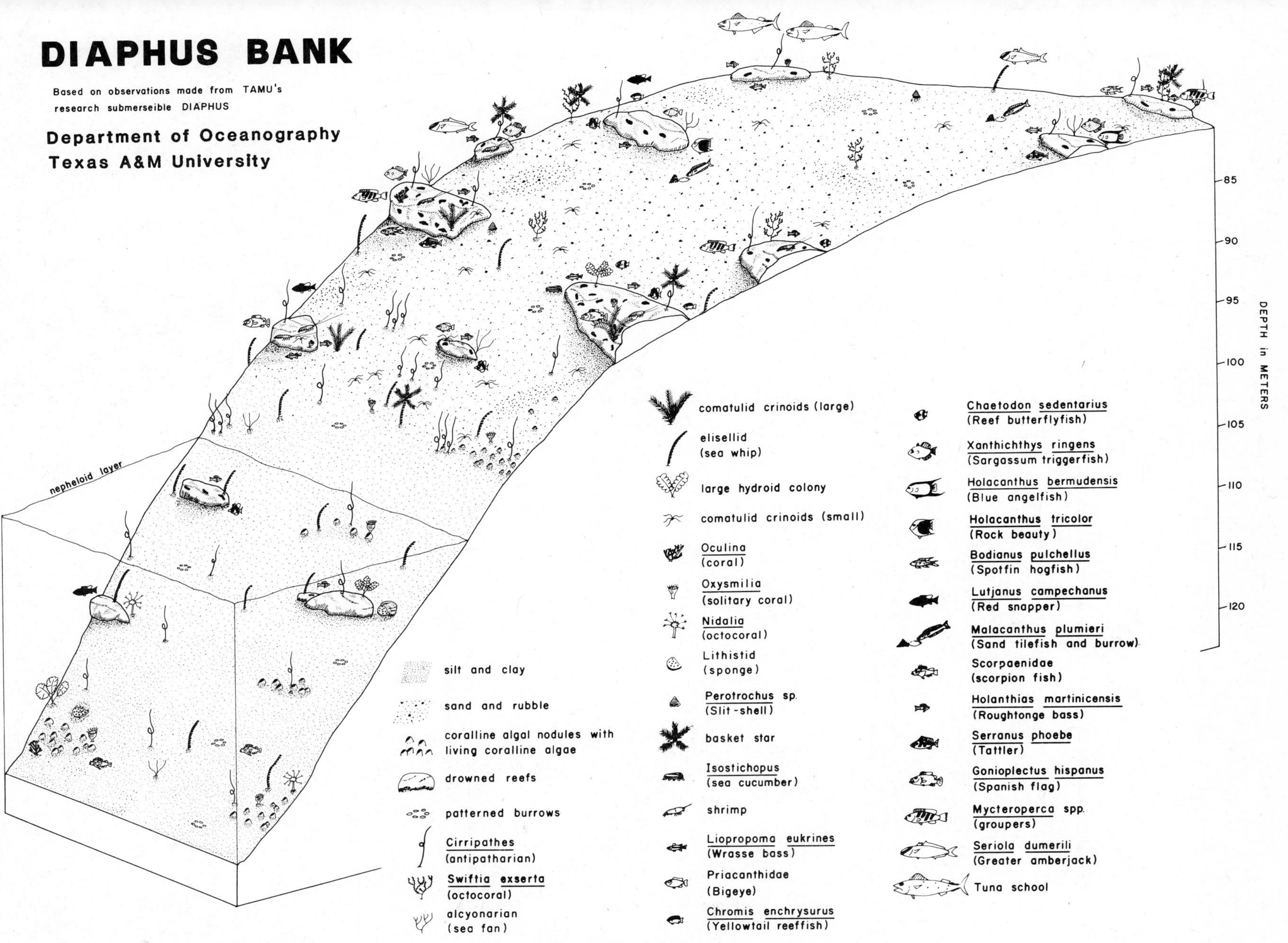

Figure 7.41. Biota of Diaphus Bank.

ment–epifauna mats and sponges with small amounts of coralline algae.

The largest drowned reefs (2.5 to 3 m high), which were seen between 85- and 95-m depths, mark the break in slope at the edge of the bank's upper platform (Figure 7.41). This was presumably the level of most active reef building in the past. As they progress from the bank edge toward the top central portion of the bank (82-m depth) the drowned reefs become increasingly less elevated above the bottom (some only 0.5 m high), but their lateral dimensions (which average 3 to 6 m across) and spacing (generally 3 to 9 m or more apart) are fairly uniform. Below 95 m the drowned reefs are smaller and more heavily laden with fine sediment.

Sediment cover on the drowned reefs is somewhat less above 88-m depth, but even at 82 m it is substantial, sometimes occurring on 90 to 95% of the rock surface. Above 95 m, at least, the sediment on rocks is most often entrapped by low-growing populations of attached epibenthic organisms which form a mat. At greater depths the fine sediment cover may be so great that most small attached benthos are obscured or precluded.

Coralline algal cover on the drowned reefs on top of the bank generally varies from nil to 10 or 15% and probably averages 3 to 5%. The greatest cover (20 to 30%) was encountered on a large reef (3 m high) at 88 m on the bank edge. Although a few living patches of coralline algae were seen down to 107-m depth, little occurred below 98 m. In general, the contemporary coralline algae population is incidental on the part of the bank surveyed and cannot be considered an effective reef-building population.

It is possible that on the shallowest peaks (74- to 78-m depth) there is some degree of expression of the Algal-Sponge Zone. Between 80 and 98 m, however, the assemblages are comparable to those of the Antipatharian Zone. The remainder of the shelf-edge carbonate banks which have crest depths above 80 m are occupied by well developed Algal-Sponge Zones (Ewing, Jakkula, Sweet, Alderdice, Parker, Rezak–Sidner, Bouma, 18 Fathom, Elvers, Geyer, Bright, 28 Fathom, East Flower Garden, West Flower Garden, and probably Applebaum). The crest of Phleger Bank is too deep (122 m), to support an Algal-Sponge Zone or Antipatharian Zone. The hard-bottom biota atop Phleger, although apparently existing in clear water (Continental Shelf Associates, 1980), are types found in turbid water (Nepheloid Zones) near the bases of the shelf-edge banks slightly farther inshore.

Clear-water assemblages of the Algal-Sponge Zone, dominated by coralline algae, extend downward to nearly 100 m on those banks that are positioned at the extreme edge of the continental shelf and surrounded by depths in excess of 180 m (Geyer, Elvers, and probably Sweet). At the other extreme, on those shelf-edge carbonate banks (like Alderdice) that are substantially inshore from the shelf break and surrounded by depths of less than 100 m, the Algal-Sponge assemblages are restricted to shallower depths (generally less than 75 m, Table 7.1).

Alderdice Bank. The 84-m depth contour can be taken as an inshore limit for the distribution of shelf-edge carbonate banks that bear reefal communities. Alderdice Bank exists at this inshore limit, and were it not for the several peaks that rise into clear water above 65-m depth, probably would not even support an Algal-Sponge Zone.

The bank is composed of four major topographic peaks on a flattish platform of about 80-m depth. In addition, a spectacular basalt outcrop juts vertically out of the bottom at 76 m, cresting at 55 m (Figure 7.42). Reconnaissance of one of the major peaks, the basalt outcrop, and a good part of the 80-m platform, revealed basic differences in the biotic communities that occupy the three structural zones.

Healthy, growing coralline algal nodules underlain by carbonate sand occur at the crest of the large southeastern peak (58- to 67-m depth). The nodules are accompanied here and there by small reefal structures and carbonate blocks covered with the dominant coralline algae. The extreme variability in size of nodules, blocks, and firmly affixed reef rock gives the substratum an irregular appearance not typical of other algal nodule zones. Contributing to the irregularity is the "lumpy" nature of the highly bioturbated sand, where it is exposed.

The most conspicuous invertebrates on the peak are *Cirrhipathes,* massive sponges of several species, and large branching bryozoan colonies, *Holoporella.* An exceptionally large number of Yellowtail reeffish, *Chromis enchrysurus,* was encountered above 67 m; the peak was literally swarming with them. Creole-fish, *Paranthias furcifer,* were numerous; groupers, *Mycteroperca,* congregated on top; schools of snappers, *Lutjanus* and *Rhomboplites aurorubens,* and Greater amberjack, *Seriola dumerili,* were seen.

The Algal-Sponge Zone is probably restricted to crests of the several peaks at Alderdice Bank; it is therefore of limited areal extent. It is nevertheless

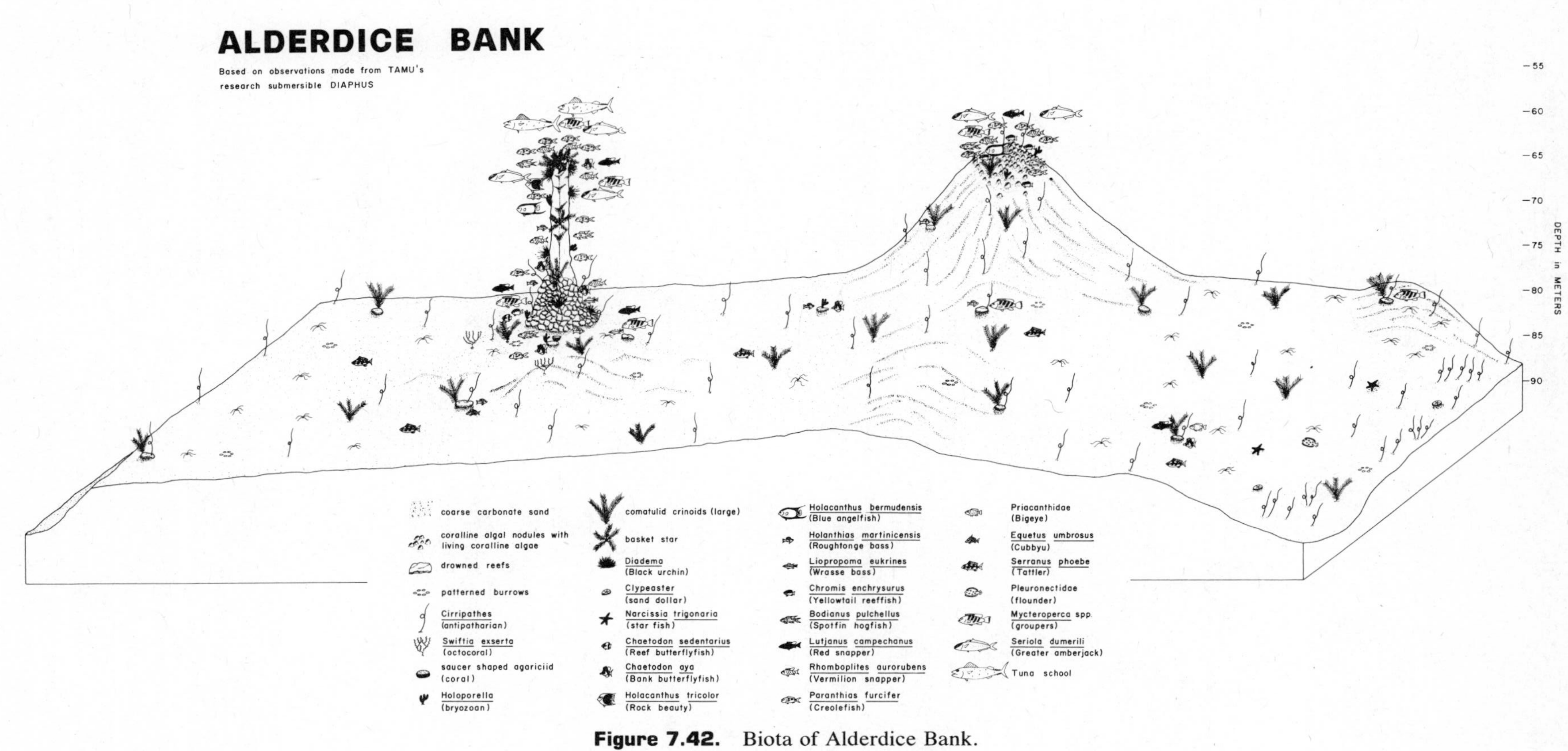

Figure 7.42. Biota of Alderdice Bank.

a zone of active reef building and carbonate substratum production.

Small "drowned" reefal structures occur on the bank down to at least 85-m depth. They are surrounded by the large expanse of unconsolidated sediment that constitutes most of the bank. With increasing depth, the sediment grades from carbonate sand, gravel, and nodular material (75 m), to mixtures of sand, silt, clay, and gravel (79 m), to soft, primarily fine sediment (82 m). Reefal structures below the Algal-Sponge Zone are laden typically with veneers of sediment entrapped by mats of low epifaunal growth. Below 82 m the drowned reefs are almost totally covered with thin layers of fine sediment. Nevertheless, small amounts of coralline algae occur on the drowned reefs down to at least 79 m (5% cover at 76 m). No algae were seen below 79 m.

Cirrhipathes is generally the most conspicuous invertebrate at all depths and is particularly abundant locally below 84 m. Between 76 and 85 m enormous populations of small comatulid crinoids cling to rocks and gravel on the unconsolidated bottom. Larger crinoids are numerous on rocks and drowned reefs between 76 and 85 m and were seen to 88 m. Branching colonies of *Holoporella* are abundant above 76 m. Various alcyonarians occur on the rocks between 76 and 79 m; the largest are white fans and the orange and white branching form *Swiftia exserta.*

The deeper, muddy bottoms below 82 m are comparatively barren but are abundantly etched with tracks, trails, and burrows, which indicate an active population of mobile benthic invertebrates. The sand dollar, *Clypeaster ravenelii,* is fairly numerous below 85 m and the starfish, *Narcissia trigonaria,* was seen between 82 and 85 m.

Two species of hermatypic corals were encountered at 76 m, saucer-shaped agariciids and a small head of what appeared to be *Stephanocoenia.* Neither was abundant, and both occurred on drowned reefs.

Holanthias martinicensis was the most frequently encountered fish around the drowned reefal structures below 76 m. Others included Yellowtail reeffish, *Chromis enchrysurus;* Spotfin hogfish, *Bodianus pulchellus;* Blue angelfish, *Holacanthus bermudensis;* Reef butterflyfish, *Chaetodon sedentarius;* Cubbyu, *Equetus umbrosus;* Spanish flag, *Gonioplectrus hispanus;* and bigeye, *Priacanthus.* A school of snappers, *Lutjanus,* and several groupers, *Mycteroperca,* were also seen. Tattlers, *Serranus phoebe,* were numerous on the unconsolidated bottom.

The most impressive feature on Alderdice Bank is the basalt outcrop, an elongated narrow ridge that extends vertically upward from the 76-m surrounding depth to 55-m crest depth. Spires examined at the crest are two or so meters across at the top and sheer cliffs extend downward to approximately 67 or 69 m, below which large blocks of basalt talus are piled around the base of the outcrop. The hard basalt is covered with crusts of coralline algae, sponges, bryozoans, and other epifauna. Near the top of the outcrop these crusts are almost total; up to 50% are coralline algae. At 69 m on the large blocks coralline algal cover is 70 to 80%, but the cover decreases with depth to small patches at 76 m.

Large basket stars, *Diadema* urchins, and branching colonies of the bryozoan *Holoporella* are particularly abundant and visible on the outcrop and talus slope. Basket stars tend to accumulate at the peak of the outcrop, and *Cirrhipathes, Antipathes,* large comatulid crinoids, and small branching alcyonarians are numerous on the talus slope surrounding it.

Fishes that swarm around the crests of the outcrops include Creole- fish, *Paranthias furcifer;* Vermilion snapper, *Rhomboplites aurorubens;* Greater amberjack, *Seriola dumerili;* tuna; large Red snapper, *Lutjanus campechanus;* and groupers, *Mycteroperca.* Creole-fish and Vermilion snapper were the most numerous large fishes on the structure at all depths, and *Holanthias martinicensis* was abundant, particularly on the talus slope. Closely associated with the outcrop and talus blocks are the Wrasse bass, *Liopropoma eukrines;* Spotfin hogfish, *Bodianus pulchellus;* Bank butterflyfish, *Chaetodon aya;* Blue angelfish, *Holacanthus bermudensis;* Rock beauty, *Holacanthus tricolor;* and a damselfish that resembles *Chromis scotti.*

ENVIRONMENTAL CONTROLS

Based on the nature, distribution, and degree of development of their epibenthic communities, hard banks on the Texas–Louisiana Outer Continental Shelf can be divided into six environmental groups (see also Table 7.1):

1. South Texas mid-shelf relict Pleistocene carbonate reefs that bear turbidity-tolerant Antipatharian Zones and Nepheloid Zones (surrounding depths of 60 to 80 m; crests 56 to 70 m): Mysterious, Small Adam, Blackfish, Big

Adam, Dream, Southern, North Hospital, Hospital, Aransas, South Baker, and Baker.

2. North Texas–Louisiana mid-shelf Tertiary outcrop banks that bear clear-water *Millepora*-Sponge Zones and turbid-water-tolerant Nepheloid Zones (surrounding depths of 50 to 62 m; crests 18 to 40 m): Stetson, Claypile, Sonnier.
3. North Texas–Louisiana mid-shelf banks that bear turbidity-tolerant assemblages approximating the Antipatharian Zone (surrounding depths of 65 to 78 m; crests 52 to 66 m): 32 Fathom, Coffee Lump, Fishnet.
4. North Texas–Louisiana shelf-edge carbonate banks that bear clear-water coral reefs, clear-water Algal Sponge Zones, transitional assemblages approximating the Antipatharian Zone, and Nepheloid Zones (surrounding depths of 84 to 200 m; crests 15 to 75 m): Appelbaum, East Flower Garden, West Flower Garden, 28 Fathom, Bright, Geyer, Elvers, 18 Fathom, Bouma, Rezak–Sidner, Parker, Sweet, Alderdice, Jakkukla, Ewing.
5. Eastern Louisiana shelf-edge carbonate banks that bear poorly developed elements of the Algal-Sponge Zone, transitional Antipatharian Zone assemblages, and Nepheloid Zones (surrounding depths of 100 to 110 m; crests 67 to 73 m): Diaphus, Sackett.
6. Extreme shelf-edge banks with crest depths too deep to permit the development of light-dependent, reef-building communities but which support elements of transitional Antipatharian Zone and Nepheloid Zone assemblages (crests deeper than 100 m, surrounding depths 200 m or more): Phleger.

The clear-water biotic zones on these banks (*Millepora*-Sponge, several coral reef zones, and Algal-Sponge Zone) are distinctly tropical in faunal and floral content. Biota of the Antipatharian and related transitional zones are composed largely of tropical species apparently more tolerant of turbidity. Environmental factors that can be correlated with and probably control regional patterns of community structure, distribution, abundance, and zonation of tropical epibenthos in the northwestern Gulf are distance from shore, regional patterns of substratum type, bottom depth, bank relief, water temperature, salinity, river runoff, turbidity, sedimentation, currents, and seasonal variation in the last six (Figure 7.43). These factors have been described substantially in Chapters 1 and 2 and in this chapter.

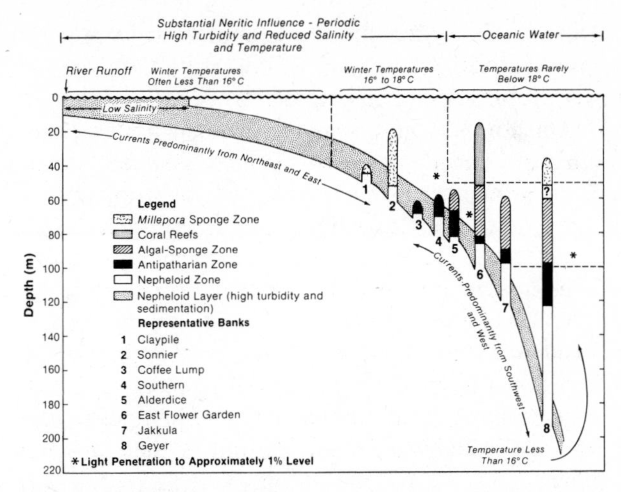

Figure 7.43. Distribution of biotic zones relative to conditions of temperature, salinity, turbidity, and light at selected banks.

Conditions at the shelf edge near and beyond the 80-m depth contour on the broad North Texas–Louisiana Shelf west of about 91° longitude are favorable to the development of tropical reef communities. Current patterns are such that shelf-edge waters come primarily from the southwest and are oceanic, with little admixture of neritic water from the Texas–Louisiana Shelf (Chapter 2, Figures 2.11, 2.12, and 2.39). These currents carry larvae, spores, and juveniles from the Gulf of Campeche, Yucatan Shelf, and the Caribbean.

There is a strong tendency for coastal water masses, highly influenced by outflow from the Mississippi and other rivers in Louisiana and North Texas, to be held onshore and shunted west most of the year (particularly during the February to May periods of peak runoff) by the general shelf circulation pattern. As a result, turbidity in the shelf-edge waters is usually nil and salinity averages 36 ppt (Figure 2.17). When high runoff combines with seasonal disruption of the typical counterclockwise current regime on the shelf (which may occur in late spring or early summer) lower salinities may occur in shelf-edge waters; however, the lowest surface salinity we have ever measured at the Flower Gardens was 32 ppt, and this was accompanied by 34 ppt at 25-m depth.

For most of the year near-surface water temperatures throughout the Gulf are tropical to subtropical (27 to 30°C) (Chapter 2, Figure 2.20). Near shore in the northern Gulf, however, temperatures become warm-temperate from December through March.

During the coldest months (January to February) temperatures grade from as low as 10°C in the estuaries to 18°C on the outer shelf edge (Figure 2.19).

Onshore–offshore seasonal movements of the 18 and 16°C surface isotherms probably have a significant influence on the distribution of tropical reef biota in the northwestern Gulf. As indicated in Chapter 6, 18°C is considered the minimum seasonal temperature limit for vigorous growth of coral reefs. The lower limit, 16°C, is stressful for most reef-building corals. Though reef-building coralline algae and other biotic elements of the tropical reef ecosystem may tolerate somewhat lower temperatures, 16°C is probably near the bottom of their optimal range.

In the northwestern Gulf the 18°C winter surface isotherm can be expected to occur somewhere between the locations of the 30- and 80-m depth contours, projected upward. The 16°C isotherm occurs between the 20- and 40-m depth contours (see Chapter 2 and Harrington, 1966). The surface isothermal layer during winter extends 50 to 75 m downward (Chapter 2, Figure 2.26), with temperatures only 1 to 3°C lower at 100 m. Thus above 50-m depth off North Texas and Louisiana and seaward of the general 80-m bottom depth contour salinities are high and temperatures range annually from approximately 18 to 30°C (Figure 7.43). Wherever suitable hard substratum exists above about 95-m depth in the absence of chronically turbid water conditions on this part of the shelf are favorable to the growth of tropical reef communities dominated by corals or coralline algae, or both.

The degree of light penetration into clear surface waters and the antagonistic effects of turbidity in bottom nepheloid layers are almost certainly the factors that control depth ranges for these communities on the various shelf-edge banks. High turbidity decreases light penetration and is inimical to the development of coral and algal reef communities. Sedimentation associated with high turbidity results in smothering of encrusting epibenthos by silt and clay veneers. In the northwestern Gulf, because of the enormous sediment load entering from the rivers, turbidity and sedimentation are major factors that limit the development of tropical reef assemblages. It is speculated that reef development at Sackett Bank (Algal-Sponge Zone) is seriously attenuated, due in part to increased turbidity in surface waters from admixed Mississippi River outflow. This influence, accompanied by somewhat reduced salinities, diminishes westward but may extend as far as Diaphus Bank (91°W) during periods of particularly high runoff (Figure 2.17).

Mid-shelf banks (Mysterious to Fishnet) rise from surrounding depths of 60 to 80 m. Their tops, which support Antipatharian-Zone type assemblages between 56 and 73 m, exist in a depth range which on shelf-edge carbonate banks (Flower Gardens to Elvers) is occupied by diverse, clear-water Algal-Sponge Zones. The lack of Algal-Sponge Zones on the mid-shelf banks and the occurrence instead of Antipatharian assemblages found typically in deeper water at the shelf edge is probably due largely to high turbidity.

The effects of bottom nepheloid layers and associated sedimentation are certainly more pronounced on the mid-shelf banks than at the shelf-edge banks (Figures 2.34–2.37 and associated text). We speculate that most or all of the mid-shelf banks are frequently totally covered by the nepheloid layer, especially during severe wave conditions. At the Flower Garden Banks a substantial nepheloid layer shallower than about 80 m has not been observed and the water is usually fairly clear, even at that depth. Fine, terrestrial sediments are not found on shelf-edge carbonate banks above the lower limit of their Algal-Sponge Zones. The mid-shelf banks, however, are generally coated with thin-to-thick layers of fine sediment, presumably derived from nepheloid layers.

Relief above the surrounding bottom is of considerable importance in alleviating the negative impacts of bottom nepheloid layers and attendant sedimentation on development of epibenthos. An Algal-Sponge Zone probably will not become established on banks that have less than about 15 m relief above a mud bottom because of the impact of nepheloid layers. At Alderdice Bank the Algal-Sponge Zone extends downward to only 67 m, which is 17 m above the surrounding mud. Farther offshore, at the East Flower Garden, the Algal-Sponge Zone extends downward to 82 m, about 18 m above the surrounding soft bottom. In even deeper water, where surrounding depths are more than 180 m (Geyer and Elvers banks), the vertical extent of the Algal-Sponge Zone is not limited by bottom nepheloid layers because of the high relief. Here, the zone extends down to more than 95-m depth and is probably limited primarily by the degree of light penetration from above (Figure 7.43).

Thus on the shelf-edge banks there is a gradual increase in the maximum depth of expression of coralline algae-dominated communities with in-

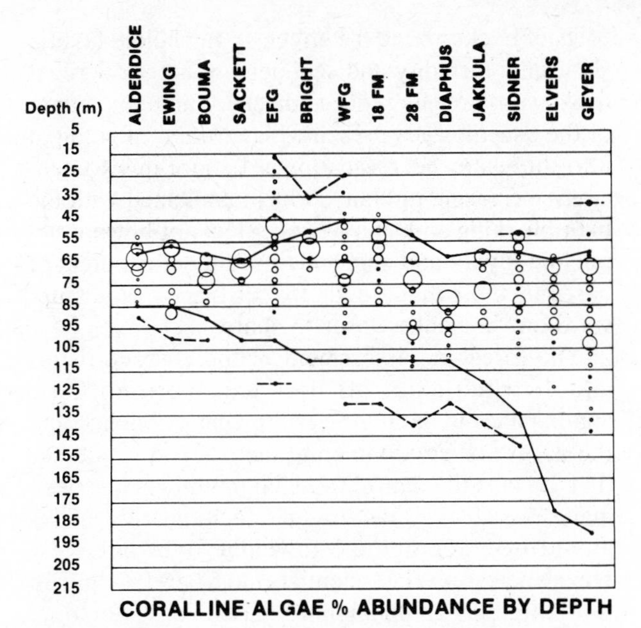

Figure 7.44. Coralline algae relative abundance by depth on shelf-edge carbonate banks in the northwestern Gulf. Sizes of circles represent percentage abundance of coralline algae at various depths within banks only (total area of all circles within one bank equals 100% of the coralline algae population for that bank). Bank-to-bank variation in abundance is not represented. Diagonal solid lines between 45- and 65-m depths represent the crest depths of the major carbonate platforms at the various banks. Dashed lines or presence of circle *centers* above the solid line represent crest depths of reefs or pinnacles. Diagonal solid lines below 85-m depth represent the general depth of the mud bottom surrounding the banks. Lower dashed lines indicate that parts of the surrounding bottom adjacent to the banks extend to the depths of the lines.

creasing surrounding depth (Figures 7.43 and 7.44; Table 7.1). A similar trend is apparent for the Antipatharian Zones on these banks (Figure 7.43; Table 7.1). These observations imply that the bottom nepheloid layers are of great importance as ecological limiting factors on the lower 15 to 20 m of the banks.

The deeper clear-water reefal communities (Algal-Sponge Zones) are therefore excluded from the mid-shelf carbonate banks by high turbidity, reduced light, and high sedimentation and are limited in downward extent on some of the shelf-edge banks by the effects of nepheloid layers. In their place is a less diverse and less abundant Antipatharian Zone assemblage made up of epibenthic forms which, though basically tropical in origin, are tolerant of the stresses imposed.

Coral reefs would not be expected on the mid-shelf carbonate banks because the depth of the crests is too great (more than 52 m). Even if shallower hard substratum were present the aforementioned conditions of mid-shelf water quality would probably preclude active coral-reef development on these banks. Hard substratum exists within suitable depths for coral-reef development in the form of mid-shelf claystone–siltstone banks rising from surrounding depths of 50 to 62 m and extending upward to 18 m (Sonnier Bank) and 20 m (Stetson). The *Millepora*-Sponge Zones of these banks are undoubtedly subject to seasonal temperatures somewhat less than the 18°C minimum for vigorous reef growth but probably not much less than 16°C (Chapter 2).

The crests of Stetson and Sonnier banks may, however, be fairly well isolated from the effects of bottom nepheloid layers due to their relief (40 and 42 m, respectively) above the surrounding mud bottom. The other mid-shelf claystone–siltstone bank, Claypile, with a crest depth of 40 m and only 10 m relief, is certainly often covered by the nepheloid layer. Consequently, the abundance of dominant epibenthos is least at Claypile (lowest relief) and greatest at Sonnier (highest relief).

Speculation on environmental factors governing the development of the *Millepora*-Sponge Zone is complicated by the fact that the zone also occurs on shelf-edge bedrock outcrops that protrude from the crest of Geyer Bank between 37- and 52-m depth. The implication here is that the development of this zone is dependent on the presence of newly exposed bedrock outcrops and vigorous development of the biota is favored by clear water and winter minimum temperatures above 16°C.

Why tropical coral reefs have not developed on the claystone–siltstone outcrops on Geyer Bank, which are exposed to the same oceanic conditions as the coral reefs at the Flower Garden Banks, is not known. Hypothetically, the claystone, which is soft and disintegrates readily on exposure to water, may be unsuitable substratum for most reef-building corals. Indeed, the epifauna that inhabit the *Millepora*-Sponge Zone obviously prefer the thin beds of rock-hard siltstone that protrude from the softer but more extensive claystone on these outcrops.

The distributions of major epibenthic biotic zones in relation to the environmental factors discussed above are summarized in Figure 7.43. Correlations exist between regional patterns of winter temperature, turbidity, and light penetration in shelf waters and distribution patterns of biotic assemblages on outer continental shelf banks. Coral reefs are restricted to clear oceanic water in which temperatures rarely drop below 18°C. Most of the other bank zones experience low winter temperatures around 16°C. Deep tropical reef zones dominated by coralline algae are restricted to the clear, oceanic, shelf-edge waters.

Neritic seasonal variability in temperature and salinity is greatest in the upper 10 m of the coastal and mid-shelf water column, diminishing offshore to roughly the 80- to 85-m regional depth contour, beyond which more stable oceanic conditions predominate. Off the mouth of the Mississippi River neritic influences extend to the edge of the narrow shelf, severely limiting reefal development on nearby banks.

Chronic turbidity of bottom water (nepheloid layers) and associated sedimentation severely limit epibenthos on the lower 15 to 20 m of most banks. Vigorous reef development is restricted to those parts of banks above the effects of nepheloid layers. Continual turbidity and sedimentation on low relief banks substantially reduces diversity and abundance of the assemblages present. Penetration of sunlight into the water decreases toward shore due to generally increasing turbidity, which, combined with the light blocking and smothering effects of bottom nepheloid layers and suspended sediment around the bases of the banks, tends to "displace" zones upward on the banks closer to shore.

The foregoing assessment of the characteristics and distribution of offshore hard-bottom biotic communities in the northwestern Gulf of Mexico and the nature of environmental factors and processes that apparently control these communities has been made possible by the gradual accumulation of regional oceanographic knowledge over decades. Nevertheless marine scientists have barely begun in their attempts to understand the ecology, geology, and hydrography of the Gulf of Mexico.

REFERENCES

Abbott, R. E., 1975. The Faunal Composition of the Algal-Sponge Zone of the Flower Garden Banks, Northwest Gulf of Mexico. M.S. Thesis. Department of Oceanography, Texas A&M University, College Station, 205 pp.

Abbott, R. E., 1979. Ecological Processes Affecting the Reef Coral Population at the East Flower Garden Bank, Northwest Gulf of Mexico. Ph.D. Thesis. Department of Oceanography, Texas A&M University, College Station, 154 pp.

Amery, G. B., 1978. Structure of continental slope, northern Gulf of Mexico. In A. H. Bouma, G. T. Moore, and J. M. Coleman, Eds., Framework, Facies, and Oil Trapping Characteristics of the Upper Continental Margin, *AAPG Stud. Geol.*, **7,** 141–153.

Angelovic, J. W., 1976. Environmental Studies of the South Texas Outer Continental Shelf, 1975. Vol. 2. *Physical Oceanography*. Final Report to the Bureau of Land Management, Contract 08550-IA5-19, U.S. National Marine Fisheries Service, Gulf Fisheries Center, Galveston, Texas, NTIS Order No. PB-283-872/AS, 290 pp.

Antoine, J. W., W. Bryant, and B. Jones, 1967. Structural features of continental shelf, slope, and scarp, northeastern Gulf of Mexico. *AAPG Bull.*, **51,** 257–262.

Avent, R. M., M. E. King, and R. H. Gore, 1977. Topographic and faunal studies of shelf-edge prominences off the central eastern Florida coast. *Int. Rev. Gesamten Hydrobiol.*, **62**(2), 185–208.

Ayala-Castanares, A., 1981. Foraminiferos Recientes de La Laguna de Tamiahua. *An. Inst. Cienc. Mar. Limnol. Univ. Nac. Auton. Mex.*, **8**(1), 103–158.

Bahr, L. M. and W. P. Lanier, 1981. *The Ecology of Intertidal Oyster Reefs of the South Atlantic Coast: A Community Profile*. FWS/OBS–81/15.

Baines, P. G., 1973. The generation of internal tides by flat-bump topography, *Deep-Sea Res.*, **20,** 179–205.

Baines P. G., 1974. The generation of internal tides over steep continental slopes, *Phil. Trans. R. Soc.*, **277**(A), 27–58.

Baines, P. G. and P. A. Davies, 1980. Laboratory studies of topographic effects in rotating and/or stratified fluids. In R. Hide and P. W. White, Eds., *Orographic Effects in Planetary Flows,* GARP Publications, Series No. 23, WMO-ICSU Joint Scientific Committee, 235–293.

Bak, R. P. M., 1977. Coral reefs and their zonation in Netherlands Antilles. *AAPG Stud. Geol.*, **4,** 3–16.

Baker, P. A. and J. N. Weber, 1975. Skeletal growth variations in the reef coral, *Montastrea annularis,* with depth. *Geol. Soc. Am., Abstracts with Programs,* **7,** 24.

Ballard, R. D. and E. Uchupi, 1970. Morphology and Quaternary history of the continental shelf of the Gulf Coast of the United States. *Bull. Mar. Sci.*, **20**(3), 547–559.

Bathurst, R. G. C., 1975. *Carbonate Sediments and Their Diagenesis,* 2nd ed. Elsevier, Amsterdam, 658 pp.

Behringer, D. W., R. L. Molinari, and J. F. Festa, 1977. The variability of anticyclonic current patterns in the Gulf of Mexico. *J. Geophys. Res.*, **82,** 5469–5476.

Berryhill, H. L., Jr., Ed., 1975. *Environmental Studies, South Texas Outer Continental Shelf, 1975: An Atlas and Integrated Summary*. Final Report to the Bureau of Land Management, Contract 08550-MU5-20, NTIS Order No. PB80-217763, 303 pp.

Berryhill, H. L., Jr., 1981. Map Showing Paleogeography of the Continental Shelf During the Low Stand of Sea Level, Wisconsin Glaciation, Corpus Christi 1° × 2° Quadrangle, Texas. *U.S. Geol. Surv.*, **1**:250,000, Map #I-1287-E.

Berryhill, H. L., Jr., G. L. Shideler, C. W. Holmes, S. S. Barnes, G. W. Hill, E. A. Martin, and C. A. Pyle, 1976. *Environmental Studies, South Texas Outer Continental Shelf, 1976.* Final Report to the Bureau of Land Managment, Contract AA550-MU6-24, NTIS Order No. PB-277-337/AS, 626 pp.

Berryhill, H. L., Jr., and A. R. Trippet, 1980. Map showing water circulation and rates of sedimentation in the Port Isabel 1° × 2° Quadrangle, Texas. *U.S. Geol. Surv.,* **1**:250,000, Map #I-1254-A.

Berryhill, H. L., Jr., and A. R. Trippet, 1981a. Map showing water circulation and rates of sedimentation in the Beeville 1° × 2° Quadrangle, Texas. *U.S. Geol. Surv.,* **1**:250,000, Map #I-1288-A.

Berryhill, H. L., Jr., and A. R. Trippet, 1981b. Map showing water circulation and rates of sedimentation in the Corpus Christi 1° × 2° Quadrangle, Texas. *U.S. Geol. Surv.,* **1**:250,000, Map #I-1287-A.

Blaha, J. P. and W. Sturges, 1981. Evidence for wind-forced circulation in the Gulf of Mexico. *J. Mar. Res.,* **39**(4), 711–733.

Blumberg, A. F. and G. L. Mellor, 1981. *A numerical calculation of the circulation in the Gulf of Mexico.* Prepared for Division of Solar Technology, U.S. Department of Energy, Contract #DE-ACO2-78ET 20612, by Dynalysis of Princeton, Princeton, New Jersey [pages not numbered].

Bosellini, A. and R. N. Ginsburg, 1971. Form and internal structure of recent algal nodules (rhodolites) from Bermuda. *J. Geol.,* **79,** 669–682.

Briggs, J. C., 1974. *Marine Zoogeography.* McGraw-Hill, New York, 475 pp.

Bright, T. J., 1977. Coral reefs, nepheloid layers, gas seeps and brine flows on hard-banks in the northwestern Gulf of Mexico. *Proc. Third Int. Coral Reef Symp.,* University of Miami, Rosenstiel School of Marine and Atmospheric Science, **1,** 39–46.

Bright, T. J. and Cara Cashman, 1974. Fishes. In T. J. Bright and L. H. Pequegnat, Eds., *Biota of the West Flower Garden Bank.* Gulf Publishing Company, Houston, pp. 339–409.

Bright, T. J., C. Combs, G. Kraemer, and G. Minnery, 1982. In *Environmental Studies at the Flower Gardens and Selected Banks,* Final Report to U.S. Department of the Interior, Minerals Management Service, Contract #AA851-CTO-25, Chapter III, NTIS Order No. PB83-101303, pp. 39–102.

Bright, T. J., G. P. Kraemer, G. A. Minnery, and S. T. Viada, 1984. Hermatypes of the Flowering Garden Banks. *Bull. Mar. Sci.* **34**(3), 461–476.

Bright, T. J., P. A. LaRock, R. D. Lauer, and J. M. Brooks, 1980a. A brine seep at the East Flower Garden Bank, northwestern Gulf of Mexico. *Int. Rev. Gesamten Hydrobiol.,* **65,** 535–549.

Bright, T. J. and L. H. Pequegnat, Eds., 1974. *Biota of the West Flower Garden Bank.* Gulf Publishing Company, Houston, Texas, 435 pp.

Bright, T. J., E. Powell, and R. Rezak, 1980b. Environmental effects of a natural brine seep at the East Flower Garden Bank, northwestern Gulf of Mexico. In R. A. Geyer, Ed., *Marine Environmental Pollution,* Elsevier Oceanography Series, 27A. Elsevier, New York, pp. 291–316.

Bright, T. J. and R. Rezak, 1976. *A Biological and Geological Reconnaissance of Selected Topographical Features on the Texas Continental Shelf.* Final Report to U.S. Department of the Interior, Bureau of Land Management, Contract #08550-CT5-4, NTIS Order No. PB80-166036, 377 pp.

Bright, T. J. and R. Rezak, 1977. Reconnaissance of reefs and fishing banks of the Texas Continental Shelf. In R. A. Geyer, Ed., *Submersibles and Their Use in Oceanography.* Elsevier, New York, pp. 113–150.

Bright, T. J. and R. Rezak, 1978a. *South Texas Topographic Features Study.* Final Report to U.S. Department of the Interior, Bureau of Land Management, Contract #AA550-CT6-18, NTIS Order No. PB-294-768/AS, 772 pp.

Bright, T. J. and R. Rezak, 1978b. *Northwestern Gulf of Mexico Topographic Features Study.* Final Report to U.S. Department of the Interior, Bureau of Land Management, Contract #AA550-CT7-15, NTIS Order No. PB-294-769/AS, 667 pp.

Bright, T. J., J. W. Tunnell, L. H. Pequegnat, T. E. Burke, C. W. Cashman, D. A. Cropper, J. P. Ray, R. C. Tresslar, J. Teerling, and J. B. Wills, 1974. Biotic zonation on the West Flower Garden Bank. In T. Bright and L. Pequegnat, Eds., *Biota of the West Flower Garden Bank.* Gulf Publishing Company, Houston, pp. 4–54.

Bright, T. J., S. Viada, C. Combs, G. Dennis, E. Powell, and G. Denoux, 1981. East Flower Garden monitoring study. In *Northern Gulf of Mexico Topographic Features Study,* Final Report to U.S. Department of the Interior, Bureau of Land Management, Contract #AA551-CT8-35, Vol. 3, Part C, NTIS Order No. PB81-248676.

Broecker, W., 1961. Radiocarbon dating of late Quaternary deposits, South Louisiana: A discussion. *Geol. Soc. Am. Bull.,* **72,** 159–162.

Brooks, D. A. and M. C. Eble, 1982. *Moored Array Observations in the Western Gulf of Mexico: Current Meter Data Report for the July 1980 to February 1981 Mooring Period.* Department of Oceanography, Texas A&M University, Tech. Rep. #82-12-T, 259 pp.

Brooks, J. M., T. J. Bright, B. B. Bernard, and C. R. Schwab, 1979. Chemical aspects of a brine pool at the East Flower Garden Bank, northwestern Gulf of Mexico. *Limnol. & Oceanogr.,* **24**(4), 735–745.

Bryant, W. R., A. A. Meyerhoff, N. K. Brown, Jr., M. Furrer, T. E. Pyle, and J. W. Antoine, 1969. Escarpments, reef trends and diapiric structures, Eastern Gulf of Mexico. *AAPG Bull.,* **53**(12), 2506–2542.

Burke, T. E., 1974a. Echinoderms. In T. J. Bright and L. H. Pequegnat, Eds., *Biota of the West Flower Garden Bank.* Gulf Publishing Company, Houston, pp. 311–332.

Burke, T. E., 1974b. Echinoderms of the West Flower Garden Reef Bank. Master's Thesis, Department of Oceanography, Texas A&M University, College Station, 165 pp.

Cairns, S. D., 1977. Stony Corals. I. *Carophylliina* and *Dendrophylliina* (Anthozoa:Scleractinia). In *Memoirs of the Hourglass Cruises,* Marine Research Laboratory, Florida Department of Natural Resources, St. Petersburg, Florida, Vol. 3, Part IV, 27 pp.

Capurro, L. R. A. and J. L. Reid, 1970. *Contributions on the Physical Oceanography of the Gulf of Mexico,* Vol. 2. Gulf Publishing Company, Houston, 288 pp.

Carsey, J. B., 1950. Geology of Gulf coastal area and continental shelf. *AAPG Bull.,* **34,** 361–385.

Cashman, Cara W., 1973. Contributions to the Ichthyofaunas of the West Flower Garden Reef and other reef sites in the Gulf of Mexico and Western Caribbean. Ph.D. Dissertation, Department of Oceanography, Texas A&M University, College Station, 248 pp.

Causey, Billy D., 1969. The Fish of Seven and One-Half Fathom Reef., Master's Thesis, Department of Biology, Texas A&I University, Kingsville, 110 pp.

Cervignon, F., 1966. *Los Peces Marinos De Venezuela*. Monographs No. 11 and 12, 2 vols., Fundacion La Salle de Ciencias Naturales, Fondo De Cultura Cientifica, Caracas, Venezuela, 1387 pp.

Chamberlain, C. K., 1966. Some Octocorallia of Isla de Lobos, Veracruz, Mexico. *Brigham Young Univ. Geol. Stud.*, **13,** 47–54.

Chavez, E. A., 1973. Observaciones generales sobre las comunidades del arrecife de Lobos, Veracruz. *An. Esc. Nac. Cienc. Biol. Mex.*, **20,** 13–21.

Chavez, E. A., E. Y. Sevilla, and M. L. Hidalgo, 1970. *Datos acerca de las comunidades bentonicas del arrecife de Lobos, Veracruz*. Revista de la Sociedad Mexicana de Historia Natural, Tomo XXXI.

Chittenden, M. E., Jr., and J. D. McEachran, 1976. Composition, Ecology and Dynamics of Demersal Fish Communities on the Northwestern Gulf of Mexico Continental Shelf, with a similar synopsis for the Entire Gulf. Department of Oceanography, Texas A&M University, Tech. Rep. TAMU-SG-76-208.

Chittenden, M. E. and D. Moore, 1977. Composition of the ichthyofauna inhabiting the 110-meter bathymetric contour of the Gulf of Mexico, Mississippi River to the Rio Grande. *Northeast Gulf Sci.*, **1,** 106–114.

Clark, S. T. and P. B. Robertson, 1982. Shallow water marine isopods of Texas. *Contrib. Mar. Sci.*, **25,** 45–59.

Cochrane, J. D. and F. J. Kelly, Jr., 1982. Proposed annual progression in the mean Texas-Louisiana shelf circulation. *Trans. Am. Geophys. Union.*, **63**(45), 1012.

Collard, S. B. and C. N. D'Asaro, 1973. *The Biological Environment: Benthic Invertebrates of the Eastern Gulf of Mexico*. In J. I. Jones, R. E. Ring, M. O. Rinkel, and R. E. Smith, Eds. *A Summary of Knowledge of the Eastern Gulf of Mexico*. State University System of Florida, Institute of Oceanography, St. Petersburg, pp. III G-1–III G-27.

Continental Shelf Associates, Inc., 1980. *Video and Photographic Reconnaissance of Phleger and Sweet Banks, Northwest Gulf of Mexico*. Prepared for Bureau of Land Management, Washington, D.C., Contract No. AA551-CT9-36, 20 pp.

Continental Shelf Associates, Inc., 1982. *Environmental Monitoring Program for Platform "A," Lease OCS-G 3061, Block A-85, Mustang Island Area, East Addition, Near Baker Bank*. Final Report to Conoco, Inc., 171 pp.

Cropper, Dennis A., 1973. Living Cheilostome Bryozoa of West Flower Garden Bank, Northwest Gulf of Mexico. Master's Thesis, Department of Oceanography, Texas A&M University, College Station, 89 pp.

Crout, R. L. and R. D. Hamiter, 1981. Response of bottom waters on the west Louisiana shelf to transient wind events and resulting sediment transport. *Trans. Gulf Coast Assoc. Geol. Soc.*, **31,** 273–278.

Curray, J. R., 1960. Sediments and history of Holocene transgression, continental shelf, northwest Gulf of Mexico. In F. P. Shepard, F. B. Phleger, and T. H. Van Andel, Eds., *Recent Sediments, Northwest Gulf of Mexico*. AAPG, Tulsa, Oklahoma, pp. 221–266.

Curray, J. R., 1965. Late Quaternary history, continental shelves of the United States. In H. E. Wright, Jr. and D. G. Frey, Eds., *The Quaternary of the United States*. Princeton University Press, Princeton, New Jersey, pp. 728–736.

Davies, D. K. and W. R. Moore, 1970. Dispersal of Mississippi sediment in the Gulf of Mexico. *J. Sediment. Petrol.*, **40**(1), 339–353.

Davis, G. E., 1982. A century of natural change in coral distribution at the Dry Tortugas: A comparison of reef maps from 1881 and 1976. *Bull. Mar. Sci.*, **32**(2), 608–623.

Defenbaugh, R. E., 1974. Hydroids. In T. J. Bright and L. H. Pequegnat, Eds., *Biota of the West Flower Garden Bank*. Gulf Publishing Company, Houston, pp. 93–114.

Defenbaugh, R. E., 1976. A Study of the Benthic Macroinvertebrates of the Continental Shelf of the Northern Gulf of Mexico. Ph.D. Thesis. Department of Oceanography, Texas A&M University, College Station, 410 pp.

Dennis, R. E. and E. E. Long, 1971. *A User's Guide to a Computer Program for Harmonic Analysis of Data at Tidal Frequencies*. NOAA Tech. Rep. NOS 41, Office of Marine Surveys and Maps, Oceanographic Division, Rockville, Maryland, 31 pp.

Dodge, R. E., A. Logan, and A. Antonius, 1982. Quantitative reef assessment studies in Bermuda: A comparison of methods and preliminary results. *Bull. Mar. Sci.*, **32**(3), 745–760.

Downey, M. E., 1973. Starfishes from the Caribbean and the Gulf of Mexico. *Smithson. Contrib. Zool.* No. 126, 158 pp.

Dubois, Random, 1975. A Comparison of the Distribution of the Echinodermata of a Coral Community with That of a Nearby Rock Outcrop on the Texas Continental Shelf. Master's Thesis, Department of Oceanography, Texas A&M University, College Station, 153 pp.

Dustan, P., 1975. Growth and form in the reef-building coral *Montastrea annularis*. *Mar. Biol.*, **33,** 101–107.

Earle, S. A., 1969. Phaeophyta of the Eastern Gulf of Mexico. *Phycologia*, **7**(2), 71–254.

Earle, S. A., 1972. Benthic algae and seagrasses. In *Atlas Folio No. 22, Chemistry, Primary Productivity, and Benthic Algae of the Gulf of Mexico*. American Geographical Society, New York, pp. 15–18, 25–29, and plate 6.

Edwards, G. S., 1971. *Geology of the West Flower Garden Bank*. Texas A&M Sea Grant Publ., TAMU-SG-71-215, 199 pp.

Edwards, P., 1976. *Illustrated Guide to the Seaweeds and Seagrasses in the Vicinity of Port Aransas, Texas*. University of Texas Press, Austin, 126 pp.

Ehrlich, R. and B. Weinberg, 1970. An exact method for characterization of grain shape. *J. Sediment. Petrol.*, **40,** 205–212.

Eiseman, N. J. and S. M. Blair, 1982. New records and range extensions of deepwater algae from East Flower Garden Bank, northwestern Gulf of Mexico, *Contrib. Mar. Sci.*, **25,** 21–26.

Elliott, B. A., 1979. Anticyclonic Rings and the Energetics of the Circulation of the Gulf of Mexico. Ph.D. Thesis. Department of Oceanography, Texas A&M University, College Station, 188 pp.

Elliott, B. A., 1982. Anticyclonic rings in the Gulf of Mexico. *J. Phys. Ocean.*, **12,** 1292–1309.

Etter, P. C. and J. D. Cochrane, 1975. *Water Temperature on the Texas–Louisiana Shelf.* Marine Advisory Bulletin, Commerce. Texas A&M Sea Grant Publ., TAMU-SG-75-604, 22 pp.

Everett, D. K., 1971. *Hydrologic and Quality Characteristics of the Lower Mississippi River.* Louisiana Department of Public Works, 48 pp.

Fandino, V. S., unpublished. *Algunos Estudies Sobre las Madreporas del Arecife "La Blanquilla," Veracruz, Mexico.* Universidad Autonoma Metropolitana, 103 pp.

Felder, D. L. and A. C. Chaney, 1979. Decapod crustacean fauna of Seven and One-Half Fathom Reef, Texas: species composition, abundance, and species diversity. *Contrib. Mar. Sci.,* **22,** 1–29.

Flint, R. W. and C. W. Griffin, Eds., 1979. *Environmental Studies, South Texas Outer Continental Shelf, Biology and Chemistry.* Final Report to U.S. Department of the Interior, Bureau of Land Management, Contract AA550-CT7-11, Executive Summary, NTIS Order No. PB81-106692, 75 pp.

Folk, R. L., 1974. *Petrology of Sedimentary Rocks,* 2nd ed. Hemphill Publication Company, Austin, 182 pp.

Forristall, G. Z., R. C. Hamilton, and V. J. Cardone, 1977. Continental shelf currents in Tropical Storm Delia: observations and theory. *J. Phys. Oceanogr.,* **7,** 532–546.

Fotheringham, N., 1980. *Beachcomber's Guide to Gulf Coast Marine Life.* Gulf Publishing Company, Houston, 124 pp.

Frohlich, C., 1982. Seismicity in the Central Gulf of Mexico. *Geology,* **10,** 103–106.

Gallaway, B. J. and G. S. Lewbel, 1982. *The Ecology of Petroleum Platforms in the Northwestern Gulf of Mexico: A Community Profile.* FWS/OBS-82/27, 91 pp.

George, R. Y. and P. J. Thomas, 1979. *Aspects of Fouling on Offshore Oil Platforms in Louisiana Shelf in Relation to Environmental Impact.* Final report, Offshore Ecological Investigations, Gulf Universities Research Consortium, 54 pp.

Giammona, Charles P., 1978. Octocorals in the Gulf of Mexico—Their Taxonomy and Distribution with Remarks on Their Paleontology. Ph.D. Dissertation, Department of Oceanography, Texas A&M University, College Station, 260 pp.

Ginsburg, R. N. and M. Stanley, 1970. *Seminar on Organism-Sediment Interrelationships.* Reports of Research, 1969, Bermuda Biological Station for Research, Spec. Publ. No. 6, St. George's West, Bermuda, 111 pp.

Gittings, S. R., 1983. Hard-Bottom Macrofauna of the East Flower Garden Brine Seep: Impact of a Long-Term, Point-Source Brine Discharge. M.S. Thesis. Department of Oceanography, Texas A&M University, College Station, 72 pp.

Goedicke, T. R., 1955. Origin of the pinnacles on the continental shelf and slope of the Gulf of Mexico. *Tex. J. Sci.,* **7,** 149–159.

Goreau, T. F., 1959. The ecology of Jamaican coral reefs: I. Species composition and zonation. *Ecology,* **40,** 67–90.

Goreau, T. F. and J. W. Wells, 1967. The shallow-water scleractinia of Jamaica: revised list of species and their vertical distribution range. *Bull. Mar. Sci.,* **17**(2), 442–453.

Gould, H. R. and R. H. Stewart, 1955. Continental terrace sediments in the northeast Gulf of Mexico. In Finding Ancient Shorelines, *Soc. Econ. Paleontol. Mineral., Spec. Publ. 3,* Tulsa, Oklahoma, pp. 3–20.

Grimm, D. E. and T. S. Hopkins, 1977. Preliminary characterization of the octocorallian and scleractinian diversity at the Florida Middle Ground. *Proc. Third Int. Coral Reef Symp.,* University of Miami, Rosenstiel School of Marine and Atmospheric Science, **1,** 135–141.

Harding, J. L., and W. D. Nowlin, Jr., 1966. Gulf of Mexico. In Rhodes W. Fairbridge, Ed., *Encyclopedia of Oceanography.* Reinhold, New York, pp. 324–331.

Harrington, D. L., 1966. Oceanographic Observations on the Northwest Continental Shelf of the Gulf of Mexico, 1963–1965. Contribution No. 329, National Marine Fisheries Service Biological Laboratory, Galveston, Texas, 25 pp.

Hawkins, J. W., 1983. Sources of Relict Sand on Northeast Texas Continental Shelf. M.S. Thesis. Department of Oceanography, Texas A&M University, College Station, 63 pp.

Hayes, S. P. and D. Halpern, 1976. Variability of semidiurnal internal tide during coastal upwelling. *Mem. Soc. R. Sci. Liège,* **10,** 175–186.

Hedgpeth, J. W., 1953. An introduction to zoogeography of the northwestern Gulf of Mexico with reference to the invertebrate fauna. *Publ. Inst. Mar. Sci. Univ. Tex.,* **3,** 107–224.

Hedgpeth, J. W., 1954. Bottom communities of the Gulf of Mexico. *Fish. Bull. 89, Fish Wildl. Serv. U.S.,* **55,** 203–214.

Heilprin, A., 1890. The corals and coral reefs of the western waters of the Gulf of Mexico. *Proc. Acad. Nat. Sci. Philadelphia,* **42,** 301–316.

Hildebrand, H. H., 1954. A study of the fauna of the brown shrimp (*Penaeus aztecus* Ives) grounds in the Western Gulf of Mexico. *Publ. Inst. Mar. Sci. Univ. Texas,* **3**(2), 233–366.

Hildebrand, H. H., 1955. A study of the fauna of the pink shrimp (*Penaeus duorarum* Burkenroad) grounds in the Gulf of Campeche. *Publ. Inst. Mar. Sci. Univ. Texas,* **4**(1), 169–232.

Hildebrand, H. H., 1957. Estudios Biologicos Preliminares Sobre La Laguna Madre De Tamaulipas. *Ciencia,* **17**(7–9), 151–173.

Hoese, H. D. and R. H. Moore, 1977. *Fishes of the Gulf of Mexico, Texas, Louisiana, and Adjacent Waters.* Texas A&M University Press, College Station, 327 pp.

Hogg, N. G., 1980. Effects of bottom topography on ocean currents. In R. Hide and P. W. White, Eds., Orographic Effects in Planetary Flows, GARP Publications Series No. 23, WMO-ICSU Joint Scientific Committee, 167–205.

Holmes, C. W., 1982. Geochemical indices of fine sediment transport, northwest Gulf of Mexico. *J. Sediment. Petrol.,* **52**(1), 307–321.

Hopkins, T. S., D. R. Blizzard, and D. K. Gilbert, 1977. The molluscan fauna of the Florida Middle Ground with comments on its zoogeographical affinities. *Northeast Gulf Sci.,* **1**(1), 39–47.

Hopkins, T. S. and W. Schroeder, 1981. *Physical and Chemical Oceanography.* In *Northern Gulf of Mexico Topographic Features Study,* Final Report to U.S. Department of the Interior, Bureau of Land Management, Contract #AA551-CT8-35, Department of Oceanography, Texas A&M University, Tech. Rep. #81-2-T, Vol. 5, 21–47.

Hudson, J. H., 1981. Growth rates in *Montastrea annularis:* A record of environmental change in Key Largo Coral Reef Marine Sanctuary, Florida. *Bull. Mar. Sci.,* **31,** 444–459.

Hudson, J. H. and D. M. Robbin, 1980. Effects of drilling mud on the growth rate of the reef building coral, *Montastrea annularis*. In R. A. Geyer, Ed., *Marine Environmental Pollution*, Elsevier Oceanography Series, 27A. Elsevier, New York, pp. 455–470.

Hudson, J. H., E. A. Shinn, R. B. Halley, and B. Lidz, 1976. Sclerochronology: A tool for interpreting past environments. *Geology*, **4**, 361–364.

Humm, H. J., 1973. *The Biological Environment; Salt Marshes, Benthic Algae of the East Gulf of Mexico, Seagrasses, Mangrove. A Summary of the Eastern Gulf of Mexico*. State University System of Florida, Institute of Oceanography, St. Petersburg, pp. 1–6.

Humm, H. J. and H. H. Hildebrand, 1962. Marine algae from the Gulf Coast of Texas and Mexico. *Publ. Inst. Mar. Sci. Univ. Texas*, **8**, 227–268.

Humphris, C. C., Jr., 1978. Salt movement on continental slope, northern Gulf of Mexico. In A. H. Bouma, G. T. Moore, and J. M. Coleman, Eds., *Framework Facies and Oil Trapping Characteristics of the Upper Continental Margin*, AAPG, Studies in Geology No. 7. Tulsa, Oklahoma, pp. 69–85.

Huntsman, G. R. and I. G. MacIntyre, 1971. Tropical coral patches in Onslow Bay. *Underwater Nat. Bull. Am. Littoral Soc.*, **7**(2), 32–34.

Hurlburt, H. E. and J. D. Thompson, 1982. The Dynamics of the Loop Current and Shed Eddies in a Numerical Model of the Gulf of Mexico. In J. C. J. Nihoul, Ed., *Hydrodynamics of Semi-Enclosed Seas*. Elsevier, Amsterdam, pp. 243–298.

Huthnance, J. M., 1981. Waves and currents near the continental shelf edge. *Prog. Oceanogr.*, **10**, 193–226.

Ichiye, T., H. Kuo, and M. Carnes, 1973. Assessment of currents and hydrography of the Eastern Gulf of Mexico. *Texas A&M Univ. Oceanogr. Stud.*, No. 601, 311 pp.

Jameson, S. C., Ed., 1981. Key Largo Coral Reef National Marine Sanctuary Deep Water Resource Survey. *NOAA Tech. Rep. #CZ/SP-1*, 144 pp.

Jordan, G. F., 1952. Reef formation in the Gulf of Mexico off Appalachicola Bay, Florida. *Geol. Soc. Am. Bull.*, **63**, 741–744.

Joyce, E. A., Jr., and J. Williams, 1969. Memoirs of the Hourglass Cruises: Rationale and Pertinent Data. Marine Research Laboratory, Florida Department of Natural Resources, St. Petersburg, Vol. 1, Part I, 50 pp.

Kim, C. S., 1964. Marine Algae of Alacran Reef, Southern Gulf of Mexico. Ph.D. Thesis. Department of Botany, Duke University, Durham, North Carolina, 212 pp.

Kornicker, L. S., F. Bonet, R. Cann, and C. M. Hooker, 1959. Alacran Reef, Campeche Bank, Mexico. *Publ. Inst. Mar. Sci. Univ. Texas*, **6**, 1–22.

Kraemer, G. P., 1982. Population Levels and Growth Rates of Scleractinian Corals Within the *Diploria-Montastrea-Porites* Zones of the East and West Flower Garden Banks. M.S. Thesis. Department of Oceanography, Texas A&M University, College Station, 139 pp.

Kundu, P. K., 1976. An analysis of inertial oscillations observed near Oregon coast. *J. Phys. Oceanogr.*, **6**, 879–893.

Kutkuhn, H., 1962. Gulf of Mexico commercial shrimp populations—trends and characteristics, 1956–59. *Fish. Bull. #212*, **62**, 401 pp.

Lang, J. C., 1973. Interspecific aggression by scleractinian corals. 2. Why the race is not only to the swift. *Bull. Mar. Sci.*, **23**(2), 260–279.

LeBlanc, R. J., 1972. Geometry of sandstone reservoir bodies. In T. D. Cook, Ed., Underground Waste Management and Environmental Implications, *AAPG Mem.* **18**, 133–190.

LeBlond, P. H. and L. Mysak, 1978. *Waves in the Ocean*. Elsevier, New York, 602 pp.

Leuterman, Arthur, 1979. The Taxonomy and Systematics of the Gymnolaemate and Stenolaemate Bryozoa of the Northwest Gulf of Mexico. Ph.D. Dissertation, Department of Oceanography, Texas A&M University, College Station, 309 pp.

Levert, C. F. and H. C. Ferguson, Jr., 1969. Geology of the Flower Garden Banks, northwest Gulf of Mexico. *Trans. Gulf Coast Assoc. Geol. Soc.*, **19**, 89–100.

Lindquist, Paul, 1978. Geology of South Texas Shelf Banks. Masters Thesis, Department of Oceanography, Texas A&M University, College Station, 138 pp.

Lipka, D. A., 1974. Mollusks. In T. J. Bright and L. H. Pequegnat, Eds., *Biota of the West Flower Garden Bank*. Gulf Publishing Company, Houston, pp. 141–198.

Logan, B. W., 1969. Carbonate Sediments and Reefs, Yucatan Shelf, Mexico: Part 2, Coral Reefs and Banks, Yucatan Shelf, Mexico (Yucatan Reef Unit). *AAPG Mem.*, **11**, 129–198.

Logan, B. W., J. L. Harding, W. M. Ahr, J. D. Williams, and R. G. Snead, 1969. Carbonate sediments and reefs, Yucatan Shelf, Mexico: Part 1, Late Quaternary carbonate sediments of Yucatan Shelf, Mexico. *AAPG Mem.*, **11**, 5–128.

Loya, Y., 1976. Effects of water turbidity and sedimentation on the community structure of Puerto Rican corals. *Bull. Mar. Sci.*, **26**, 450–466.

Loya, Y., 1978. Plotless and transect methods. In D. R. Stoddart and R. E. Johannes, Eds., *Coral Reefs: Research Methods*. UNESCO, Paris, pp. 197–217.

Lyons, W. G. and S. B. Collard, 1974. Benthic Invertebrate Communities of the Eastern Gulf of Mexico. Proceedings of Marine Environmental Implications of Offshore Drilling Eastern Gulf of Mexico: 1974. Contributions No. 233, FDNR/MRL, 157–167.

Macdonald, K., 1982. Introductory Overview: Southwest Florida Shelf Ecosystems Study. Proceedings Third Annual Gulf of Mexico Information Transfer Meeting, Minerals Management Service, New Orleans, pp. 31–35.

MacIntyre, I. G., 1972. Submerged reefs of the Eastern Caribbean. *AAPG Bull.*, **56**, 720–738.

Maddocks, R. F., 1974. Ostracodes. In T. J. Bright and L. H. Pequegnat, Eds., *Biota of the West Flower Garden Bank*. Gulf Publishing Company, Houston, pp. 199–230.

Marmer, H. A., 1954. Tides and sea level in the Gulf of Mexico. *Fish. Bull. Fish Wildl. Serv. U.S.*, **55**, 101–118.

Marmorino, G. O., 1982. Summertime Coastal Currents in the Northeastern Gulf of Mexico. Unpublished manuscript, Department of Oceanography, Florida State University, Tallahassee, 30 pp.

Martin, R. G., 1978. Northern and eastern Gulf of Mexico continental margin: Stratigraphic and structural framework. In A. H. Bouma, G. T. Moore, and J. M. Coleman, Eds., *Framework, Facies, and Oil-Trapping Characteristics of the Upper Continental Margin*. AAPG, Tulsa, Oklahoma, pp. 21–42.

Martin, R. G. and A. H. Bouma, 1978. Physiography of Gulf of Mexico. In A. H. Bouma, G. T. Moore, and J. M. Coleman, Eds., *Framework, Facies, and Oil-Trapping Characteristics of the Upper Continental Margin.* AAPG, Tulsa, Oklahoma, pp. 3–19.

Mayor, A. G., 1915. The lower temperature at which reef corals lose their ability to capture food. *Yearbook, Carnegie Inst. Pub.* **183,** 6, 1–24.

McConnaughey, T. and C. McRoy, 1979. Food-web structure and the fractionation of carbon isotopes in the Bering Sea. *Mar. Biol. Berlin,* **53,** 257–262.

McFarlan, E., Jr., 1961. Radiocarbon dating of late Quaternary deposits, South Louisiana. *Geol. Soc. Am. Bull.,* **72,** 129–158.

McGrail, D. W., 1977. Shelf edge currents and sediment transport in the northwestern Gulf of Mexico. (Abs.) *Trans. Am. Geophys. Union,* **58,** 1160.

McGrail, D. W., 1978. Boundary layer processes, mixed bottom layers, and turbid layers on continental shelves. (Abs.) *Trans. Am. Geophys. Union,* **59,** 1110.

McGrail, D. W., 1979. The role of air–sea interaction in sediment dynamics. *Abstracts of Papers at the 145th National Meeting,* American Association for the Advancement of Science, 45.

McGrail, D. W., 1982. Anomalous flow on the Outer Continental Shelf in the Gulf of Mexico and its effect on sediment transport. (Abs.) *Trans. Am. Geophys. Union,* **63**(3), 65.

McGrail, D. W. and M. R. Carnes, 1983. Shelfedge dynamics and the nepheloid layer. In D. J. Stanley and G. T. Moore, Eds., Shelf Break: Critical Interface on Continental Margins, *Soc. Econ. Paleontol. Mineral., Special Pub. No. 33,* 251–264.

McGrail, D. W., M. Carnes, D. Horne, T. Cecil, J. Hawkins, and F. Halper, 1982a. Water and sediment dynamics at the Flower Garden Banks. In *Environmental Studies at the Flower Gardens and Selected Banks.* Final Report to Minerals Management Service, Contract #AA851-CT0-25, NTIS Order No. PB83-101303, pp. 103–226.

McGrail, D. W., M. Carnes, D. Horne, and J. Hawkins, 1982b. *Hydrographic Data Report. Northern Gulf of Mexico Topographic Features Study.* U.S. Department of the Interior, Bureau of Land Management, Contract #AA851-CT0-25. Department of Oceanography, Texas A&M University. Tech. Rep. #82-4-T, 516 pp.

McGrail, D. W., T. M. Cecil, and F. B. Halper, 1982e. Stacking of nepheloid and boundary layers at the shelf edge in the Gulf of Mexico. (Abs.) *Trans. Am. Geophys. Union,* **63,** 988.

McGrail, D. W., F. Halper, D. Horne, T. Cecil, M. Carnes, 1982c. *Time Series Data Report. Northern Gulf of Mexico Topographic Features Study.* U.S. Department of the Interior, Bureau of Land Management, Contract #AA851-CT0-25. Department of Oceanography, Texas A&M University, Tech. Rep. #82-5-T, 155 pp.

McGrail, D. W. and D. Horne, 1979. Currents, thermal structure and suspended sediment distribution induced by internal tides on the Texas Continental Shelf. Paper presented at the spring meeting of the American Geophysical Union SANDS Symposium.

McGrail, D. W. and D. Horne, 1981. Water and sediment dynamics [Flower Garden Banks]. In *Northern Gulf of Mexico Topographic Features Study,* Final Report to U.S. Department of the Interior, Bureau of Land Management, Contract #AA551-CT8-35, Vol. 3, NTIS Order No. PB81-248676, Part B, 9-45.

McGrail, D. W. and D. W. Huff, 1978. Shelf sediment and local flow phenomena: *in situ* observations. (Abs.) *Program, AAPG-SEPM Annual Convention,* 93.

McGrail, D. W., D. W. Huff, and S. Jenkins, 1978. Current measurements and dye diffusion studies. In T. Bright and R. Rezak, Eds., *Northwestern Gulf of Mexico Topographic Features Study.* Final Report to U.S. Department of the Interior, Bureau of Land Management, Contract #AA550-CT7-15, NTIS Order No. PB-294-769/AS, pp. III-3 to III-72.

McGrail, D. W. and R. Rezak, 1977. Internal waves and the nepheloid layer on continental shelf in the Gulf of Mexico. *Trans. Gulf Coast Assoc. Geol. Soc.,* **27,** 123–124.

McGrail, D. W., R. Rezak, and T. J. Bright, 1982d. *Environmental Studies at the Flower Gardens and Selected Banks: Northwestern Gulf of Mexico, 1979–1981.* Final Report to Minerals Management Service, Contract #AA851-CT0-25, NTIS Order No. PB83-101303, 314 pp.

Merrell, W. J. and J. M. Morrison, 1981. On the circulation of the Western Gulf of Mexico with observations from April 1978. *J. Geophys. Res.,* **86**(C5), 4181–4185.

Minnery, Gregory, 1984. Distribution, Growth Rates and Diagenesis of Coralline Algal Structures on the Flower Garden Banks, Northwestern Gulf of Mexico. Ph.D. dissertation, Department of Oceanography, Texas A&M University, College Station, 174 pp.

Moore, D. R., 1958. Notes on Blanquilla Reef, the most northerly coral formation in the western Gulf of Mexico. *Publ. Inst. Mar. Sci. Univ. Tex.,* **5,** 151–155.

Moore, D. R. and H. R. Bullis, Jr., 1960. A deep water coral reef in the Gulf of Mexico. *Bull. Mar. Sci. Gulf and Caribb.,* **10,** 125–128.

Mosley, F., 1966. Notes on fishes from the snapper banks off Port Aransas, Texas, *Texas J. Sci.* **18,** (1) 75–79.

Murray, G. E., 1961. *Geology of the Atlantic and Gulf Coastal Province of North America.* Harper, New York, 692 pp.

Nakai, N., 1960. Carbon isotope fractionation of natural gas in Japan. *J. Earth Sci.,* **8,** 174–180.

Nettleton, L. L., 1957. Gravity survey over a Gulf coast continental shelf mound. *Geophysics,* **22,** 630.

Nichols, J. A., 1976. The effect of stable dissolved-oxygen stress on marine benthic invertebrate community diversity. *Int. Rev. Gesamten Hydrobiol.,* **61,** 747–760.

Niiler, P. P., 1976. Observations of low-frequency currents on the West Florida continental shelf. *Mem. R. Soc. Sci. Liege,* Ser. 6, **10,** 331–358.

NOAA, 1980. *Draft Environmental Impact Statement on the Proposed Gray's Reef Marine Sanctuary.* Office of Coastal Zone Management, National Oceanic and Atmospheric Administration, 193 pp.

Nowlin, W. D., Jr., 1972. Winter circulation patterns and property distributions. In L. R. A. Capurro and J. L. Reid, Eds., *Texas A&M University Oceanographic Studies,* Vol. 2. Gulf Publishing Company, Houston, pp. 3–53.

Nowlin, W. D., Jr. and H. J. McLellan, 1967. A characterization of the Gulf of Mexico waters in winter. *J. Mar. Res.,* **25**(1), 29–59.

Nowlin, W. D., Jr., and C. A. Parker, 1974. Effects of a cold-air outbreak on shelf waters of the Gulf of Mexico. *J. Phys. Oceanogr.* **4**(3), 467–486.

Oppenheimer, C. H. and K. G. Gordon, 1972. *Texas Coastal Zone Biotopes: An Ecography.* Interim Report for the Bay and Estuary Management Program (CRMP). University of Texas Marine Science Institute, Port Aransas, Texas [pages unnumbered].

Parker, R. H. and J. R. Curray, 1956. Fauna and bathymetry of banks on continental shelf, Northwestern Gulf of Mexico. *Bull. Am. Assoc. Petrol. Geol.*, **40**, 2428–2439.

Parker, R. P., J. M. Morrison, and W. D. Nowlin, Jr., 1979. Surface Drifter Data from the Caribbean Sea and Gulf of Mexico, 1975–1978. ONR #N00014-75-C-0537, Department of Oceanography, Texas A&M University, Tech Rep. #79-8-T. 157 pp.

Pequegnat, L. H. and J. P. Ray, 1974. Crustaceans and other Arthropods. In T. J. Bright and L. H. Pequegnat, Eds., *Biota of the West Flower Garden Bank.* Gulf Publishing Company, Houston, pp. 231–290.

Pequegnat, W. E., 1970. Deep-water brachyuran crabs. In W. E. Pequegnat and F. A. Chace, Jr., Eds., *Contributions on the Biology of the Gulf of Mexico,* Vol. 1, Texas A&M University, College Station, 171-204.

Pielou, E. C., 1966. The measurement of diversity in different types of biological collections. *J. Theor. Biol.*, **13**, 131–144.

Pionetti, J. and A. Toulmond, 1980. Tide-related changes of volatile fatty acids in the blood of the lugworm *Arenicola marina* (L.). *Can. J. Zool.*, **58**, 1723–1727.

Poag, C. W., 1973. Late Quaternary sea levels in the Gulf of Mexico. *Trans. Gulf Coast Assoc. Geol. Soc.*, **23**, 394–400.

Poole, R. W., 1974. *An Introduction to Quantitative Ecology.* McGraw-Hill, New York, 532 pp.

Porter, J. W., 1972. Ecology and species diversity of coral reefs on opposite sides of the Isthmus of Panama. In M. L. Jones, Ed., The Panamic Biota: Some Observations Prior to the Sea Level Canal, *Bull. Biol. Soc. Wash.*, **2**, 89–116.

Powell, E. N. and T. J. Bright, 1981. A thiobios does exist—Gnathostomulid domination of the canyon community at the East Flower Garden brine seep. *Int. Rev. Gesamten Hydrobiol.*, **66**(5), 675–683.

Powell, E. N., T. J. Bright, A. Woods, and S. Gittings, 1983. Meiofauna and the thiobios in the East Flower Garden brine seep. *Mar. Biol.*, **73**, 269–283.

Prinsenberg, S. J. and M. Rattray, Jr., 1975. Effects of continental slope and variable Brunt–Vaisala frequency on the coastal generation of internal tides. *Deep-Sea Res.*, **22**, 251–263.

Prinsenberg, S. J., W. Wilmot, and M. Rattray, Jr., 1974. Generation and dissipation of coastal internal tides. *Deep-Sea Res.*, **16**, 179–195.

Pulley, T. E., 1952. A zoogeographic study based on the bivalves of the Gulf of Mexico. Ph.D. Thesis. Harvard University, 215 pp.

Pulley, T. E., 1963. Texas to the tropics. *Houston Geol. Soc. Bull.*, **6**, 13–19.

Pyle, C. A., 1977. Late Quaternary Geologic History of the South Texas Continental Shelf. M.S. Thesis. Department of Oceanography, Texas A&M University, College Station, 72 pp.

Randall, J., 1965. Grazing effects on sea grasses by herbivorous reef fishes of the West Indies. *Ecology,* **46**, 255–260.

Rannefeld, J. W., 1972. The Stony Corals of Enmedio Reef Off Veracruz, Mexico. M.S. Thesis. Department of Oceanography, Texas A&M University, College Station, 104 pp.

Rattray, M., Jr., 1960. On the coastal generation of internal tides. *Tellus,* **12**, 54–62.

Rattray, M., Jr., J. G. Dworski, and P. E. Kovala, 1969. Generation of long internal waves at the continental slope. *Deep-Sea Res.*, **16**, 179–195.

Ray, J. P., 1974. A Study of the Coral Reef Crustaceans (Decapoda and Stomatopoda) of Two Gulf of Mexico Reef Systems: West Flower Garden, Texas and Isla De Lobos, Veracruz, Mexico. Ph.D. Dissertation, Department of Oceanography, Texas A&M University, College Station, 323 pp.

Reed, J. H., 1977. *Barium and Vanadium Determination by Neutron Activation Analysis—A Study of Procedures.* Report to U.S. Department of the Interior, Bureau of Land Management, Contract #AA550-CT7-38, 84 pp.

Reed, J. K., 1980. Distribution and structure of deep-water *Oculina varicosa* coral reefs off central eastern Florida. *Bull. Mar. Sci.*, **30**(3), 667–677.

Reid, R. O. and R. Whitaker, in press. Numerical Model for Astronomical Tides in the Gulf of Mexico. U.S. Army Engineer Waterways Experiment Station, Vicksburg, Mississippi.

Rezak, R., 1977. West Flower Garden Bank, Gulf of Mexico. *Stud. Geol.*, **4**, 27–35.

Rezak, R., 1981. Geology. In *Northern Gulf of Mexico Topographic Features Study.* Final Report to U.S. Department of the Interior, Bureau of Land Management, Contract #AA551-CT8-35, Vol. 1, NTIS Order No. PB81-248650, 23–59.

Rezak, R., 1982a. Geology of the Flower Garden Banks. In *Environmental Studies at the Flower Gardens and Selected Banks.* Final Report to Minerals Management Service, Contract #AA851-CT0-25, Chapter II, NTIS Order No. PB83-101303, 19–37.

Rezak, R., 1982b. Geology of selected banks. In *Environmental Studies at the Flower Gardens and Selected Banks.* Final Report to Minerals Management Service, Contract #AA851-CT0-25, Chapter VI, NTIS Order No. PB83-101303, 253–300.

Rezak, R. and T. J. Bright, 1981a. *Northern Gulf of Mexico Topographic Features Study.* Final Report to U.S. Department of the Interior, Bureau of Land Management, Contract #AA551-CT8-35. 5 vols., 995 pp. NTIS Order Nos.: Vol. I, PB81-248650; Vol. II, PB81-248668; Vol. III, PB81-248676; Vol. IV, PB81-248684; Vol. V, PB81-248692.

Rezak, R. and T. J. Bright, 1981b. Seafloor instability at East Flower Garden Bank, northwest Gulf of Mexico. *Geo-Marine Lett.*, **1**(2), 97–103.

Rezak, R. and W. R. Bryant, 1973. West Flower Garden Bank. *Trans. Gulf Coast Assoc. Geol. Soc. 23rd Annual Conv.* (Oct. 24–26), 377–382.

Rezak, R. and G. S. Edwards, 1972. Carbonate sediments of the Gulf of Mexico. *Texas A&M Univ. Ocean. Stud.*, **3**, 263–280.

Rezak, R. and T. T. Tieh, 1980. Basalt on Louisiana Outer Continental Shelf. EOS, *Trans. Am. Geophys. Union,* **61**(46), 989.

Rhoads, D. C., 1974. Organism-sediment relations on the muddy sea floor. *Oceanogr. Mar. Biol. Ann. Rev.*, **12,** 263–300.

Rigby, J. K. and W. G. McIntyre, 1966. The Isla de Lobos and associated reefs, Veracruz, Mexico. *Brigham Young Univ. Geol. Stud.*, **13,** 1–46.

Roberts, H. H., L. J. Rause, Jr., N. D. Walker, and J. H. Hudson, 1982. Cold water stress in Florida Bay and northern Bahamas: A product of winter air outbreaks. *J. Sediment. Petrol.*, **52**(1), 145–155.

Roos, P. J., 1964. The distribution of reef corals in Curacao. *Stud. Fauna Curacao,* **20,** 1–51.

Roos, P. J., 1971. The shallow water stony corals of the Netherlands Antilles. *Vitgaven "Natuurwetenschappelijke Studiekring voor Suriname en de Nederlandse Antillen,"* **37**(64), 1–108.

Sahl, L. E., W. J. Merrell, and D. W. McGrail, 1982. Currents and property distribution on the Texas shelf and slope in March 1981 and 1982. (Abs.) *Trans. Am. Geophys. Union,* **63,** 1012.

Salvador, A., 1980. Late Triassic–Jurassic Paleogeography and the Gulf of Mexico. In R. H. Pilger, Jr., Ed., Proceedings of a Symposium, the Origin of the Gulf of Mexico and the Early Opening of the Central North Atlantic Ocean, School of Geoscience, Louisiana State University, Baton Rouge, 101 pp.

Sanchez-Gil, P., A. Yanez-Arancibia, and F. A. Linares, 1981. Diversidad, Distribucion y Abundancia de las Especies y Poblaciones de peces Demersales de la Sonda De Campeche. *An. Inst. Cienc. Del Mar. Limnol.*, **8**(1), 209–240.

Schott, F., 1977. On the energetics of baroclinic tides in the North Atlantic, *Ann. Geophys.*, **33,** 41–62.

Schroeder, W. W. and D. W. McGrail, 1982. Shelf current observations during tropical storms. (Abs.) *Trans. Am. Geophys. Union,* **63**(3), 66.

Shepard, F. P., 1937. Salt domes related to Mississippi submarine trough. *Geol. Soc. Am. Bull.*, **48,** 1349–1361.

Shepard, F. P., 1960. Rise of sea level along northwest Gulf of Mexico. In F. P. Shepard, F. B. Phleger, and T. H. Van Andel, *Recent Sediments, Northwest Gulf of Mexico.* AAPG, Tulsa, Oklahoma, pp. 338–344.

Shepard, F. P., 1963. *Submarine Geology,* 2nd ed., Harper and Row, New York, 557 pp.

Shideler, G. L., 1977. Late Holocene sedimentary provinces, South Texas Outer Continental Shelf. *AAPG Bull.*, **61**(5), 708–722.

Shideler, G. L., 1978. A sediment-dispersal model for the South Texas continental shelf, northwest Gulf of Mexico. *Mar. Geol.*, **26,** 289–313.

Shideler, G. L., 1981. Development of the benthic nepheloid layer on the South Texas continental shelf, western Gulf of Mexico. *Mar. Geol.*, **41,** 37–61.

Shinn, E. A., 1974. Oil structures as artificial reefs. In Colunga, L. and R. Stone, Eds., *Proc. Int. Conf. Artificial Reefs,* Center for Marine Resources, Texas A&M University, College Station, 91–96.

Shinn, E. A., 1975. Coral reef recovery in Florida and in the Persian Gulf. Shell Oil Environmental Conservation Department, *Environ. Geol.*, **1,** 241–254.

Shinn, E. A., J. H. Hudson, R. V. Halley, and B. Lidy, 1977. Topographic control and accretion rate of some coral reefs: South Florida and Dry Tortugas. In *Proc. Third Int. Coral Reef Symp.*, University of Miami, Miami, Florida, 1–7.

Shipp, R. L. and T. S. Hopkins, 1978. Physical and biological observations of the northern rim of the DeSoto Canyon made from a research submersible. *Northeast Gulf Sci.*, **2**(2), 113–121.

Smith, F. G. W., 1971. *Atlantic Reef Corals: A Handbook of the Common Reef and Shallow-water Corals of Bermuda, Florida, the West Indies, and Brazil,* rev. ed. University of Miami Press, Coral Gables, Florida, 164 pp.

Smith, G. B., 1976. Ecology and distribution of eastern Gulf of Mexico reef fishes. *Fla. Mar. Res. Publ.*, **19,** 78 pp.

Smith, N. P., 1980. Temporal and spatial variability in longshore motion along the Texas Gulf coast. *J. Geophys. Res.*, 85(C3), 1531–1536.

Sonnier, F., J. Teerling and H. D. Hoese, 1976. Observations on the Offshore Reef and Platform Fish Fauna of Louisiana. *Copeia,* **1,** 105–111.

Sorenson, L. O., 1979. *A Guide to the Seaweeds of South Padre Island, Texas.* Gorsuch Scarisbrick Publishing Company, Dubuque, Iowa, 123 pp.

Stafford, J. M., 1982. An Evaluation of the Carbonate Cements and Their Diagenesis on Selected Banks, Outer Continental Shelf, Northern Gulf of Mexico. M.S. Thesis. Department of Oceanography, Texas A&M University, College Station, 78 pp.

Stearn, C. W. and C. Colassin, 1979. A simple underwater pneumatic hand drill. *J. Paleontol.*, **53**(5), 1257–1259.

Steel, R. G. D. and J. H. Torrie, 1960. *Principles and Procedures of Statistics.* McGraw-Hill, New York, 481 pp.

Stetson, H. C., 1953. The sediments of the western Gulf of Mexico, Part I—The continental terraces of the western Gulf of Mexico: Its surface sediments, origin, and development. Papers in *Phys. Oceanogr. Meteorol.*, M.I.T./W.H.O.I., **12**(4), 1–45.

Steward, R. G., 1981. Light Attenuance Measurements of Suspended Particulate Matter: Northeastern Gulf of Mexico. M.S. Thesis. Dept. of Marine Science, University of South Florida, St. Petersburg, 140 pp.

Stoddart, D. R., 1969. Ecology and morphology of recent coral reefs. *Biol. Rev. Cambridge Philos. Soc.*, **44,** 433–498.

Sverdrup, H. U., M. W. Johnson, and R. Fleming, 1942. *The Oceans, Their Physics, Chemistry and General Biology.* Prentice-Hall, Englewood Cliffs, New Jersey, 1087 pp.

Taylor, J. L., L. Feignenbaum, and M. L. Stursa, 1973. *Utilization of Marine and Coastal Resources.* In J. I. Jones, R. E. Ring, M. O. Rinkel, and R. E. Smith, Eds. *A Summary of Knowledge of the Eastern Gulf of Mexico.* State University System of Florida, Institute of Oceanography, St. Petersburg, pp. IV-1–IV-63.

Teerling, Joyce, 1975. A Survey of Sponges from the Northwestern Gulf of Mexico. Ph.D. Dissertation, Department of Biology, University of Southwestern Louisiana, Lafayette, 186 pp.

Temple, R. F., D. S. Harrington, and J. A. Martin, 1977. *Monthly Temperature and Salinity Measurements of Continental Shelf Waters of the Northwestern Gulf of Mexico, 1963–1965.* U.S.

Department of Commerce, National Oceanic and Atmospheric Administration and National Marine Fisheries Service, Tech. Rept. #SSRF-707, 26 pp.

Thierstein, H. and W. Berger, 1978. Injection events in ocean history. *Nature,* **276,** 461–466.

Thistle, D. and F. Lewis, III, 1979. *Literature Search on the Soft-bottom Benthos of the Open Waters of the Gulf of Mexico.* Ocean Chemistry Laboratory, Atlantic Oceanographic and Meteorological Laboratory. NOAA 20 September 1979.

Torgrimson, G. M. and B. M. Hickey, 1979. Barotropic and baroclinic tides of the continental slope and shelf off Oregon. *J. Phys. Oceanogr.,* **9,** 945–961.

Tresslar, R. C., 1974a. The Living Benthonic Foraminiferal Fauna of the West Flower Garden Bank Coral Reef and Biostrome. Master's Thesis, Department of Oceanography, Texas A&M University, College Station, 239 pp.

Tresslar, R. C., 1974b. Foraminifers. In T. J. Bright and L. H. Pequegnat, Eds., *Biota of the West Flower Garden Bank.* Gulf Publishing Company, Houston, pp. 67–92.

Tresslar, R. C., 1974c. Corals. In T. J. Bright and L. H. Pequegnat, Eds., *Biota of the West Flower Garden Bank,* Gulf Publishing Company, Houston, pp. 116–139.

Trippett, A. R. and H. Berryhill, 1982. *Geology of the continental shelf edge and upper continental slope off southwest Louisiana.* U.S. Department of the Interior, Minerals Management Service. Open File Report #82-02, Sheet VI.

Trowbridge, A. C., 1930. Building of Mississippi Delta. *AAPG Bull.,* **14,** 867–901.

Tunnell, J. W., Jr., 1973. Molluscan population of a submerged reef off Padre Island, Texas. *Bulletin of the American Malacological Union, Inc.,* 25–26.

Tunnell, J. W. and A. H. Chaney, 1970. A checklist of the mollusks of Seven and One Half Fathom Reef, Northwestern Gulf of Mexico. *Contrib. Mar. Sci.,* **15,** 194–203.

Uchupi, E., and K. O. Emery, 1968. Structure of continental margin off Gulf Coast of United States. *AAPG Bull.,* **52,** 1162–1193.

U.S. Naval Weather Service Command, 1975. *Summary of Synoptic Meteorological Observations.* Vol. 4, May 1975, Area 28, Galveston, Texas.

Van Andel, T. H., 1960. Sources and dispersion of Holocene sediments, northern Gulf of Mexico. In F. P. Shepard, F. B. Phleger, and T. H. Van Andel, Eds., *Recent Sediments, Northwest Gulf of Mexico, AAPG Bull.,* Tulsa, Oklahoma, pp. 34–55.

Van Andel, T. H. and J. R. Curray, 1960. Regional aspects of modern sedimentation in northern Gulf of Mexico and similar basins, and paleogeographic significance. In F. P. Shepard, F. B. Phleger, and T. H. Van Andel, Eds., *Recent Sediments, Northwest Gulf of Mexico, AAPG Bull.,* Tulsa, Oklahoma, 345–364.

Van Andel, T. H. and D. M. Poole, 1960. Sources of recent sediments in the northern Gulf of Mexico. *J. Sediment. Petrol.,* **30**(1), 91–122.

Vaughan, T. W., 1916. On recent Madreporaria of Florida, the Bahamas, and the West Indies and on collections from Murray Island, Australia. *Carnegie Inst. Wash. Year Book,* 220–231.

Viada, S. T., 1980. Species Composition and Population Levels of Scleractinian Corals Within the *Diploria-Montastrea-Porites* Zone of the East Flower Garden Bank, Northwest Gulf of Mexico. M.S. Thesis. Department of Oceanography, Texas A&M University, College Station, 96 pp.

Villalobos, A., 1971. *Estudios Ecologicos en un Arrecife Coralino en Veracruz, Mexico.* Symposium on Investigations and Resources of the Caribbean Sea and Adjacent Regions, UNESCO and FAO, 531–545.

Watkins, J. S., J. W. Ladd, R. T. Buffler, F. J. Shaub, M. H. Houston, and J. L. Warzel, 1978. Occurrence and evaluation of salt in deep Gulf of Mexico. In A. H. Bouma, G. T. Moore, and J. M. Coleman, Eds., *Framework, Facies, and Oil-Trapping Characteristics of the Upper Continental Margin.* AAPG, Tulsa, Oklahoma, pp. 43–65.

Weatherly, G. L., 1972. A study of the bottom boundary layer of the Florida current. *J. Phys. Oceanogr.,* **2,** 54–72.

Weatherly, G. L., 1975. A numerical study of time-dependent turbulent Ekman layers over horizontal and sloping bottoms. *J. Phys. Oceanogr.,* **5,** 288–299.

Weatherly, G. L., S. L. Blumsack, and A. A. Bird, 1980. On the effect of diurnal tidal currents in determining the thickness of the turbulent Ekman bottom boundary layer. *J. Phys. Oceanogr.,* **10,** 297–300.

Weigand, J. G., H. Farmer, S. Prinsenberg, and M. Rattray, Jr., 1969. Effects of friction and surface tide angle of incidence on the coastal generation of internal tides. *J. Mar. Res.,* **27,** 241–259.

Wells, J. W., 1957. Coral reefs. In J. W. Hedgpeth, Ed., Treatise on Marine Ecology and Paleoecology. *Geol. Soc. Amer. Mem.* **67,** 1, 609–631.

Wells, J. W., 1973. New and old Scleractinian corals from Jamaica. *Bull. Mar. Sci.,* **23**(1), 16–58.

Wentworth, C. K., 1922. A scale of grade and class terms for clastic sediments. *J. Geol.,***30,** 377–392.

Wills, J. B., 1976. Benthonic Polychaeta of the West Flower Garden Bank. Master's Thesis, University of Houston, Department of Biology, Houston, 181 pp.

Wills, J. B. and T. J. Bright, 1974. Worms. In T. J. Bright and L. H. Pequegnat, Eds., *Biota of the West Flower Garden Bank.* Gulf Publishing Company, Houston, pp. 291–310.

Wilson, O., 1969. Three coral reefs of Bermuda's North Lagoon: Physiography and distribution of corals and calcareous algae. In R. N. Ginsburg and P. Garrett, Eds., *Seminar on Organism-Sediment Interrelationships, Reports of Research,* Bermuda Biological Station for Research, Spec. Publ. No. 2, St. George's West, Bermuda, 51, 64.

Winker, C. D., 1982. Cenozoic shelf margins, northwest Gulf of Mexico. *Trans. Gulf Coast Assoc. Geol. Soc.,* **32,** 427–448.

Wiseman, W. J., Jr., D. W. McGrail, and L. J. Rouse, Jr., 1982. Shelf break exchange driven by cold fronts. (Abs.) *Trans. Am. Geophys. Union,* **63,** 1012.

Withjack, M. O. and C. Scheiner, 1982. Fault patterns associated with domes: An experimental and analytical study. *AAPG Bull.,* **66,** 302–316.

Yañez-Arancibia, A., F. A. Linares, and J. W. Day, Jr., 1980. *Fish Community Structure and Function in Terminos Lagoon, A Tropical Estuary in the Southern Gulf of Mexico.* In Estuarine Perspectives. Academic Press, New York, pp. 465–482.

APPENDIX I

Plants and Animals of the Outer Continental Shelf Hard Banks in the Northwestern Gulf of Mexico

This list of organisms is derived from final Topographic Features project reports submitted to the U.S. Bureau of Land Management (now U.S. Minerals Management Service) between 1976 and 1983 (Bright and Rezak, 1976); Bright and Rezak, 1978a,b; Rezak and Bright, 1981a; McGrail, Rezak, and Bright, 1982d). The necessary information was compiled largely through the efforts of Dr. Linda H. Pequegnat, who, with the help of Dr. Richard Titgen, managed sorting and distribution of specimens to systematic specialists for identification. The cooperation of the following specialists is greatly appreciated:

ALGAE	Mr. George Dennis, Texas A&M University Dr. Nat Eiseman, Harbor Branch Foundation, Fort Pierce, Florida Ms. Elaine Stamman, Washington, D.C.
PORIFERA	Dr. Joyce Teerling, Lafayette, Louisiana
COELENTERATA	
HYDROZOA	
HYDROIDA	Dr. Dale Calder, So. Carolina Wildlife & Marine Resources Department
HYDROCORALS	Mr. Walter Jaap, Florida Department of Natural Resources

Taxon	Authority
ANTHOZOA	
ALCYONARIA	Ms. Jenifer Wheaton Lowry, Florida Department of Natural Resources
	Dr. Charles Giammona, Texas A&M University
SCLERACTINIA	Mr. Walter Jaap, Florida Department of Natural Resources
	Dr. Stephen Cairns, Smithsonian Institution
ANTIPATHARIA	Dr. Dennis Opresko, Oak Ridge, Tennessee
ANNELIDA	
POLYCHAETA	Dr. Barry Vittor, Mobile, Alabama
	Mr. Fain Hubbard, TerEco Corporation
MOLLUSCA	Dr. William Lyons, Florida Department of Natural Resources
BRYOZOA	Dr. Arthur J. Leuterman, Houston, Texas
CRUSTACEA	
AMPHIPODA	Dr. Larry McKinney, Moody College of Marine Sciences Galveston, Texas
CIRRIPEDIA	Mr. Stephen Gittings, Texas A&M University
DECAPODA	Dr. James Ray, Shell Oil Co., Houston, Texas.
	Dr. Linda Pequegnat, Texas A&M University
	Dr. Patsy McLaughlin, Florida International University
	Dr. Darryl Felder, University of Southwestern Louisiana
ISOPODA	Mr. Scott Clark, Oklahoma State University
TANAIDACEA	Dr. Richard Heard, Ocean Springs, Mississippi
	Mr. John Ogle, Gulf Coast Research Laboratory
BRACHIOPODA	Dr. G. Arthur Cooper, Smithsonian Institution
ECHINODERMATA	
ASTEROIDEA	Ms. Maureen Downey, Smithsonian Institution
CRINOIDEA	Dr. Charles Messing, Smithsonian Institution
ECHINOIDEA	Dr. Dave Pawson, Smithsonian Institution
HOLOTHUROIDEA	Dr. Robert Carney, Smithsonian Institution
OPHIUROIDEA	Dr. Gordon Hendler, Smithsonian Institution
CHORDATA	Dr. Cara Cashman, Mahopac, New York
	Mr. George Dennis, Texas A&M University
	Dr. Thomas Bright, Texas A&M University
	Dr. William Anderson, Grice Marine Biol. Lab., Charleston, SC

Additional works pertaining to the systematics of biota occupying outer continental shelf banks in the northwestern Gulf are the following

Taxon	Publication
ALGAE	Eiseman and Blair, 1982
FORAMINIFERS	Tresslar 1974a,b
SPONGES	Teerling 1975
HYDROIDS	Defenbaugh 1974
CORALS	Tresslar 1974c Giammona 1978
MOLLUSKS	Lipka 1974
ANTHROPODS	Maddocks 1974, Ray 1974 Pequegnat & Ray 1974 Clark & Robertson 1982
WORMS	Wills & Bright 1974 Wills 1976
BRYOZOA	Cropper 1973 Leuterman 1979
ECHINODERMS	Burke 1974a, 1974b DuBois 1975
FISHES	Cashman 1973 Moseley 1966 Sonnier et al. 1976 Bright & Cashman 1974 Hoese & Moore 1977 Causey 1969

In the following list, the depths given represent "known" depths. Many, if not most, of the species certainly occur shallower and deeper than indicated. When uncertainty exists concerning depth of collection the species presence is indicated by an "x." The absence of any indication of depth or presence means that the species has not been collected by us from that particular bank group.

The specific banks were grouped as follows for purposes of tabulation (see Fig. 7.1 for locations):

Mid-Shelf Claystone/Siltstone (MS): Sonnier, Stetson, Claypile

Flower Gardens (FG): East Flower Garden, West Flower Garden

Other Shelf Edge Banks (SEB): Alderice, Ewing, Bouma, Sackett, Applebaum, Bright, Diaphus, 18 Fathom, 28 Fathom, Jakkula, Rezak-Sidner, Elvers, Geyer, Parker

Transitional Mid-Shelf (TRAN): Fishnet, Coffee Lump, 32 Fathom

South Texas Mid-Shelf (STMS): Blackfish, North Hospital, Big Adam, Hospital Rock, Aransas, Baker, South Baker, Southern, Dream, Mysterious, Small Adam

TAXA	Bank Group and Depths of Collection (M)				
	MS	FG	SEB	TRAN	STMS
***CHLOROPHYTA* (green algae)**					
Acetabularia crenulata		54			
Anadyomene stellata		54	46–61		
Caulerpa microphysa		31–55	x		
Caulerpa peltata		31	46–61		
Caulerpa prolifera			x		
Caulerpa racemosa v. *macrophysa*		24–55	x		
Caulerpa sp.		24–54	54		
Chaetomorpha sp.			x		
Cladophora corallicola			46–79		
Cladophora cf. *repens*		47			
Cladophora sp.		x	x		
Codium decorticatum		x			
Codium taylori		x	x		
Codium isthmocladum		47			
Codium sp.			54		
Derbesia sp.		54			
Halimeda discoidea		x	x		
Halimeda gracilis		x			
Halimeda opuntia			x		
Halimeda tuna f. *platydisca*		48–61	76		
Halimeda sp.		21–91	61–108		

TAXA	Bank Group and Depths of Collection (M)				
	MS	FG	SEB	TRAN	STMS
Microdictyon boergesenii		55	61–69		
Palmellaceae		x			
Pseudocodium floridanum			x		
Struvea sp.		x			
Udotea cyathiformis		58	x		
Udotea flabellum		x			
Udotea spp.		40–64			
Ulva lactuca		x			
Ulva sp.		54	94		
Valonia macrophysa		31	x		
Valonia ventricosa		31–55	46–76		
Valonia sp.		24–55	52–76		
***PHAEOPHYTA* (brown algae)**					
Dictyopteris delicatula			x		
Dictyopteris justii		x			
Dictyota bartayresii		31–50	55		
Dictyota cervicornis			x		
Dictyota dichotoma		31			
Dictyota divaricata			x		
Dictyota linearis			x		
Dictyota spp.		x	x		
Lobophora (Pocockiella) variegata		40–69	55–78		
Sargassum hystrix		31	46–61		
Sargassum sp.			44–58		
Spatoglossum schroederi		54			
Stypopodium zonale		31–84	46–67		
***RHODOPHYTA* (red algae)**			55–58		
Acrochaetium sp.		54			
Amphiroa rigida v. *antillana*		x			
Amphiroa tribulus		30–31			
Apoglossum ruscifolium		x			
Botryocladia occidentalis		x			64–70
Botryocladia pyriformis		x		x	
Callithamnion cf. *halliae*		x			
Champia parvula		31			
Chondria cnicophylla		33–55			
Chondria floridana		54			
Chrysymenia enteromorpha		x	x		
Chrysymenia halymenioides		24–53	x		
Chrysymenia ventricosa			x		
Coelarthrum albertisii		x	x		
Compsothamnion thujoides		x			
Cryptonemia luxurians		x			
Cryptonemia sp.		x			
Dasya corymbifera		x			
Erythrocladia subintegra		54	64		
Fauchea hassleri		x	x		
Fauchea peltata		33–50			

TAXA	Bank Group and Depths of Collection (M)				
	MS	FG	SEB	TRAN	STMS
Galaxaura cylindrica		x	x		
Galaxaura oblongata		x			
Galaxaura obtusata		x	55		
Galaxaura obtusata v. *major*		50			
Gelidium sp.		33–55			
Gloiophlaea halliae		24			
Goniotrichum alsidii		54	76		
Gracilaria cf. *mammillaris*			x		
Gracilaria sp.		31	46–61		
Grateloupia filicina		x			
Halymenia floresia		x			
Halymenia floridana		50			
Halymenia cf. *hancockii*			x		
Halymenia sp.		x	x		
Hypoglossum involvens		54			
Hypoglossum tenuifolium		x			
Jania capillacea		54			
Jania sp.		31			
Kallymenia westii		x	x		
Laurencia sp.		x			
Leptofauchea rhodymenioides			x		
Nemalion schrammi		x			
Nemalion sp.		x			
Nitophyllum punctatum		54			
Peyssonnelia rubra		31			
Peyssonnelia simulans		x	x		
Peyssonnelia sp.		24–62	46–90		
Polysiphonia gorgoniae		x	x		
Polysiphonia cf. *tepida*		x			
Pterocladia bartlettii		x			
Pterocladia sp.		54	64		
Rhodymenia occidentalis		x			
Rhodymenia pseudopalmata		33–50	76		
Rhodymenia sp.		x			
Scinaia complanata v. *intermedia*		x			
Searlesia subtropica		x			
Spyridia filamentosa		x			
Titanophora incrustans		x			
CORALLINACEAE (coralline algae)					
Archaeolithothamnion spp.		23–72			
Fosliella spp.		23			
Hydrolithon spp.		23–65			
Lithophyllum bermudense		x			
Lithophyllum spp.		23–80			58
Lithoporella spp.		23–85			
Lithothamnium cf. *sejunctum*		x			
Lithothamnium spp.		23–90			
Mesophyllum spp.		15–86			
Tenarea spp.		23–90			

TAXA	Bank Group and Depths of Collection (M)				
	MS	FG	SEB	TRAN	STMS
***PORIFERA* (sponges)**					
Adocia albifragilis					67–70
Adocid	23–46		56		61–71
Agelas sp.	37–52		44–67		60–64
Agelas conifera			115		
Agelas dispar	43–52	18–57	56–120		18–71
Alcyospongia sp.			76		
Ancorinidae			55		
Anthoarcuata sp.			56		59
Auletta sp.			73–115	x	58–73
Axinella sp.			56		61–71
Axinellidae			67	x	61–71
Callyspongia armigera	43–46		46–56		59
Callyspongia vaginalis	24		24		
Callyspongia sp.	38–44		54	x	
Chelotropella sp.			115		
Chondrosid			56		
Choristidae			x		
Cinachyra sp.			56		
Cliona celata					62–73
Cliona delitrix		22			61–71
Cliona lampa			56		
Cliona schmidti		22–28	120		64–70
Cliona vastifica					59
Cliona sp.					59–71
Corallistes sp.			76–125		
Cyamon vickersi					69
Desmacidon sp.					64–70
Desmanthus sp.			x		
Didiscus sp.			46–120		
Diplastrella sp.			55		
Ectoforcepia sp.					61–71
Epallax sp.			55–62		61–71
Epipolasidae		x	56–120		61
Epipolasis sp.		23			
Erylus alleni		x	55		
Eurypon clavatum			x		69
Gelliodes areolata			x		
Gelliodes ramosa	27				59
Gelliodes sp.			56–76		61–71
Geodia gibberosa			62–115		
Geodia sp.			73–113		
Hadromeridae			115		
Halichondridae			120	x	61–71
Haliclona sp.		20–23			58
Haliclona palmata		81			
Haliclonidae		x	115		59
Halisarca sp.	21–27		56		69

TAXA	Bank Group and Depths of Collection (M)				
	MS	FG	SEB	TRAN	STMS
Haploscleridae		81	73–88		
Higginsia sp.			x		
Iotrochota birotulata					59–70
Ircinia campana	27–54		62	x	58–70
Ircinia fasciculata			56	x	59–61
Ircinia spp.	21–46	55–81	58–81		57–72
Ircinia strobilina	27	18–58	46–120		x
Jaspidae			x		
Jaspis sp.			x		
Keratose sponge		x	56–115	x	67–70
Leuconia aspera		x			67–70
Leuconia sp.			x		
Lissodendoryx isodictyalis					59
Lissodendoryx sp.					61
Merriamium tortugasensis			56		
Microciona rarispinosa					59
Microciona sp.			56		69
Microcionidae			x		61–71
Microscleritoderma sp.			46–62		
Mycale angulosa		18–52	55–56		
Mycale laevis					69
Mycale sp.			56		59–69
Mycalidae			55		
Myriastrea fibrosa		x	x		
Myriastrea sp.			46–55		
Neofibularia sp.	18–56	24–61	52–83		60–69
Neofibularia nolitangere oxeata		24–57		x	56–58
Phorbasid					61
Placospongia carinata			55		
Placospongia melobesioides	43–46				
Placospongia sp.	40–46	52			
Plakinidae		x	56		
Plakortis zyggompha					58–73
Poecillastra sp.			62–115		
Poecilosceridae			128	x	61–71
Prianos sp.	21–24		67		
Pseudaxinella rosacea				x	
Pseudosuberites sp.	21–24				
Rhabdodictyon sp.			115		
Rhizochalina sp.		19–25	73–88		59–70
Sigmadocia caerulea					69
Sigmadocia sp.			67–115		
Siphonodictyon coralliphagum		18–24	56		59–71
Siphonidium sp.			120		
Sollasellid					59
Spirogastrella coccinee			56		59
Spirastrella sp.		24–81			
Spongia sp.		x			59–63

TAXA	Bank Group and Depths of Collection (M)				
	MS	FG	SEB	TRAN	STMS
Spongia barbara	27	18–52	58–61		58–72
Spongiidae			46		
Stellata sp.			62		
Strongylophora sp.					61
Suberites sp.					58
Suberitidae			115		61–71
Tedania ignis					61
Tedania sp.					67–70
Terpios fugax	52				61
Tethya sp.					61–71
Tethyidae			44–62		
Thalyseurypon sp.					61
Thoosa sp.			56		
Timea					x
Timeinid				x	
Trachygellius sp.					61–71
Tretodictyum sp.			115–120		
Tricheurypon viride					61
Ulosa sp.			62		
Verongia spp.			69		58
Verongia cauliformis		18–55			62
Xestospongia halichondrioides				x	
Xestospongia muta		20			
Xystopsene sigmatum					61–72
***COELENTERATA* (coelenterates)**					
HYDROIDS					
Acryptolaria rectangularis			115–120		x
Aglaophenia elongata			120		
Cryptolaria pectinata			x		
Dynamena dalmasi			120		
Halecium sp.			120	x	
Halopterus catharina		x			
Lafoea sp.			x		
Monostaechas quadridens			x		
Plumularia sp.		x	127–128		
Plumularidae		55			
Sertularia distans				x	
Sertularidae					73
Thyroscyphus marginatus		x			
Zygophylax convallaria		x	126		
HYDROZOAN CORALS					
Millepora sp.		18–52	20–54	43–54	
Millepora alcicornis		24	15–55		
ANEMONES					
Condylactis gigantea		43–55	52–78		
Lebrunia danae		x	51–67		

TAXA	Bank Group and Depths of Collection (M)				
	MS	FG	SEB	TRAN	STMS
ANTHOZOAN STONY CORALS					
Agariciidae	x	x	x	x	x
Agariciid (saucer-shaped)	52	18–73	47–71		61–69
Agaricia sp.		15–76	44–76		58–61
Agaricia agaricites agaricites		20			
Agaricia agaricites purpurea			76		
Agaricia fragilis	24		75		
Agaricia undata		x			
Caryophyllia parvula			120		
Caryophyllia sp.			100		
Coenocyathus n. sp.			100		
Coenosmilia arbuscula			x		
Colpophyllia amaranthus		21–26			
Colpophyllia sp.		21–47			
Colpophyllia natans		21–26			
Diploria strigosa		15–55			
Eusmilia sp.		24			
Guynia annulata			91		
Helioseris cucullata		20–84	20–128		
Javania cailleti			115		
Madracis asperula		47–84	56–128	x	61
Madracis sp.	41–52	15–92	52–73		58–61
Madracis brueggemanni		62			58–70
Madracis decactis	24	15–41			
Madracis cf. *formosa*		62	72		
Madracis mirabilis	43–46	23–40	56		
Madracis myriaster		113	100–115		69
Madrepora carolina			100		
Madrepora sp.			72		
Montastrea annularis	18–43				
Montastrea cavernosa		15–60	44–54		
Mussa angulosa		21–54			
Oculina diffusa			x		
Oculina varicosa			100		
Oculina sp.	24		85		
Oxysmilia sp.		82–101	76–102		
Oxysmilia rotundifolia			91–115		
Paracyathus pulchellus	43–46		100–128	x	60–70
Porites astreoides		21–40			
Porites furcata		21			
Scolymia cubensis		21–27			
Scolymia sp.		18–46			
Siderastrea radians	24				
Siderastrea siderea		21–50			
Siderastrea sp.	34–44				
Stephanocoenia michelini	27–41	38–49	43–54		

TAXA	Bank Group and Depths of Collection (M)				
	MS	FG	SEB	TRAN	STMS
ALCYONARIAN CORALS					
Alcyonarians	40	45	58–146		60–68
Bebryce cinerea			64–120		
Bebryce sp.			30–83		62
Bellonella spp.			120–128		
Caliacis nutans			120		
Callogorgia verticillata			127–128		
Ellisella atlantica			x		
Ellisella barbadensis			x		
Ellisella elongata			126–128		
Ellisella funiculina			115		
Ellisellidae			62–73		
Neospongodes sp.			30		
Nicella guadalupensis			120–128		
Nicella flagellum			64–91		
Nicella schmitti			55		
Nidalia occidentalis			30–115		
Nidalia sp.			80–105		
Paramuriceidae			76–87	x	62
Placogorgia rudis			115–128		
Riisea paniculata			115–120		
Scleracis guadalupensis			115–120		
Scleracis sp.			73		63–69
Siphonogorgia agassizii			115–120		
Swiftia exserta			127–128		
Swiftia sp.			73		
Thesea grandiflora			x		
Thesea granulosa			x	x	
Thesea guadalupensis			115		
Thesea rugosa			x		
Thesea spp.					62–76
ANTIPATHARIA					
Antipathes atlantica			115		
Antipathes barbadensis			x		
Antipathes furcata			115–120		
Antipathes pedata			115		
Antipathes tanacetum			120		
Antipathes sp.	47–55		30–83		58–73
Aphanipathes abietina			115–120		
Cirrhipathes lutkeni			x		
Cirrhipathes spp.	55–58	44–104	56–129	x	52–78
POLYCHAETA					
Acanthicolepis sp.			120		
Aglaophamus circinata			x		
Ampharetidae			56		
Anaitides madeirensis			46–61		
Arabella iricolor				x	58

TAXA	Bank Group and Depths of Collection (M)				
	MS	FG	SEB	TRAN	STMS
Arabella mutans			x		
Armandia maculata			x		
Autolytus prolifer			55–61		
Ceratonereis irritabilis			115	x	
Ceratonereis mirabilis		x	46–115		59
Chloeia viridis			46–61		59
Chloenopsis sp.			115		
Dasybranchus sp.				x	
Dorvillea rubrovittata			76		
Dorvillea sociabilis			56	x	
Eunice antennata			56–115	x	
Eunice aphroditois		x	56		
Eunice filamentosa					59
Eunice vittata		x	46–115		
Eunice websteri			120		
Eupholoe sp.			x		
Eupolymnia sp.			x		
Eurysyllis sp.			x		
Eurythoe complanata		x		x	
Exogone dispar			x		
Filograna implexa			55–61		
Filograna sp.			120		
Gyptis brevipalpa			x		
Glycera papillosa			x		
Glyceridae			x		
Haplosyllis spongicola			46–67		
Harmothoe aculeata			56		
Harmothoe spp.	52	x	55–115	x	59
Hermodice carunculata	21–24		56		
Hermodice sp.		90–93	64		62–72
Hesionidae		x		x	
Hesione sp.			x		
Hydroides norvegica			56–61		
Hypsicomus sp.		x	x	x	
Inermonephtys inermis				x	
Kefersteinia cirrata		x	x		
Laonice cirrata				x	
Leocratides filamentosa			55–61		
Lumbrineris coccinea		62			
Lumbrineris heteropoda				x	
Lumbrineris cf. *papillifera*				x	
Lysidice ninetta			x	x	
Marphysa sanguinea				x	
Micropodarke dubia			67		
Neanthes acuminata			62		
Nematonereis unicornis		x			
Nematonereis sp.				x	
Nereidae			55–88		59

TAXA	Bank Group and Depths of Collection (M)				
	MS	FG	SEB	TRAN	STMS
Nereis riisei			115	x	
Nicon sp.		x	x	x	
Nothria sp.			x		
Onuphis cf. *quadricuspis*				x	
Ophiodromus obscurus			74–88		
Opisthodonta spp.			x		
Opisthosyllis cf. *ankylochaeta*			56		
Opisthosyllis sp.			74–88		
Paleanotus debilis			x	x	
Paleanotus heteroseta			56		
Palmyridae			x		
Paramarphysa sp.			120		
Paratyposyllis sp.			120	x	
Pareurythoe americana			x		
Pherusa cf. *parmata*				x	
Pholoe dorsipapillata			x		
Phyllodoce arenae			115	x	
Phyllodoce (Anaitides) madeirensis			74–88		
Phyllodocidae			x		x
Pionosyllis spp.			x		
Pista sp.			x	x	
Platynereis dumerilii			74–88		
Podarke sp.			115		
Podarkeopsis galagui			55–61		
Poecilochaetus johnsoni				x	
Polycirrus eximius dubius			x		
Polydora armata			120	x	
Polydora sp.				x	
Polyphthalmus pictus			55–61		
Polynoidae			46–88		
Pomatoceros americanus				x	
Pomatoceros sp.			62		
Pomatostegus sp.				x	
Pontogenia chrysocoma			126		
Potamilla cf. *reniformis*			67–88		
Potamilla cf. *torelli*			120		
Pseudovermilia sp.			120	x	
Sabellidae		43–89	56–92		
Salmacina sp.				x	
Serpula vermicularis		81			
Sphaerosyllis bulbosa				x	
Sphaerosyllis piriferopsis			67		
Spiophanes bombyx				x	
Spiophanes wigleyi				x	
Spirobranchus giganteus	27–56	17–55		x	61
Steninonereis martini			62–120		
Sthenolepis cf. *japonica*				x	
Subadyte pellucida					59–70

TAXA	Bank Group and Depths of Collection (M)				
	MS	FG	SEB	TRAN	STMS
Subprotula cf. *longiseta*		x			
Syllidae					58
Syllis gracilis	21–24	x	120	x	67–70
Syllis (Haplosyllis) spongicola	43–46		56		59–70
Syllis (Langerhansia) cornuta			56		
Syllis sp.	21–24				
Synelmis sp.			120		
Tachytrypane jeffreysii			55–120		
Terebella pterochaeta			46		
Terebellidae			56		
Terebellides stroemi				x	
Trypanosyllis zebra			55–152		67–70
Trypanosyllis sp.		x	56	x	
Typosyllis alternata			56–120	x	
Typosyllis hyalina			46–61		59–70
Typosyllis prolifera			56–120		
Typosyllis regulata carolinae		x	120	x	
Typosyllis cf. *taprobanensis*				x	
Typosyllis variegata			55–61		
Typosyllis sp.		x	46		
Vermiliopsis annulata			55–120		64–70
Vermiliopsis cf. *bermudensis*			56		
***MOLLUSCA* (Molluscs)**					
CHITONS					
Ischnochiton sp.			67		
Chiton spp.					54–70
GASTROPODS					
Acmaea pustulata		x			
Anachis lafresnayi			x		
Antillophos candei			x		
Astraea phoebia		x	115		
Astraea tecta	23–26				
Astraea sp.			55		
Bursa finlayi			x		
Bursa granularis cubaniana	x				
Busycon spp.			64–72		64
Busycon egg case			82		
Calliostoma roseolum			x	x	
Cantharus multangulus		x			
Cassis flammea	x				
Cassis sp.			64		62
Cerithium litteratum	46–52				
Cerithiopsis sp.		x	x		
Compsodrillia haliostrephis				x	
Conus daucus				x	
Conus testudinarius		x			

TAXA	Bank Group and Depths of Collection (M)				
	MS	FG	SEB	TRAN	STMS
Conus sp.	46	76	64–80		62–69
Coralliophila abbreviata		x			
Coralliophila caribaea		21			
Costellaria sykesi		x			
Crepidula plana		x			
Cuvierina columnella			x		
Cymatium pileare	x				
Cypraea sp.		20			61–62
Cypraea spurca acicularis			126		
Cyprea zebra	x				
Diodora sp.				x	
Emarginula phrixodes			x		
Emarginula pumila			56		
Emarginula sicula		x			
Fasciolariidae				x	
Fissurellidae			67	x	
Glyophostoma dentifera			x		
Haliotis pourtalesii			76		
Latiaxis mansfieldi			115		
Limpet			55–61		
Lobiger sp.			120		
Mitra barbadensis			x		
Murex beauii			115–126		
Murex sp.				x	
Murexiella hidalgoi			115		
Nassarius sp.			x	x	
Nesta atlantica			67		
Perotrochus adansonianus			115		
Pleurotomaria sp.			98		
Polinices sp.		23			
Polystira vibex			x		
Psarostola minor				x	
Siliquaria modesta			115		
Siliquaria squamata			46–67		
Strombus alatus			x		
Strombus costatus			126		
Strombus raninus	x				
Strombus sp.			52		
Terebra limatula			x		
Triphora sp.			x		
Turbo cailletti		x			
Turbo castanea			126		
Turritella exoleta				x	
"worm shells"			62–64		
Vermicularia fargoi			x		
PELECYPODS					
Aequipecten muscosus				x	
Americardia media			126	x	

TAXA	Bank Group and Depths of Collection (M)				
	MS	FG	SEB	TRAN	STMS
Anadara sp.			52		
Arca imbricata	46–52				67–70
Arca zebra		49–88	55–126	x	69
Arcidae				x	
Arcinella cornuta				x	
Arcopsis adamsi				x	69
Argopecten gibbus				x	
Atrina sp.			62–115		
Barbatia cancellaria			56		
Barbatia candida			55–76	x	64–70
Barbatia domingensis	46–52		55–120		x
Barbatia sp.	46–52				
Botula fusca			67–79		
Cardiidae		50			
Chama macerophylla					58
Chama sinuosa				x	
Chama sp.		x	120		
Chlamys benedicti			56–76	x	
Corbula sp.				x	
Glycymeris pectinata			126		
Gouldia cerina				x	
Gregariella coralliophaga		x			
Gregariella opifex	21–24	88	55–88		
Hiatella arctica					58
Hiatella azaria				x	
Jouannetia quillingi			127–128		
Laevicardium pictum				x	
Lima scabra			76		59
Lima sp.		43			67
Lithophaga bisulcata			56		
Lyropecten nodosus	38–48	90		x	
Lyropecten sp.			59–67		
Musculus lateralis				x	
Nemocardium transversum			x		
Nemocardium tinctum				x	
Papyridea soleniformis			128	x	
Pecten raveneli				x	
Pinctada radiata	21–24				
Pinna sp.		x			
Pinnidae		x			
Plicatula gibbosa			67		
Propeamussium sp.				x	
Pteria sp.			120	x	68–76
Pycnodonte hyotis			115		
Spondylus americanus	37–58	37–78	43–122	x	60–72
Trachycardium magnum			120		
Varicorbula operculata				x	

TAXA	Bank Group and Depths of Collection (M)				
	MS	FG	SEB	TRAN	STMS
CEPHALOPODS					
Nautilus sp.			75		
Octopus macropus		55			
Octopus sp.	58	62			
SCAPHOPODS					
Dentalium laqueatum			x		
BRACHIOPODA **(brachiopods)**					
Argyrotheca barrettiana			73–100		55–73
ARTHROPODA **(Arthropods)**					
CRUSTACEA—LEPTOSTRACA					
Paranebelia sp.	43–46		55–67		67–70
CRUSTACEA–CIRRIPEDIA					
Acasta cyathus		23–26			
Ceratoconcha floridana		x			
Conopea galeata		101			
Arcoscalpellum diceratum		101			
Lepas anatifera		87			
Lithotrya dorsalis		23–24			
CRUSTACEA—DECAPODA, NATANTIA					
Alpheidae					67–70
Alpheopsis labis			128		
Alpheopsis sp.			55–67		
Alpheus amblyonyx			126		
Alpheus beanii		x	128		
Alpheus sp.			x		
Discias sp.			x		
Hippolysmata grabhami	37		57–72		76
Hippolytidae		x	55–128		
Leptochela bermudensis			x		
Lysmata sp.			64		
Palaemonid		x	126–128	x	
Penaeopsis goodei				x	
Periclimenaeus bermudensis			46		
Periclimenaeus bredini			x	x	
Periclimenaeus caraibicus					67–70
Periclimenaeus cf. *perlatus*		x			
Periclimenaeus wilsoni			120		
Periclimenes americanus			56		
Periclimenes sp.				59	
Processa tenuipes			126		
Pseudocoutiera sp.			128		
Salmoneus ortmanni		x	120		
Stenopus sp.					67–70
Synalpheus cf. *agelas*			120		
Synalpheus barahonensis			x		
Synalpheus bousfieldi					67–70

TAXA	Bank Group and Depths of Collection (M)				
	MS	FG	SEB	TRAN	STMS
Synalpheus fritzmuelleri	24				
Synalpheus minus	27				
Synalpheus pandionis	24		55–129		
Synalpheus tanneri			73–88		
Synalpheus townsendi	43–54	x	46–120	x	58–70
Synalpheus spp.	47–54		46–120		67–70
Thor sp.			56		
Tuleariocaris neglecta		23–24			
CRUSTACEA—DECAPODA, REPTANTIA					
Actaea sp.		x			
Anomuran			78		
Arachnopsis filipes				x	
Axiopsis sp.			120		
Calappa angusta			126		
Carpilius corallinus		18			
Clibanarius anomalus			115		
Collodes cf. *trispinosus*				x	
Dardanus insignis			x		
Dardanus sp.		x			
Dromidia antillensis			x		
Euphrosynoplax clausa			115		
Galathea rostrata	27–46		76	x	58–69
Goneplacidae	54		46–67		
Gonodactylus bredini	43–46				
Gonodactylus torus			76		
Hexapanopeus lobipes			128	x	
Homola barbata		x	115		
Hyas sp.			46		
Iridopagurus sp.			x		
Leptodius agassizii			x		
Lobopilimnus agassizii	24				
Macrocoeloma concavum			128		
Majidae			55		
Melybia thalamita	54	x	55–79		
Microcassiope granulimanus			x	x	
Micropanope lobifronss				x	
Micropanope sculptipes		x	126		
Micropanope urinator			115		
Micropanope sp.			120		x
Mithrax acuticornis			120	x	64–70
Munida angulata			115–128		
Munida flinti			62		
Munida nuda			120		
Munida sp.		92–104			
Munida pusilla			x		
Munida simplex		58	46–126		
Munidopsis squamosa			115		
Myropsis quinquespinosa			126		

TAXA	Bank Group and Depths of Collection (M)				
	MS	FG	SEB	TRAN	STMS
Nibilia antilocapra			46		
Pagurus pygmaeus			56		
Pagurus brevidactylus		x	120		
Pagurus piercei			120	x	
Paguridae	46–58	x	55–120	x	58–62
Paguristes cf. *oxyophthalmus*			x		
Paguristes sp.			120		
Palicus alternatus				x	
Panulirus sp.	27	x	45–53		
Panulirus argus		28			
Panulirus guttatus		27			
Parapinnixa sp.		x			
Parthenope pourtalesii			x		
Petrochirus diogenes					59
Phimochirus holthuisi			120		
Phimochirus sp.			128		
Pilumnoides nudifrons			115		
Pilumnus floridanus		x	46–52	x	
Pilumnus sayi	43–46	18–52	56	x	
Podochela gracilipes				x	
Porcellana sp.				x	
Portunus spinicarpus				x	
Pseudomedaeus distinctus			115	x	
Pylopagurus sp.			126		
Raninoides lousianensis					76
Scyllarides nodifer	24				
Scyllarides aequinoctialis		18			
Scyllarides sp.					67
Sphenocarcinus corrosus			126		
Stenocionops furcata			120		
Stenorynchus seticornis		49–83	62	x	
Stenorynchus sp.	38	58–83	52–134	x	60–62
Thalassinid				x	
Xanthidae	24			x	58–70
CRUSTACEA—ISOPODA					
Aega antillensis			126		
Cirolana mayana		x	115–120	x	
Cirolana parva			126–128	x	
Eurydice littoralis			x		
Excorallana sp.			x		
Excorallana tricornis			126		
Gnathia sp.			x		
Jaeropsis sp.			x		
Stenetrium occidentale			x		
Isopoda		88	55–76	x	58–70
CRUSTACEA—MYSIDACEA					
Mysidae	54		62–128		

TAXA	Bank Group and Depths of Collection (M)				
	MS	FG	SEB	TRAN	STMS
CRUSTACEA—TANAIDACEA					
Apseudes propinquus			x		x
Apseudes sp.			x		x
Leptochelia sp.		x	x	x	x
Paratanaidae		x	126	x	x
Pseudotanais sp.			x		
Synapseudes sp.			x		
CRUSTACEA—AMPHIPODA					
Ampelisca abdita				x	
Ampelisca cristata			x		
Ampelisca cristoides				x	
Ampelisca schellenbergi			120–128	x	
Ampelisca sp.				x	59–70
Ampelisca venetensis				x	
Amphithoidae			120		
Ampithoe sp.		x	56–128		
Anamixis hanseni				x	
Anamixidae			56		
Carinobatea carinata			120		
Ceradocus sp.			x	x	
Ceradocus sheardi		x	126–128	x	
Chevalia aviculae			120		
Colomastix pusilla			120	x	
Colomastix sp.			56		
Corophiidae		x	126	x	59–70
Elasmopus rapax		x	46–120		
Erichtonius brasiliensis					59
Gammaridae	54				
Gammaropsis sp.				x	
Grandidierella bonneroides			x		
Heterophoxus oculatus			x		
Leucothoe spinicarpa	43–46	x	56–128	x	59–70
Liljeborgia bousfieldi		x	115–128	x	
Lysianassa alba		x	115–120		
Maera sp.			115–128	x	
Melita appendiculata			56–120	x	59
Melita dentata			73–88		
Stenothoe sp.					67–70
Unicola dissimilis			128		
Unicola laminosa			x		
***PYCNOGONIDA* (sea spiders)**					
Pyncnogonid			46–55		
***BRYOZOA* (moss animals)**					
Aimulosia uvulifera			67–120		61–73
Aetea truncata	21–46		55–91		61–73
Alderina smitti	27	x	55–91		61–73

TAXA	Bank Group and Depths of Collection (M)				
	MS	FG	SEB	TRAN	STMS
Antropora tincta	22–46		67–120	x	73
Aplousina filum	27	x	67–128	x	61–71
Arthropoma cecilii		x	73–100	x	61–72
Arthropoma circinatum			72–91	x	
Beania mirabilis	21–46				
Bellulopora bellula	43–46		67–91		61–72
Bracebridgia subsulcata	43–46	x	67–128	x	
Buskea dichotoma			76–100	x	59–73
Caberea boryi	43–46		56–120		72
Canda retiformis		x			69
Cauloramphus brunea		x	73–120		
Cellaria irregularis				x	61–73
Celleporaria albirostris	27–46	x	46–126	x	58–70
Celleporaria magnifica		x	67–100		73
Celleporaria mordax	21–27	x	67–128		72–73
Celleporaria tubulosa	43–46	x	67–120	x	61–73
Celleporaria spp.			115–126		
Chaperia patula			100–115		
Chorizopora brogniarti			55–61		
Cigclisula pertusa			120		73
Cigclisula serrulata			x		
Cigclisula turrita			67–91	x	60–73
Cigclisula protecta		x	60–91		67–73
Cleidochasma contractum	27	x	46–128	x	61–83
Cleidochasma porcellanum	27–46	x	56–128	x	60–73
Codonellina montferrandii			100–120		67–70
Colletosia radiata	27–46	x	55–128	x	61–83
Colletosia sp.			x	x	
Crepidacantha longiseta		x	76–126		59–64
Crepidacantha poissonii	27	x	67–91	x	61–83
Crepidacantha setigera			73		61
Crepidacantha sp.			120	x	
Cribellopora trichotoma	27		55–120	x	73
Crisia cf. *eburnea*	43–46	x		x	61–72
Crisia elongata			x		61–72
Crisia sp.	43–46	x	62		59–72
Cupuladria biporosa	60		100	x	73–83
Cupuladria canariensis	51–60		100	x	61–83
Cupuladria doma	60–67		100	x	64–73
Dakaria biserialis			x	x	61–72
Desmeplagioecia sp.	43–46			x	
Diaperoecia floridana	27–60		62	x	58–73
Discoporella umbellata	60–62			x	61–83
Disporella fimbriata	27		60–126		61–72
Disporella hispida			x		61–72
Disporella sp.	27		67–91		
Drepanophora tuberculatum		x	60–91	x	61–73
Entalophora proboscideoides	43–46		46–120	x	59–73

TAXA	Bank Group and Depths of Collection (M)				
	MS	FG	SEB	TRAN	STMS
Escharina pesanseris	27	x	67–120	x	58–70
Escharipora stellata			67		
Fenestrulina malusii	60		100		61–73
Floridina antiqua			62–128		61–73
Gemelliporidra aculeata			100–120		61–73
Hippaliosina rostrigera	27		x	x	68
Hippopetraliella bisinuata				x	
Hippopetraliella lanceolata			x		
Hippopodina bernardi			100–115		
Hippoporidra edax				x	65–73
Hippoporina americana			76–100	x	
Hippoporella gorgonensis	27	x	67–91		
Hippothoa distans	21–46		56–91	x	59–70
Holoporella albirostris	43–46				59
Holoporella sp.			58–80	x	60–73
Idmidronea atlantica	27–60		120	x	58–72
Idmidronea flexuosa			100	x	59–83
Labioporella granulosa		x	x	x	59–73
Labioporella sinuosa	43–46	x	60–91		73
Lekythopora longicollis	43–46		x		61
Lichenopora radiata			56–120		61–73
Lichenopora sp.			x		
Mastigophora porosa	27	x	62–120	x	59–83
Mecynoecia delicatula	43–46		x	x	59–72
Membranipora savartii				x	
Membraniporella aragoi	27	x	x	x	61–73
Microporella ciliata			62–120	x	61–83
Microporella cf. *coronata*	27		x		59–73
Microporella marsupiata			67–120		73
Microporella pontifica			67–91		60–73
Microporella sp.			x		
Mollia patellaria		x	55–120		73
Monoporella divae			115–128		
Parasmittina mildredae			60–104		59–72
Parasmittina nitida			67–91	x	59–83
Parasmittina signata				x	73
Parasmittina spathulata	43–46	x	67–120	x	64–73
Parasmittina sp.		x	55–128	x	73
Parasmittina trispinosa	21–24		67–91	x	64–73
Parellisina curvirostris				x	58
Parellisina latirostris	27		55–120	x	
Parellisina sp.				x	
Plagioecia sarniensis			76–91		61–83
Plagioecia sp.	27	x			
Proboscina sp.		x			73
Reptadeonella violacea	62		62–115	x	59–73
Retevirgula flectospinata		x	55–91		61–72
Retevirgula tubulata	27	x	60–128	x	61

TAXA	Bank Group and Depths of Collection (M)				
	MS	FG	SEB	TRAN	STMS
Rhynchozoon bispinosum		x	46–91		
Rhynchozoon spicatum	27	x	91		61–72
Rhynchozoon verruculatum	27	x	67–120		73
Rhynchozoon sp.			55–94		
Schizomavella sp.			x		
Schizomopora sp.			62		
Schizoporella trichotoma			55–61		
Schizoporella unicornis			67–91	x	61–73
Schizoporella sp.				x	
Scrupocellaria harmeri			67–91		58–73
Scrupocellaria regularis		x	62		
Scrupocellaria spp.		x	x		
Setosella vulnerata			73–120		
Smittina nitidissima			56–91	x	59–73
Smittipora levinseni	43–46	x	55–91	x	61–73
Smittipora nitidissima			60		
Steganoporella magnilabris		x	55–94		
Stenopsella fenestrata			67–120	x	
Stephanosella sp.			115		
Stylopoma sp.			67–73		
Stylopoma spongites	27–62		46–100	x	60–83
Tetraplaria dichotoma				x	
Trematooecia turrita			46–67		
Tremogasterina lanceolata			100	x	69–73
Tremogasterina mucronata			100	x	
Tremogasterina sp.			x		
Tremoschizodina lata	27–62	x	67–91	x	61–73
Triporula stellata		x	67–120		
Trypostega venusta	43–46		67–91	x	73
Tubulipora sp.		x	67–120	x	58–72
***ECHINODERMATA* (echinoderms)**					
ASTEROIDS					
Anthenoides piercei			115		
Asterinopsis lymani		x	x		
Asterinopsis pilosa			115–128		
Astropecten sp.		82	98	x	
Astropecten comptus		x	115		
Chaetaster nodosus		85	126		
Coronaster briareus		x			
Goniaster tessellatus		x		x	
Goniaster sp.				x	
Hacelia superba			115		
Linckia nodosa		60			
Narcissia trigonaria	48	82–113	75–83	x	
Narcissia sp.	40–56	75–84			
Ophidiasteridae			64–80		
Ophidiaster sp.		88	73–88		
Ophidiaster guildingii		82			
Tosia parva				x	

TAXA	Bank Group and Depths of Collection (M)				
	MS	FG	SEB	TRAN	STMS
OPHIUROIDS					
Amphiodia sp.		88	x		
Amphiodia pulchella		x	55–128	x	
Amphipholis squamata	54		55		
Amphioplus sepultus				x	
Amphioplus sp.				x	69
Amphiura sp.			126–128		
Asteroschema intectum			115		
Asteroschema oligactes			115		
Astrocyclus caecilia		x			61
Astrospartus mucronatus			120		
Astrophytum muricatum			67		
Astropora sp.					59
Astroschema sp.			120		
Axiognathus squamatus			115–128	x	
Gorgonocephalidae			58–134		58–70
Micropholis sp.				x	
Ophiactidae			73–88		59
Ophiactis sp.		18–52	46–120		59
Ophiactis algicola		x	115–128	x	
Ophiactis quinqueradia		x	120–126		
Ophiactis savignyi	47–54	18–52	46–128		58
Ophiocantha sp.			120–128		
Ophiacanthella sp.			120		
Ophiocoma wendti		x			
Ophioderma phoenium			67		
Ophioderma sp.			79		
Ophiodermatidae			76		
Ophioplax ljungmani			126		
Ophiostigma isacanthum		x	115–128		
Ophiostigma sp.			120		73
Ophiothrix angulata		x	46–128	x	59–73
Ophiothrix suensoni		67	126		
Ophiothrix sp.			56		
Ophiura acervata			126		
Ophiurochaeta littoralis		x	127–128		
ECHINOIDS					
Arbacia punctulata		x	x		
Arbacia sp.				x	
Astropyga magnifica			72–115	x	
Brissus unicolor			115		
Calocidaris micans			120		
Centrostephanus rubricingulus			120		
Cidaroid				x	
Clypeaster sp.	43–58	67–107	56–126		
Clypeaster ravenelli		82	126	x	
Coelopleurus floridanus			115		
Conolampus sigsbei			115		
Diadema sp.	21–52	48	43–72	x	64

TAXA	Bank Group and Depths of Collection (M)				
	MS	FG	SEB	TRAN	STMS
Diadema antillarum		18–67			
Eucidaris tribuloides		x	x	x	
Genocidaris maculata			x		
Pseudoboletia maculata maculata		60			
Clypeasteroidea		73–90	80		
Salenia goesiana			x		
Stylocidaris affinis		x	x	x	
Stylocidaris sp.			80–95	x	
Tretocidans bartletti			126		
HOLOTHUROIDS					
Isostichopus sp.	37–54	55	62–85		61
Isostichopus badionotus		55		x	
Psolas tuberculosis			x		
CRINOIDS					
Comactinia meridionalis	x		x		
Comatulid	55–58	76–96	58–146	x	58–96
Crinometra brevipinna			115		
Hypalometra defecta		x	125		
Leptonemaster venustus			115		
Nemaster discoidea			128		
CHORODATA PISCES **(fishes)**					
ORECTOLOBIDAE					
Ginglymostoma cirratum	x	24–49	47		
RHINCODONTIDAE					
Rhincodon typus	sfc				
CARCHARHINIDAE				x	58
Carcharhinus falciformis	0–60				56–60
Carcharhinus leucas	9–25				
Galeocerdo cuvieri	0–60				
Mustelus norrisi					56
Rhizoprionodon terraenovae	53	x			x
SPHYRNIDAE	x				
Sphyrna cf. *lewini*		9–20			
SQUATINIDAE					
Squatina dumerili		42			
PRISTIDAE					
Pristis sp.		26			
DASYATIDAE					57
Dasyatis americana	x	x			
MYLIOBATIDAE					
Aetobatus narinari		0–12			
MOBULIDAE					
Manta birostris		9–43			

	Bank Group and Depths of Collection (M)				
TAXA	MS	FG	SEB	TRAN	STMS
MORINGUIDAE					
Moringua edwardsi		28			
XENOCONGRIDAE					
Kaupichthys nuchalis		28			
MURAENIDAE	53				
Enchelycore sp.		28			
Enchelycore nigricans		28			
Gymnothorax moringa	26–56	21–70	55–85		58
Gymnothorax nigromarginatus		sfc			
CONGRIDAE					
Uroconger syringus		sfc			
OPHICHTHIDAE				63–178	
Ophichthus rex	0–58	81–120	146		82
Myrichthys acuminatus		56			
SYNODONTIDAE		45–125	52–136	67	67
Synodus foetens		x	x		
Synodus intermedius		82–88	76–79		
Synodus synodus	x	28			
ANTENNARIIDAE					
Antennarius radiosus		x			
OGCOCEPHALIDAE		49–148	76–145		
Ogcocephalus corniger			78–146		
OPHIDIIDAE			149–165		
EXOCOETIDAE	sfc				
Cypselurus cyanopterus		sfc			
Cypselurus melanurus		sfc			
Euleptorhamphus velox		sfc			
Hemiramphus balao		sfc			
Hemiramphus brasiliensis		sfc			
Hirundichthys rondeleti		sfc			
Parexocoetus brachypterus		sfc			
BELONIDAE					
Ablennes hians		sfc			
Platybelone argalus		sfc			
Tylosurus acus		sfc			
HOLOCENTRIDAE	21–56	18–125	42–87	59–82	58–72
Holocentrus ascensionis	x	x	x	x	56–59
Holocentrus marianus		x	x		
Holocentrus poco		28			
Holocentrus rufus	24	x	37	x	x
Holocentrus vexillarius		x			
Myripristis jacobus	21–43	22–49	47–79		
Plectrypops retrospinis		28			62

TAXA	Bank Group and Depths of Collection (M)				
	MS	FG	SEB	TRAN	STMS
AULOSTOMIDAE					
Aulostomus maculatus		22–40			60–61
CENTRISCIDAE					
Macrohamphosus scolopax			146–213		
SERRANIDAE					
Epinephelus spp.	37	22–76	52–122	70	
Epinephelus adscensionis	21–55	20–61	43–53		
Epinephelus cruentatus	x	20–49	52–53		
Epinephelus fulvus		20–21	50		
Epinephelus guttatus		28–49			
Epinephelus inermis		21–82	37–91		
Epinephelus niveatus		91	76–177		
Epinephelus striatus			47		
Gonioplectrus hispanus	x	50–91	76–127		
Hemanthias leptus		x	x		
Hemanthias vivanus		x			
Holanthias martinicensis		61–140	62–127	66–82	57–76
Liopropoma eukrines	28–56	30–88	52–96	61–82	58–72
Liopropoma rubre		30			
Mycteroperca bonaci		x			
Mycteroperca interstitialis	x	x	x	x	x
Mycteroperca microlepis		x			
Mycteroperca phenax	27–54	21–88	44–95	61–82	55–70
Mycteroperca tigris	27	20–23	59		
Mycteroperca venenosa		x			
Paranthias furcifer	18–56	12–88	30–98	67	59–61
Serranus annularis		47–79	42–78		59–61
Serranus phoebe	38–56	61–99	64–175	59–70	58–72
Serranus subligarius		x			
GRAMMISTIDAE	31–43				
Rypticus maculatus	56				
Rypticus subbifrenatus		20–30			
PRIACANTHIDAE					
Priacanthus arenatus	28–56	21–126	53–104	59–81	56–75
Priacanthus cruentatus		21–53			
Pristigenys alata		x			x
APOGONIDAE	24–58	43–89	52–82	68	60–63
Apogon maculatus	24–58	43–58	53–64		x
Apogon pseudomaculatus			37		
Apogon townsendi		26			
Phaeoptyx conklini		26–28			
Phaeoptyx pigmentaria		26–28			
MALACANTHIDAE					
Caulolatilus sp.		64	128–206		
Caulolatilus cyanops		91			
Malacanthus plumieri		21–79	30–87		58–61

TAXA	Bank Group and Depths of Collection (M)				
	MS	FG	SEB	TRAN	STMS
POMATOMIDAE					
Pomatomus saltatrix	60				
ECHENEIDAE				x	0–37
Echeneis naucrates		sfc			56
CARANGIDAE					
Alectis ciliaris	x	x			
Caranx bartholomaei		x			
Caranx crysos	22–56	0–125	9–54		30–73
Caranx hippos	x	10–21			
Caranx latus	x	10–40			
Caranx lugubris		21–25			
Caranx ruber		10–51	46		
Elagatis bipinnulata	x	0–30	12		
Selene setapinnis					64
Selene vomer	x				52
Seriola dumerili	0–55	0–88	9–110	61–82	0–69
Seriola rivoliana	x	0–28	x		x
Trachurus lathami		sfc	128		
CORYPHAENIDAE					
Coryphaena hippurus		sfc			
LUTJANIDAE					
Lutjanus apodus	37–43				61
Lutjanus buccanella		x			
Lutjanus campechanus	44–52	31–90	46–103	61–82	56–74
Lutjanus cyanopterus					x
Lutjanus griseus	x	21–46			
Lutjanus jocu		x			
Lutjanus synagris					59
Lutjanus vivanus		x			
Ocyurus chrysurus		29–46			
Rhomboplites aurorubens	27–43	49–91	55–104	62–73	57–62
HAEMULIDAE					
Haemulon aurolineatum	x				58–64
Haemulon melanurum	24–56	21–87	79	67–73	59
Orthopristis chrysoptera	44			x	
INERMIIDAE		x	x		
SPARIDAE					
Calamus leucosteus	x	x			
Calamus nodosus	28–56	21–87	79	59–66	58–62
Lagodon rhomboides				65	64
Pagrus pagrus		x	x		
SCIAENIDAE					
Equetus lanceolatus		21–91	50–98	61–64	61
Equetus punctatus	46	x	76–94		62
Equetus umbrosus	28–56	49–86	52–93	62–73	58–69

TAXA	Bank Group and Depths of Collection (M)				
	MS	FG	SEB	TRAN	STMS
MULLIDAE					
Mulloidichthys martinicus	28–56	21–75	64–69		
Pseudupeneus maculatus	21–44				
KYPHOSIDAE					
Kyphosus incisor	x	0–22			
Kyphosus sectatrix	x	0–22			
CHAETODONTIDAE					
Chaetodon aculeatus		21–80	47–81		
Chaetodon aya		61–125	55–123	70–75	61
Chaetodon capistratus		x			
Chaetodon ocellatus	24–41	18–55	52–76		60
Chaetodon sedentarius	21–56	20–98	37–96	59–82	58–73
Chaetodon striatus	x	21–40			
POMACANTHIDAE					
Centropyge argi	x	21–79	37–77		59–60
Holacanthus bermudensis	22–56	21–91	37–83	64–81	60–69
Holacanthus ciliaris	24–53	21–64	44–59	73	58–60
Holacanthus tricolor	27–37	21–88	37–84	61–62	
Pomacanthus arcuatus	27–29	18–37			
Pomacanthus paru	22–56	18–70	47–67		
POMACENTRIDAE					
Abudefduf saxatilis		x			
Chromis cyaneus	24	21–49	44–47		
Chromis enchrysurus	38–58	21–107	37–91	59–72	43–72
Chromis insolatus	x	x	x		
Chromis multilineatus	21–38	21–46	30		
Chromis scotti	18–53	21–46	47–76		67
Microspathodon chrysurus		20–21			
Pomacentrus partitus	37–50	18–53	46–53		x
Pomacentrus planifrons		21–28			
Pomacentrus variabilis	18–46	21–49	52		x
CIRRHITIDAE					
Amblycirrhitus pinos	x	21–55	47–75		
LABRIDAE					
Bodianus pulchellus	25–54	21–91	44–96	59–82	43–73
Bodianus rufus	21–46	18–50	46–53		
Clepticus parrai	27–47	15–52			
Halichoeres bivittatus	38–41	x	66–76	x	59–69
Halichoeres caudalis	x	27–75	52–67		
Halichoeres garnoti		21–46	47–53		
Halichoeres maculipinna		21			
Halichoeres radiatus		21–46			61
Hemipteronotus sp.		21			60
Lachnolaimus maximus	x	x			
Thalassoma bifasciatum	22–50	18–53	46–52		

TAXA	Bank Group and Depths of Collection (M)				
	MS	FG	SEB	TRAN	STMS
SCARIDAE		21–58			59
Scarus taeniopterus		21–49	51–76		
Scarus vetula		21–28			
Sparisoma aurofrenatum	21–71	52–76			60
Sparisoma viride		20–52			
SPHYRAENIDAE					
Sphyraena barracuda	1–30	0–67	6–62		0–67
OPISTOGNATHIDAE		49	55–81		
Opistognathus aurifrons		21–26			
Opistognathus lonchurus				81	
CLINIDAE					
Emblemaria atlantica		61			
Emblemaria pandionis		28			
Starksia ocellata		21–28			
Starksia sp. (nov.)		80–120			
BLENNIIDAE					
Hypleurochilus bermudensis		28			
Ophioblennius atlanticus	24–40	21–28			
Parablennius marmoreus		62			
GOBIIDAE					
Coryphopterus thrix		20–28			
Gnatholepis thompsoni		22–28			
Gobiosoma (*Elacatinus*) sp.	40	46–78	52–78		61–70
Gobiosoma oceanops	24–32	21–52			
Ioglossus calliurus	28–58	55	61–79	70–73	64–66
Lythrypnus nesiotes		28			
Lythrypnus phorellus		46			
Lythrypnus spilus		20–67			
Quisquilius hipoliti		20–26			
Risor ruber		20–28			
ACANTHURIDAE	21–37	40–70	55–58		
Acanthurus bahianus	27–38	21–49			
Acanthurus chirurgus	40	21–29			
Acanthurus coeruleus	24–34	21–46	55		
SCOMBRIDAE				0–30	
Euthynnus alletteratus	x	x			
Scomberomorus cavalla	24	0–60	0–9		30–46
SCORPAENIDAE		46–148	56–135	73	78
Scorpaena spp.				x	x
Scorpaena plumieri	x				
Scorpaenodes caribbaeus		28	x		
TRIGLIDAE		82	64–213	67	
BOTHIDAE	39	94–125	85–210	67	x
Cyclopsetta fimbriata			203–210		

TAXA	Bank Group and Depths of Collection (M)				
	MS	FG	SEB	TRAN	STMS
CYNOGLOSSIDAE			146–151		
BALISTIDAE					
Aluterus scriptus		21–53			
Balistes capriscus	46	61–79	52–77	61–73	
Balistes vetula	24–44	15–70	44–67		52–61
Cantherhines macrocerus		21–50			
Cantherhines pullus	x	15–21			
Canthidermis sufflamen	18–40	21–52	12–75		
Melichthys niger		18–43	37		
Xanthichthys ringens		sfc	46–79		
OSTRACIIDAE					
Lactophrys quardricornis	25	21–50	52–85	59–67	
Lactophrys triqueter	x	21–43	52		
TETRAODONTIDAE					
Canthigaster rostrata	24–53	19–85	47–91	64	61–67
Sphoeroides sp.		40–43	79	65–67	x
DIODONTIDAE					
Diodon holacanthus		0–28			
Diodon hystrix		21–85	56–82		
MOLIDAE					
Mola mola		sfc			
***CHORDATA REPTILIA* (turtles)**					
Cheloniidae	sfc	sfc-20	x		
Caretta caretta			60		

APPENDIX II

LOCATIONS OF OUTER CONTINENTAL SHELF REEFS AND HARD BANKS IN THE NORTHWESTERN GULF OF MEXICO

Classification		Bank	Latitude	Longitude
South Texas Relict Carbonate Shelf Reefs	South Texas Fishing Banks	Small Adam	26°56.7′N	96°49.8′W
		Big Adam	26°57.2′N	96°49.0′W
		North Hospital	27°34.5′N	96°28.5′W
		Aransas	27°35.5′N	96°27.0′W
		Baker	27°45.0′N	96°14.0′W
		Blackfish	26°52.6′N	96°46.6′W
		Hospital Rock	27°32.5′N	96°28.5′W
		Mysterious	26°46.0′N	96°42.0′W
		Southern	27°26.5′N	96°31.5′W
		Dream	27°02.5′N	96°42.5′W
		South Baker	27°40.5′N	96°16.4′W

Continued

Classification		Bank	Latitude	Longitude
North Texas–Louisiana Reefs and Banks on Diapiric Structures	Mid-shelf Banks	Claypile	28°19.5′N	94°09.0′W
		Sonnier	28°20.0′N	92°27.0′W
		Stetson	28°10.0′N	94°17. ′W
		32 Fathom	28°03.8′N	94°31.3′W
		Coffee Lump	28°01.5′N	93°54. ′W
		Fishnet	28°09.0′N	91°48.5′W
	Shelf Edge Banks	Alderdice	28°05.0′N	91°59.5′W
		Ewing	28°05.2′N	91°00.0′W
		Bouma	28°02.5′N	92°27.0′W
		Parker	27°57.0′N	92°01.0′W
		Sackett	28°38.0′N	89°32.8′W
		East Flower Garden	27°54.5′N	93°36.0′W
		Applebaum	27°52.0′N	94°15.0′W
		Bright	27°53.2′N	93°18.0′W
		West Flower Garden	27°52.5′N	93°49.0′W
		Diaphus	28°05.0′N	90°42.5′W
		18 Fathom	27°58.0′N	92°36.0′W
		28 Fathom	27°55.0′N	93°27.0′W
		Jakkula	27°59.0′N	91°39.5′W
		Rezak-Sidner	27°57.0′N	92°23.0′W
		Sweet	27°51.0′N	91°49.0′W
		Elvers	27°49.0′N	92°53.5′W
		Geyer	27°51.6′N	93°04.0′W
		Phleger	27°50.0′N	91°53.5′W

AUTHOR INDEX

Subject Index